G. K. Vemulapalli

PHYSICS: The Behavior of Particles

HARPER'S PHYSICS SERIES
Frederick Seitz, Editor

PHYSICS: The Behavior of Particles

PHILIP STEHLE
University of Pittsburgh

HARPER & ROW, PUBLISHERS
New York, Evanston,
San Francisco, London

PHYSICS: THE BEHAVIOR OF PARTICLES

Standard Book Number: 06-046411-9

Library of Congress Catalog Card Number: 78-141172

Contents

Preface

The study of physics should have a broad appeal in a society that is as dominated by technology as ours is. The public is surrounded by the most complicated and sophisticated devices whose construction and operation depend on the existence of laws of nature. One would expect a large part of this public to have some curiosity about these laws of nature even though a detailed knowledge of the devices themselves may be more than it cares to have. In spite of this, physics courses are not filled with the curious but more with students fulfilling a requirement. What is the reason for this?

From the end of World War II until very recently it has been widely believed that the survival of the United States depended on having a large number of technically highly trained people. Consequently, education in the humanities and social sciences was considered to be of lesser importance. This view was naturally highly acceptable to physicists, and it became their goal to teach physics very thoroughly to a lot of students so that they would acquire a mastery of the subject early and would then be able to apply their knowledge to all sorts of problems. As a consequence, there developed a type of beginning physics course and a type of physics text which set out to produce technicians in one stroke. The introductory course became longer and more mathematical until two year introductions were not uncommon, and two volume texts the rule. The courses for students not knowing the calculus were considered inferior, and not nearly as much thought and time were spent on them.

This trend has recently been checked and even reversed. Many teachers have concluded that a massive first dose does not produce a mastery of subject matter in proportion to the time and effort involved. Furthermore it makes physics a formidable subject, not to be taken on the chance that it could be fun, because the investment required in time, effort, and mathematical preparation is well beyond what can be justified by a rather casual curiosity such as a history or philosophy student might have. The emphasis now is on making the introduction to physics on the college level a survey of physics as it is seen by contemporary physicists. Such an introduction, many feel, can serve both the future professional physicists and also the students who see physics as one of the influences which form modern life by giving them an insight into its aims and methods.

The present book is designed to serve as a text in an introductory physics course of this survey kind. It is not a survey of all of physics, but a sampling of the topics which play a leading role in contemporary physics. The book is organized around the concept of *particle*. The first few sections are quite conventional and are in conventional sequence; in them the ideas of Newtonian mechanics of particles are developed using vectors but not calculus, although a close approach to calculus is unavoidable by the nature of the subject. What is new, or at least less common, is the pursuit of the idea of particle and particle motion immediately to the area of microscopic phenomena. The experiments of J. J. Thomson and R. A. Millikan are analyzed in detail, and the proof of the existence of the electron is explained by these experiments in a way which is not at all mysterious even though the particles involved are truly microscopic and not directly observable. In similar ways the experiments which led physicists to the concepts of atomic nuclei, of the constituent particles of the nucleus, and of the "fundamental" particles of high energy physics are described and analyzed. All this can be done in essentially classical terms because only large scale motions of these particles are involved in these experiments. Atomic structure and the quantum modifications of the laws of motion are reserved until much later, but relativistic modifications of Newtonian mechanics are needed and are described and analyzed. This is first done in connection with particle mechanics, and only later is the usual approach to relativity through the study of light made.

All the subjects mentioned so far involve the study of individual particles. A large part of physics involves the description of large scale phenomena in terms of microscopic particles. This aspect of physics is developed in the second part of the book, after a short digression on fields and light waves. Here again there is a quite conventional sequence of subjects related to the kinetic theory of gases. Unusual stress is placed on the idea of momentum and momentum transport here because it shows the all-pervading character of the mechanical idea of momentum. Nearly at the end, unfortunately the usual place for it, comes the idea of the

Bohr atom and its extension to solids where again the large scale phenomena of electrical conduction are described statistically in terms of electrons in energy bands. Even if the course ends before this is reached, and one should try to avoid this, the student will have become acquainted with a large part of modern physics.

In any introductory course, the emphasis I believe should be on a qualitative understanding of the concepts involved and not on a mastery of details. In using this book as a text I believe it would be a mistake to linger over the sections on vectors or even over those on Newton's laws longer than is necessary to give some intuitive feel for their meaning. The use of these ideas in the later sections starting with Section 19 ought to make them clear and relevant, because these sections show applications where the results matter. No one really cares how fast you must shoot a bullet to have it go 1000 yards and drop only so much, but one might well care how one measures the charge and the mass of an electron or how one knows the size of an atomic nucleus and the mean life of a pion. The later sections provide worked-out examples of the laws of motion and the motion of charged particles in fields, so only a few simple examples of the traditional kind are worked out in the early sections. The places to linger are the places where actual experiments are described; if these are made clear, the abstract laws will take care of themselves.

Obviously certain sections are less essential than others. If time presses, I would suggest the omission of Sections 12, 21, 24, and 25 in the first half, and of Sections 35, 36, 37, and 42 in the second. These omissions will not make the later sections much more difficult; the greatest loss is the intrinsic interest of the subject matter.

It is possible to use the first five chapters as the text for a one-semester course. If this is done, the advice not to linger over the first three chapters becomes even more urgent because there is no second term from which to borrow time, and omission of Chapter 5 would render the course one without a climax. I have covered this material in one term with considerable success, but it required self-discipline at the start.

A word is needed about the problems. They are meant for the most part to illustrate important features of the sections they follow. The numbers that occur in many of them are real, physical values of quantities of interest, and these numbers have the awkward property of not dividing exactly into one another and of not having self-evident square roots. Usually the answer is interesting, for the purpose of this book, only to one or two significant figures; sometimes only the position of the decimal point really matters so that the order of magnitude of the answer is sufficient. Students who have slide rules and who know how to use them should do so, and those who do not probably should do some careful guessing. In no case is a lot of pure arithmetic worthwhile.

The value of the book has been increased by the cooperation of many of my colleagues and students at the University of Pittsburgh who endured courses taught from typewritten versions of preliminary drafts. I especially thank Joseph Eberly of the University of Rochester, who reviewed the manuscript in detail and made pungent remarks leading to improvements. The mistakes and obscurities that remain are mine.

PHILIP STEHLE

PHYSICS: The Behavior of Particles

An Attempt to Define Physics

The word *physics,* like many words, has a variety of meanings. We might say that physics is what academic physicists do when they consider themselves to be working. The adjective *academic* is supposed to mean that the physicist so described is free to work on problems of his own choosing. A physicist who is academic in this sense may work in industry or in a government laboratory, but most such physicists are associated with colleges and universities. This kind of definition is much better than one given in terms of specific problems studied, because the problems change as time goes on. New problems whose existence was unsuspected a few years ago now occupy the attention of many physicists, and old problems that may have been solved to a certain extent have been dropped by physicists and taken up by others whose interests are of a different kind.

An example of change in the problems studied by physicists will illustrate this point. Acoustics was once an important part of physics. Today few academic physicists are concerned with this subject even though not all acoustical problems have been solved. The design of a good auditorium for music or drama involves acoustical problems that still must be solved to a large extent by trial and error. There are military applications of sound, especially underwater sound, that are important and the object of much work. These problems are considered by academic physicists to be more in the field of engineering than physics. The part of acoustics of most interest to them now is the extremely high frequency range where sound, like light, can excite individual atoms. Many of the ideas developed in connection with acoustics when it was a vital part of physics are still important ideas in contemporary physics. They are used in other contexts now, and their origin in the study of sound is a historical accident. The application of concepts changes, but a good idea is a rare thing and we cannot afford to discard one when its original cause or stimulus is no longer of prime interest.

While some subjects have moved out of physics, some new subjects have appeared. Perhaps the most spectacular example is the subject of "elementary" particles or high energy physics. It has been discovered that the proton and the neutron, particles considered to be "elementary" 25 years ago, possess elaborate internal structures and that many previously unknown particles exist. The study of this new field of phenomena has just begun, but already it involves enormous experimental establishments such as the National Laboratories at Brookhaven, Berkeley, and Batavia. At the other end of the scale of size, a new class of very large, very bright objects has been identified in the sky. They are called "quasi-stellar objects" or quasars, and the source of energy that keeps them going is not yet understood. The mechanism must somehow involve the transformation of elementary particles on a grand scale. Other less spectacular but still important and interesting new topics that could not have been identified as parts of physics 10 or 20 years ago because they did not exist at that time are masers and lasers, many phenomena connected with semiconductors and superconductors, highly ionized plasmas, and many more.

Why are these things the objects of academic physicists' attention? Largely, I believe, because they are not understood but at the same time are not so hopelessly complicated that a detailed understanding is out of the question. This leads to another possible definition of physics. Physics is that branch of science which deals with phenomena so simple that a relatively complete description of them can be given. The objects of study are the simplest conceivable: a single proton, a perfectly pure crystal of silicon, or a perfectly parallel beam of light striking a single free electron. In the end the subject gets complicated because one wants a *complete* description of the system with no approximation that is not understood. An atom is simple when it is considered to be an indivisible unit of an element. An atom of carbon does not seem quite so simple when it is found to consist of a concentrated nucleus with 6 planetary electrons. It is even less simple when the nucleus is seen to consist of 12 particles, 6 protons and 6 neutrons. Now the proton and neutron are known to have a complex structure. A present-day physicist tends to leave the carbon atom to the chemist and to look at something he still hopes can be completely understood, such as the proton. In the past the result of looking for complete descriptions of simple things has been to alter man's view of the world and to make more advanced technology possible. The discoveries of the laws of motion of a particle, of the theory of relativity, and of quantum mechanics have all affected our conception of nature and ourselves in profound ways. These discoveries also have given us means to control nature to an unprecedented extent. Yet they all came from considering the most basic problems known at the time and from giving what was for their time a nearly complete account of what is relevant and how things fit together.

Clearly there is no sharp division between physics and chemistry or physics and biology. There is no unique virtue in seeking the complete description of phenomena, as the price to be paid for completeness of description is the simplicity of the thing described. It is as important and as difficult to give less complete descriptions of more complicated phenomena. It just is not physics.

CHAPTER 1
Particles and the Geometry of Motion

1. DIRECTLY OBSERVABLE PARTICLES

Physics is an experimental science; it is based on observation. In the beginning its object was to describe phenomena that can be observed with the unaided senses, but as the understanding of nature has increased it has become possible to use more and more indirect methods of observation until now there is a vast amount of experimental equipment between the observer and the things observed. In order to understand what physics is about today, one must understand something of the development of the concepts used because in the end the results of physics are stated in words whose meaning is established by direct experience.

A central role in physics, both of the past and of the present, is played by the *particle*. The modern particles of high energy physics are lineal descendants of the particles such as a falling stone or the moon considered by Galileo and Newton when the laws of motion of macroscopic particles were being developed. It is necessary to have a clear understanding of directly observable, macroscopic particles before the changes in the concept of particle forced on us by modern experiments can be appreciated.

The first part of this book is devoted to a description of particles that are directly observable, macroscopic particles. There are several aspects to this subject. One of the main things a particle can do is move, and both the motion itself and the external influences that affect the motion are important. The study of motion without reference to the causes of the motion is called *kinematics,* while the study of how a particle's motion depends on its environment is called *dynamics.* Some kinematics must precede the study of dynamics because we must be able to describe motion before we can describe its dependence on something else. Kinematics is little more than geometry, but it is the geometry of a moving point rather than of a fixed figure so that it requires the idea of time as well as of space. When some kinematics has been developed, the dynamics of particle motion is looked at carefully and regularities in the motion of particles under given circumstances are sought for. In this way we learn to recognize certain "laws of nature" such as Newton's laws of motion and the laws of conservation of energy, of momentum, and of angular momentum.

Next various ways in which the environment influences particles are studied. There are three principal kinds of influence or force that concern us here: the forces of gravity, of electricity, and of magnetism. The force of gravity is the most familiar from daily experience, but the other two kinds of force also can be observed to act on directly observable particles. They are especially important for physics because they provide the only controllable forces that can be exerted on microscopic particles, so that nearly all our knowledge of the properties of microscopic particles is obtained either directly or indirectly through the use of electric and magnetic forces. This is typical of the leapfrog method of progress in science. The existence and characteristics of electric and magnetic forces are established by observing their effects on visible particles; then, as in the later parts of this book, the properties of particles not directly visible can be inferred by their behavior when acted on by electric and magnetic forces. Once the characteristics of the particles are established, they can be used to discover new effects, and so on. The final test of the system of concepts that arises in this way is its ability to give a relatively simple and consistent picture that is in accurate accord with what has been observed so far.

Sometimes a law of nature takes the simple form of a conservation law, a statement that the value of some quantity is unchanged no matter what the physical system involved does. The best known of such laws is the law of energy conservation. This is a statement that the energy of a system does not change unless that system is in contact with another system from which it can acquire or to which it can give energy. There are many such laws involving, for example, momentum, angular momentum, and electric charge. These laws are simple to state and in many cases have been verified without exception up to now. When an observation is reported that indicates the violation of an often verified conservation law, the report is generally regarded with some skepticism, the attitude being that the violation of such a law is to be recognized only when all interpretations of what is observed that are consistent with the law can be ruled out. Thus some apparent violations of energy conservation can be made consistent with the law by widening the bounds of what is meant by energy to include something not included before, while others are due simply to erroneous observation. On rare occasions a conservation law that was thought to apply universally has to be recognized as applying only to a restricted domain of events; the discovery that "parity" is not always conserved is the best example of this that comes to mind. The meaning of this law and the limitation of its application are discussed in Section 28. In general, then, conservation laws are regarded as difficult to establish, but once established they are not to be declared invalid lightly. The demonstration that a conservation law is invalid must be made unambiguously to be taken seriously. This is the reason that proposed perpetual motion machines are not taken seriously by most scientists; energy conservation rules them out, and they feel that it is up to the inventor to demonstrate that energy is not conserved rather than for the scientist to find exactly where in the proposed device the difficulty in its operation will appear. So far no perpetual motion machine has been invented.

The subject matter of the first two chapters is very directly related to everyday experience, and it has many applications to things around us, a few of which will be pointed out. Our purpose here is not to give an extensive account of the world of direct observation but to give a starting place for progress into the microscopic world. For this reason the topics included are limited to those needed later, and many interesting and important applications are omitted altogether so that the study of the microscopic world can start as soon as possible.

2. PARTICLES

The objects that are directly visible are all built up out of smaller objects, and these smaller objects are built up out of still smaller ones, say atoms. At one time an atom was defined as an "indivisible" unit of a chemical element. Now we know that an atom consists of a nucleus and electrons, and we know that the nucleus consists of protons and neutrons. It is not yet clear whether this is an unending chain or whether there comes a stage beyond which it does not pay to consider one thing as a compound structure made out of more elementary objects. It is not always necessary to take the details of the structure of an object into account in describing its behavior under a wide range of circumstances. To the extent that this structure has effects that can be summarized briefly, the object can be treated as a *particle;* it is an object whose complete description involves giving its position and the values of a few other variables. Let us take a golf ball as an example. A standard golf ball consists of a rather elaborate core wrapped with many yards of rubber ribbon and covered with a tough, elastic layer colored white. The behavior of the golf ball in normal circumstances is governed by its spherical shape and size, its weight, and its elasticity. The structure that produces this shape, size, weight, and bounce is almost if not completely irrelevant to its behavior in a golf game. Under normal circumstances a golf ball is well represented by a particle. If it is hit so that it spins fast or if it strikes a sharp stone so that the cover is cut, then the structure of the ball matters and its representation by a particle is not so good.

Much of physics is concerned with determining how one kind of object is constructed out of smaller objects that for the purpose at hand can be treated as structureless. A gas can be pictured as made up of freely moving molecules without involving a detailed accounting of what a molecule is like. If one delves deeper into the behavior of the system, making very detailed observations, then one may have to ascribe some structure to the molecules. Now we can make detailed enough observations to learn about atomic structure, nuclear structure, and even the structure of the proton, but even so in describing most properties of gases all this knowledge is irrelevant and we may take the molecules of the gas to act as particles.

The position of a particle can be described only if some starting place for this description is agreed upon. If the particle is in a room, two intersecting walls and the floor provide a convenient reference frame or coordinate system. The point at which these three surfaces meet is called the *origin* of the coordinate system, and the lines in which pairs of surfaces meet are the coordinate *axes* of the system. This is illustrated in Figure 2.1. The origin is usually labeled O, and three coordinate axes are conventionally called the x-, y-, and z-axes.

Let the particle be at P inside the room as in Figure 2.1. We may construct a rectangular box with sides parallel to the walls and floor of

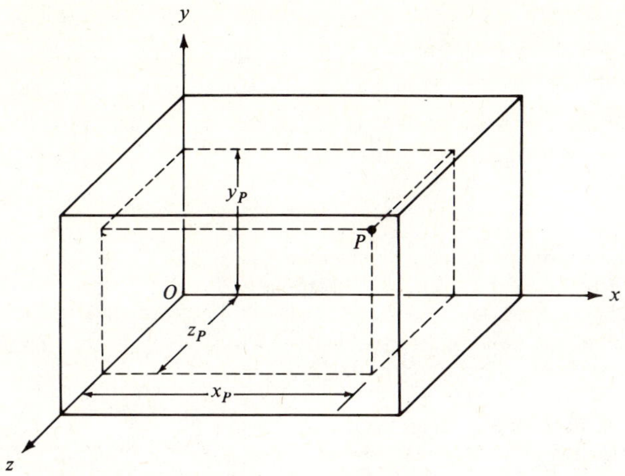

FIGURE 2.1 Particle P is located with respect to two walls and the floor of a room by the three coordinates x_P, y_P, z_P. Three edges of the room serve as the coordinate axes. The positions of the other two walls and of the ceiling do not matter; they serve only to tell which sides of the reference surfaces are "inside" surfaces.

the room, so that the origin O and the point P are at diagonally opposite corners. The lengths of the edges of the box parallel to the x-, y-, and z-axes, respectively, are then called the *coordinates* of the particle P in this coordinate system and are denoted by x_P, y_P, and z_P. If there is no possible doubt which particle is being described, the subscript P is often omitted. The sign of x_P is taken to be positive whenever an observer at P would see the inside of the reference wall containing the y- and z-axes. (The position of the other wall parallel to this one is irrelevant, so it does not matter whether P is inside the room or is so far from the reference wall that it is outside the room as long as an observer at P would see the inside of the chosen reference wall.) If an observer at P would see the outside of the reference wall, x_P is taken to be negative. The signs of y_P and z_P are chosen in a similar way. Giving the values of the three coordinates of a particle completely specifies its position with respect to the coordinate system.

It is often useful to put one more specification on the coordinate system. On looking from the origin outward along the positive z-axis, it is possible that the positive x-axis will be made to lie on top of the positive y-axis by turning it 90° in the clockwise direction. If this is so, the coordinate system is called "right-handed." If a counterclockwise rotation is needed to make the positive x-axis lie on top of the positive y-axis, the system is called "left-handed." The reader should convince himself with a couple of sketches that a left-handed system is changed into a right-handed one by interchanging the names of the x- and the y-axes. It is customary to use right-handed systems.

Problems

1. A point P lies on the positive x-axis of a coordinate system 6 cm from the origin, so that it has coordinates (6, 0, 0).
   (a) Find the coordinates of two other points on the x-axis that are 10 cm away from P. How far are they from each other?
   (b) Find the coordinates of two points on the y-axis that also are 10 cm away from P. How far are they from each other?
2. A mirror hangs on the wall of a room that is a feet long, b feet wide, and c feet high. A point P in the room has coordinates (x, y, z) relative to a coordinate system using the floor as the xy-plane and the wall on which the mirror hangs as the xz-plane. What are the coordinates of the mirror image of P? If the mirror were hung directly across the room instead, what would the coordinates of the image of P then be? Make a sketch of the room, the mirror, and the point P and its images.

3. VECTORS

The position of a particle is always specified relative to a reference frame. The values of three coordinates x, y, and z are needed to determine this position even though the *position* of a particle is a single idea. This is what is meant by saying that ordinary space is three-dimensional. It is very useful to have a single name for this position, just as it is useful to have a name for the collection of parts we call an automobile or a piano. The name given to the quantity that specifies the position of a particle is *position vector*. It is a typical vector quantity about which we shall have a lot to say throughout our discussion of physics. Vectors are useful because they represent a single idea with a single mathematical symbol, making it easier to express physical ideas in mathematical language and to give a physical interpretation to the results of mathematical manipulations. Many physical statements can be made very simply using vectors, even when these statements look complicated and are hard to write down and to read if written using the coordinates explicitly. For this reason we devote this section to vectors and introduce here most of the rules for their use that will be needed in the rest of the book.

The position of a particle can be given by specifying the length and the direction of the line segment that leads from the origin of the coordinate system to the particle. The situation is illustrated in Figure 3.1. The line segment from O to P is denoted by $\mathbf{r}_{OP}$ and is called the position vector of P in this coordinate system. (In general we use boldface letters

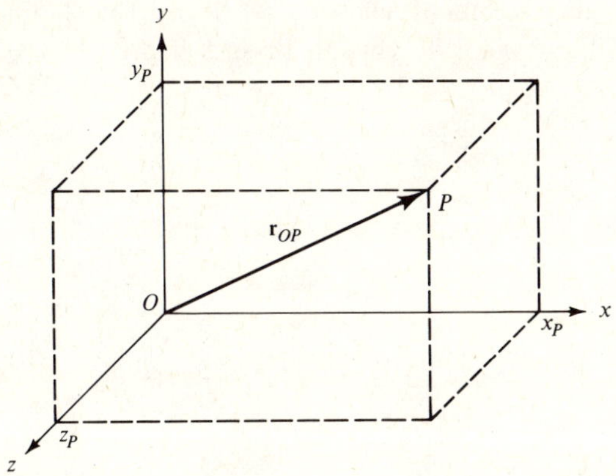

FIGURE 3.1 The position vector of the point P is the line segment from the origin O to the point P. The coordinates x_P and so on are the components of this position vector $\mathbf{r}_{OP}$ in the directions of the three coordinate axes.

to represent vectors.) The length of the segment is the distance of the particle from the origin and is denoted by r or by $|\mathbf{r}|$. It is related to the coordinates of P by the three-dimensional version of the Pythagorean theorem, so that

$$r^2 = x_P^2 + y_P^2 + z_P^2$$

$$r = \sqrt{x_P^2 + y_P^2 + z_P^2} \tag{3.1}$$

(If $z = 0$, P lies in the xy-plane and r is the hypotenuse of a right angle triangle whose sides are of length x and y. We then have the usual theorem.) The coordinates x_P, y_P, z_P of P are called the *components* of $\mathbf{r}_{OP}$ along the coordinate directions. The coordinates of a particle are usually measured in centimeters, a unit of length defined in Section 5.

The position vector $\mathbf{r}_{OP}$ may be thought of as the *displacement* that carries the particle from the origin to the point P. We shall introduce other vectors later that do not represent displacements but do represent physical quantities characterized by a magnitude and a direction. The rules used to manipulate these vectors are just those that are natural for displacements, so we develop them here in connection with displacement vectors.

A particle may be displaced from the origin O to the point P, this displacement being specified by the vector $\mathbf{r}_{OP}$. From P it may be displaced to another point Q, the displacement vector describing this being written $\mathbf{r}_{PQ}$. The particle could have been displaced directly from O to Q,

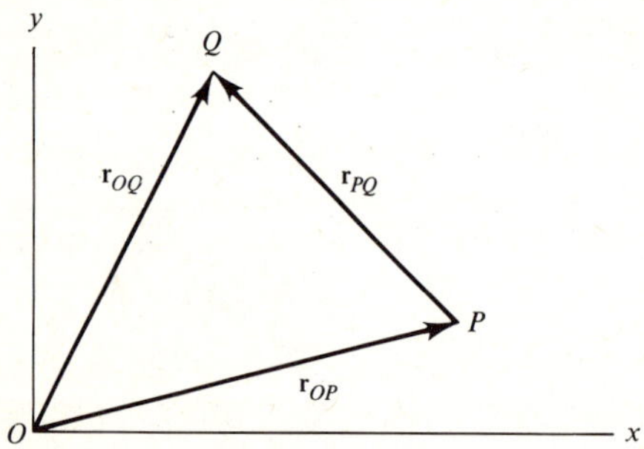

FIGURE 3.2 The sum of the two displacements $\mathbf{r}_{OP}$ and $\mathbf{r}_{PQ}$ is the displacement $\mathbf{r}_{OQ}$.

this displacement being denoted by $\mathbf{r}_{OQ}$. We say that the vector $\mathbf{r}_{OQ}$ is the *sum* or *resultant* of the vectors $\mathbf{r}_{OP}$ and $\mathbf{r}_{PQ}$, writing

$$\mathbf{r}_{OQ} = \mathbf{r}_{OP} + \mathbf{r}_{PQ} \tag{3.2}$$

The geometrical addition meant in this equation is illustrated in Figure 3.2.

In the figure all displacements are chosen to lie in the xy-plane to simplify the drawing. The reader may see what is involved here by considering the displacement necessary for his hand to reach the nearest light switch and how this displacement could be made in two stages, one vertical and one horizontal. It is seen that the order in which these displacements are made does not affect the end result.

The length of $\mathbf{r}_{OQ}$, written $|\mathbf{r}_{OQ}|$, cannot be greater than the sum of the lengths of $\mathbf{r}_{OP}$ and $\mathbf{r}_{PQ}$, and it is this large only when $\mathbf{r}_{OP}$ and $\mathbf{r}_{PQ}$ are parallel. The length of $\mathbf{r}_{OQ}$ is least when $\mathbf{r}_{OP}$ and $\mathbf{r}_{PQ}$ are in opposite directions. Vectors in opposite directions to each other are called *anti-parallel* vectors. For example, if $|\mathbf{r}_{OP}| = 10$ cm and $|\mathbf{r}_{PQ}| = 1$ cm, the maximum length the sum of these two vectors can have is 11 cm and the minimum length is 9 cm, as the reader should verify with sketches. If the two vectors are at right angles, according to the Pythagorean theorem the length of their sum is

$$\begin{aligned} |\mathbf{r}_{OQ}| &= \sqrt{|\mathbf{r}_{OP}|^2 + |\mathbf{r}_{PQ}|^2} \\ &= \sqrt{100 + 1} \\ &= \sqrt{101} = 10.05 \quad \text{(approximately)} \end{aligned}$$

Adding a short vector to a long vector when the two are perpendicular to each other results in a vector that is very little longer than the longer vector. The change in direction is more important in such a case than the change in length. The reader should repeat this example when $|\mathbf{r}_{PQ}| = 0.1$ cm and $|\mathbf{r}_{OP}|$ is still 10 cm to see how small the change in length gets when the short vector gets even shorter.

The rule for vector addition can be easily extended to any number of vectors. If from Q the particle is displaced to R, then to S, and so on, ending up at W, we write

$$\mathbf{r}_{OW} = \mathbf{r}_{OP} + \mathbf{r}_{PQ} + \mathbf{r}_{QR} + \cdots + \mathbf{r}_{VW} \tag{3.3}$$

as an algebraic form of the picture shown in Figure 3.3. The rule for drawing the picture is to place the tail of one vector on the head of the preceding vector, the sum then being the vector from the tail of

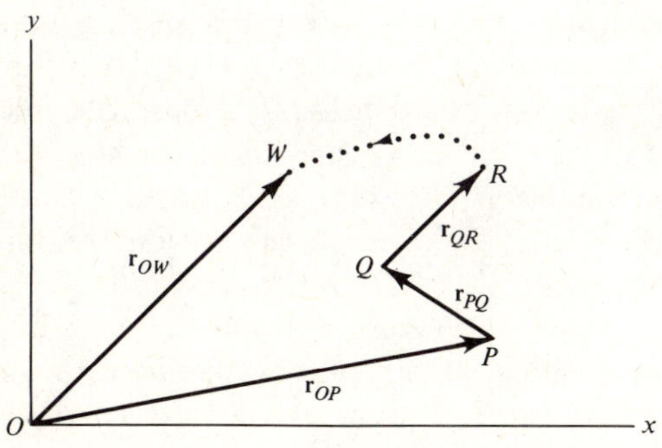

FIGURE 3.3 The sum of the sequence of displacements from O to P, from P to Q, from Q to R, and from R to $\cdots$ to W is the displacement $\mathbf{r}_{OW}$.

the first vector to the head of the last vector. When only two vectors are involved, there is another useful way of constructing the sum. This involves drawing both vectors starting from the same point and completing the parallelogram from the heads of the two vectors, as shown in Figure 3.4. The equivalence of Figures 3.2 and 3.4 follows from the geometrical fact that opposite sides of a parallelogram are equal in both magnitude and direction.

An example of the vector addition of displacements is provided by tracing one's actions in going up two floors on a common kind of stair-

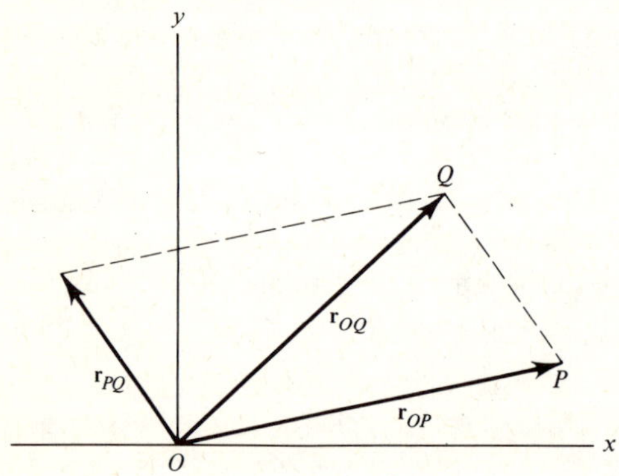

FIGURE 3.4 The displacement $\mathbf{r}_{OQ}$, the sum of $\mathbf{r}_{OP}$ and $\mathbf{r}_{PQ}$, is the diagonal of the parallelogram with the latter two vectors as sides.

case, shown in Figure 3.5. There are alternate upwardly sloping and horizontal displacements that add up to an essentially vertical sum or resultant displacement. A spiral staircase would present a more difficult problem from the vector point of view because there no long straight displacements occur, the climber's motion being along a curved path. We shall see later how this can be done; there is no conceptual trouble here, only a technical one.

The definition of the sum of a set of vectors is easy to visualize, but when the vectors are actually given, it is not obvious how to go about adding them. If all the vectors lie in a plane, one can draw them to scale on paper and evaluate the sum graphically, but this process is awkward and at best is not very accurate. An algebraic method is called for. The simplest algebraic method is one using the components of the vectors being added. Let **V** be any vector as shown in Figure

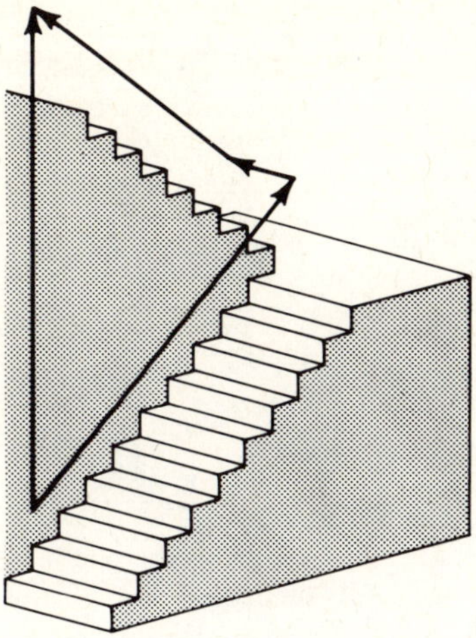

FIGURE 3.5 The almost vertical displacement made by a person climbing two floors on a staircase is composed of upward sloping and horizontal displacements whose horizontal components cancel.

3.6, taken to lie in the xy-plane for simplicity. From the head and tail of **V** we drop perpendicular lines onto the x- and y-axes. The length of the intercept between the perpendiculars that strike the x-axis is the magnitude of the x-component of **V**, V_x, while the length of the intercept of the y-axis between the other pair of perpendiculars is the magnitude of the y-component of **V**, V_y. If the projection of the head of the vector lies in the direction of increasing coordinate values from the projection of the tail, the component is positive; if it lies in the direction of decreasing coordinate values from the projection of the tail, the component is negative. Thus in Figure 3.6, V_x is positive and V_y is negative. Note that neither the magnitude nor the sign of the components of **V** in a given coordinate system depends on the location of **V** but only on the length and direction of **V**.

By dropping perpendiculars on the x-axis from the heads and tails of all the vectors in Figure 3.3, it becomes clear that the x-component of the resultant vector, $\mathbf{r}_{OW}$, is the algebraic sum of the x-components of the vectors on the right-hand side of Eq. 3.3. The same holds for the y-components and for the z-components if these are present. Equa-

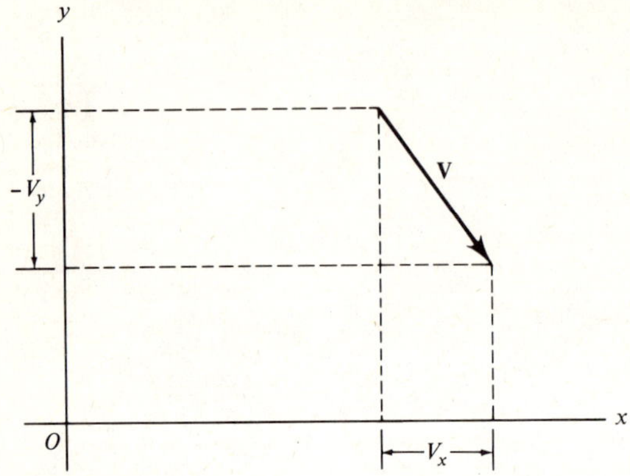

FIGURE 3.6 When perpendiculars are dropped onto the coordinate axes from the head and tail of a vector, the lengths of the intercepts give the magnitude of the components. Here V_x is positive because the vector **V** carries one from smaller to larger x-values; V_y is negative because the vector **V** carries one from larger to smaller y-values.

tion 3.3 can be regarded, then, as a quick way of writing three equations, one for the x-components, a second for the y-components, and a third for the z-components.

$$(r_{OW})_x = (r_{OP})_x + (r_{PQ})_x + (r_{QR})_x + \cdots + (r_{VW})_x \tag{3.4a}$$

$$(r_{OW})_y = (r_{OP})_y + (r_{PQ})_y + (r_{QR})_y + \cdots + (r_{VW})_y \tag{3.4b}$$

$$(r_{OW})_z = (r_{OP})_z + (r_{PQ})_z + (r_{QR})_z + \cdots + (r_{VW})_z \tag{3.4c}$$

It is much more convenient to write the single equation, Eq. 3.3, than the three equations, Eqs. 3.4, and it is much easier to read the single equation. The presence of three equations in Eqs. 3.4 tends to disguise the fact that a single concept is being described. The displacement of a particle from one place to another is a single concept, and the combination of displacements is a geometrical process that can be much more clearly thought of geometrically than by decomposing each displacement into components.

A vector equation can be reduced to equations between components in any coordinate system, and great simplification sometimes can be achieved by introducing an appropriately chosen coordinate system. For

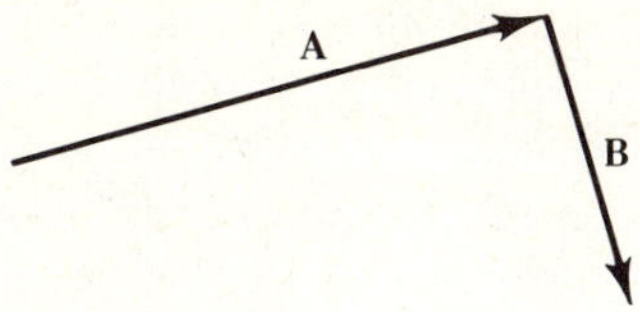

FIGURE 3.7 When two vectors **A** and **B** enter a problem, it is possible to choose the coordinate axes so that both vectors lie in the xy-plane. Both then have zero z-components.

example, suppose one has to add two vectors that are at right angles to each other, **A** and **B** in Figure 3.7. The plane of the two vectors can be chosen to be the xy-plane of a coordinate system. Then A_z and B_z both are zero. If horizontal and vertical x- and y-axes are introduced, then both **A** and **B** have both x- and y-components, as shown in Figure 3.8(a), but if the x-axis is chosen parallel to **A** and the y-axis parallel to **B**, as in Figure 3.8(b), each vector has only one nonvanishing component. This kind of simplification often is made in the later parts of this book. If the two vectors are not at right angles, then the reader should verify by a sketch that either of them can be made to have a single nonzero component while the other must have two.

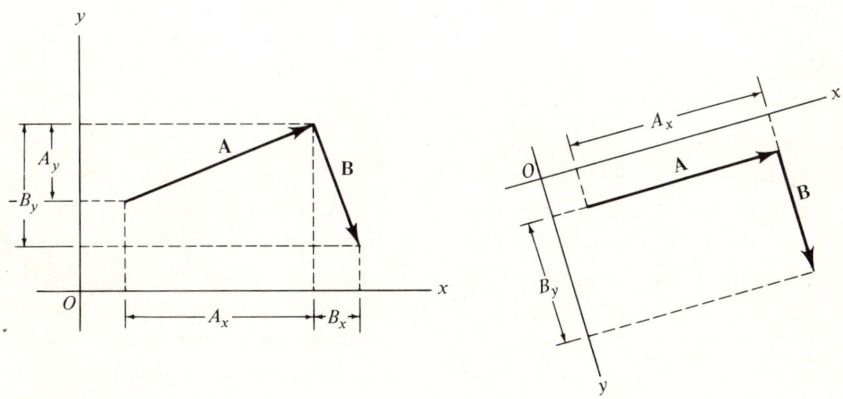

FIGURE 3.8 By proper choice of the directions of the coordinate axes, two vectors at right angles to each other can be made to have only one component apiece.

It is usually advantageous to postpone the introduction of a specific coordinate system as long as possible. At the end of the problem in question it may be evident that some particular coordinate system reduces the number of components that must be considered, making calculations easier and reducing the chance for making errors. The introduction of components reduces the geometry of vectors to ordinary algebra where symbols stand for numbers. Still later, to evaluate algebraic expressions it is necessary to reduce algebra to arithmetic by introducing numerical values for the symbols. This, too, should be put off as long as possible because an algebraic expression has meaning for many values of the symbols it contains while a numerical value has meaning only for one particular situation.

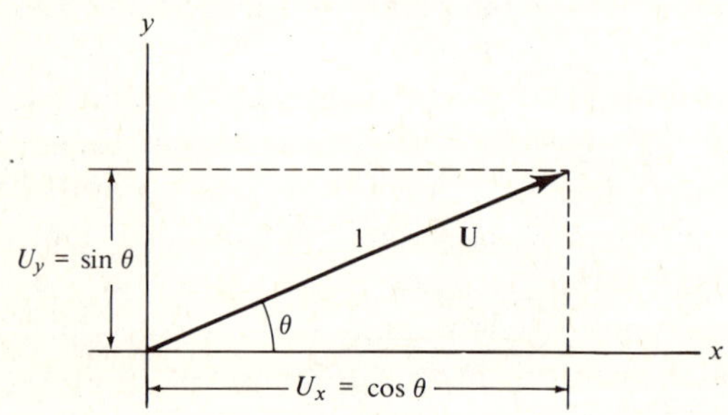

FIGURE 3.9 The x-component of a vector **U** of unit length that makes the angle θ with the x-axis is the cosine of θ by definition. If **U** lies in the xy-plane, then its y-component is the sine of θ by definition.

A method is still needed to find the magnitude of the component of a vector along any given axis. There is no simple way to calculate this, and in fact what is done is to make a table of these values for a particular vector and then the component of any other vector can be found with the help of the table. If **U** is a vector of unit length, $|\mathbf{U}| = 1$, it has a component U_x in the direction of the x-axis, as shown in Figure 3.9. The magnitude of this component depends on the angle θ between **U** and the positive x-axis and is called the *cosine* of the angle θ, written $\cos\theta$ for short. The numerical value of $\cos\theta$ can be computed for various values of θ by a numerical method we do not need to go into here, and the results of such a computation are given in the tables starting on page 420.

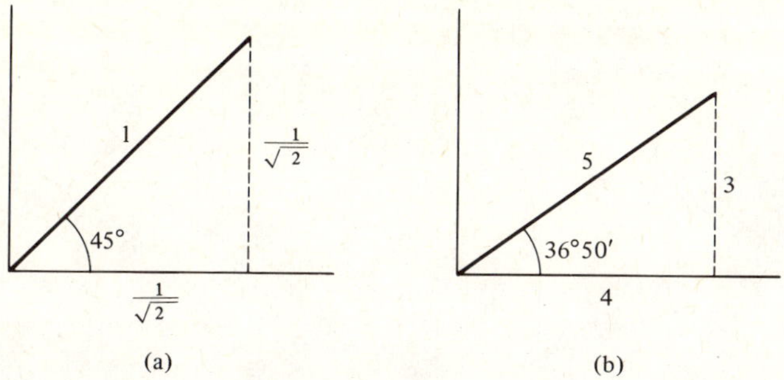

FIGURE 3.11 Two especially simple right triangles are those (a) with two equal angles, necessarily 45°, and (b) with sides whose lengths are in the ratio 3:4:5.

Then for this particular angle

$$\sin \theta = 0.60, \qquad \cos \theta = 0.80$$

According to the table of cosines used in reverse, one can see that the angle whose cosine is 0.80 is about 36°50′.

The idea of components can be used to solve practical problems. Suppose one wants to be able to ski down a trail for 5 miles with a slope of 15° to the horizontal. How high must the start of the trail be above the end? The angle between any displacement along the trail and the vertical is 90° — 15° = 75°. This angle does not depend on how the trail winds but only on its steepness. The total vertical component h is, therefore,

$$\begin{aligned} h &= 5 \text{ miles} \times \cos 75° \\ &= 5 \times 0.2588 \text{ miles} \\ &= 1.294 \text{ miles} \end{aligned}$$

Figure 3.12 shows what happens when **V** does not lie in the xy-plane. Here the x-axis, the y-axis, and the vector **V** itself form the edges of a triangular pyramid; θ is the angle between **V** and the x-axis while φ is the angle between **V** and the y-axis. Inspection of the figure shows that $\theta + \varphi$ is greater than 90°. In this situation it is still true that $V_x = V \cos \theta$ and that $V_y = V \cos \varphi$. What is not true here is that $\cos \varphi$ is equal to $\sin \theta$, although this is true when **V** has zero z-component and lies, therefore, in the plane of the x- and y-axes.

To summarize, the component of **V** in the direction of any axis, say

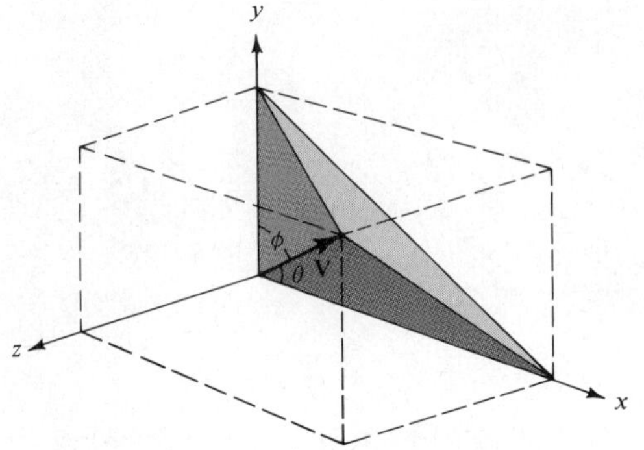

FIGURE 3.12 When a vector **V** sticks out of the xy-plane, the angles θ and ϕ which it makes with the x- and y-axes, respectively, do not add up to 180° but to a larger angle.

the w-axis, is the product of the length of **V** and the cosine of the angle between **V** and the positive direction of the w-axis. If the angle between **V** and the positive w-axis is less than 90°, this component is positive. If the angle is 90°, the component vanishes, and if it is greater than 90°, the component is negative. When it reaches the value 180°, the vector is pointing in the negative w-direction, so that angles greater than 180° never occur in this connection. In Figure 3.13 the component of **V** in the w-direction is shown for various directions of the w-axis.

We shall need one more definition in connection with vectors to make them useful: the definition of the product of a number and a vector.

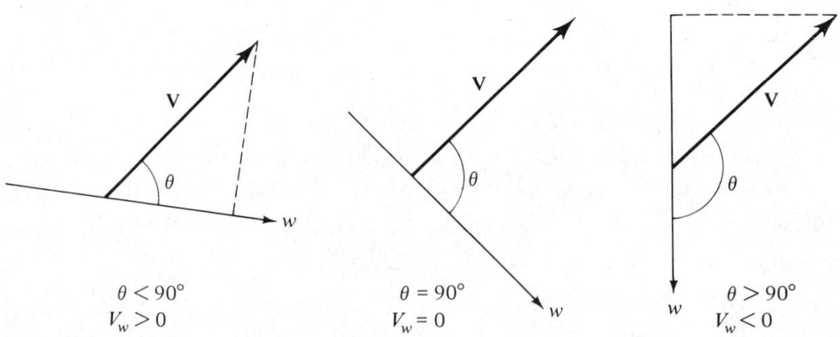

FIGURE 3.13 The sign of the component of a vector **V** in the w-direction depends on the size of the angle θ between **V** and the positive w-axis.

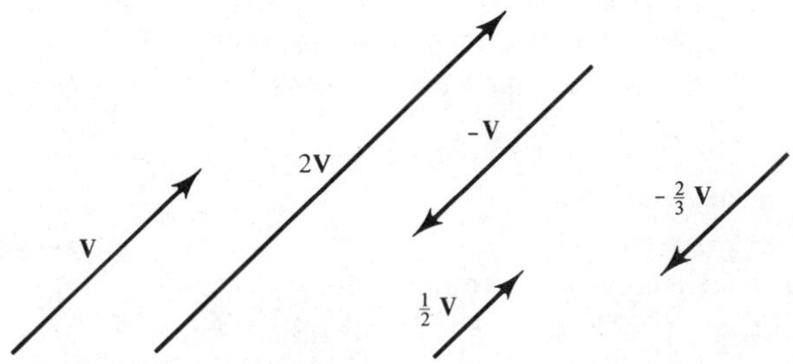

FIGURE 3.14 Several multiples of the vector **V**.

If **V** is a vector and a is a number, the quantity $a\mathbf{V}$ is a vector. The vector $a\mathbf{V}$ is parallel to **V** if a is positive and is antiparallel to **V**, that is, in the direction opposite to that of **V**, if a is negative. The length of the vector $a\mathbf{V}$ is the length of **V** multiplied by the magnitude of a. Some examples are illustrated in Figure 3.14.

If there are two vectors **A** and **B**, we can multiply **B** by —1 and add the result to **A**. This amounts to subtracting **B** from **A**. The corresponding pictures are shown in Figure 3.15. We shall often need to find the difference between two vectors. With the definitions of vector addition and subtraction given, vectors can be transposed from one side of an equation to the other with a change of sign, just as is done in ordinary algebra. If, for example,

$$\mathbf{A} + \mathbf{B} = \mathbf{C} \tag{3.9}$$

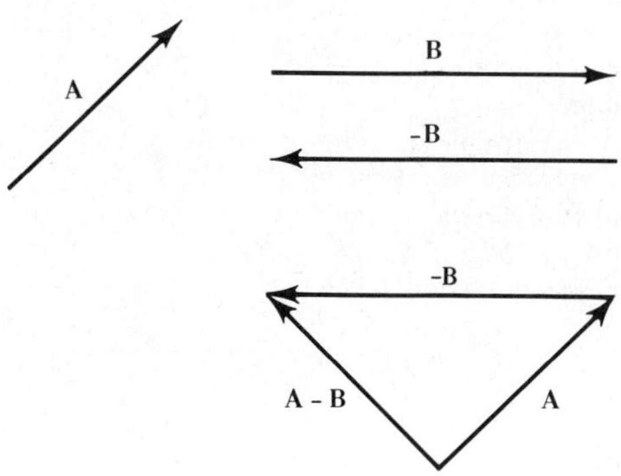

FIGURE 3.15 Steps in forming the difference of two vectors **A** and **B**.

then

$$\mathbf{A} = \mathbf{C} - \mathbf{B} \tag{3.10}$$

One cannot divide by vectors, however, so a vector equation is not altogether like a single algebraic equation.

Physical quantities that are completely specified by their magnitude alone are called *scalars* to distinguish them from vectors. The length of a vector is a scalar. The length of a vector $\mathbf{A}$ is written A, where no confusion results, or as $|\mathbf{A}|$ if the other notation is ambiguous. For example, the length of $\mathbf{A} + \mathbf{B}$ cannot be written as $A + B$ unless $\mathbf{A}$ and $\mathbf{B}$ are parallel but must be written as $|\mathbf{A} + \mathbf{B}|$.

Problems

1. A particle P lies on the x-axis of a coordinate system having coordinates (4, 0, 0). Another particle Q lies on the z-axis of this system, having coordinates (0, 0, 3). How far are these particles from each other? What are the components of the vector from P to Q?
2. Give a general formula for the distance between a point on the x-axis with coordinates $(x, 0, 0)$ and another on the y-axis with coordinates $(0, y, 0)$. Draw a sketch illustrating your solution.
3. A rectangular room is a feet long, b feet wide, and c feet high. Using the a-foot-long wall as the xz-plane and the floor as the xy-plane, find the following in terms of a, b, c:
   (a) The coordinates of the midpoint of the room.
   (b) The distance of the midpoint from the origin.
   (c) The distance of the midpoint from each of the coordinate axes. Show all these quantities on a sketch.
4. What would change if in Problem 3 the xy-plane were taken as the ceiling rather than the floor?
5. A vector $\mathbf{V}$ lies in the xy-plane of a coordinate system, making an angle of 30° with the x-axis. What is its x-component? (Of course, this can be given only in terms of the length V of the vector and not numerically.) Draw the two possible vectors $\mathbf{V}$ with this x-component and find their y-components.
6. Numbering the quadrants as in the figure, identify the quadrants in which
   (a) $\sin\theta > 0$, $\cos\theta > 0$
   (b) $\sin\theta > 0$, $\cos\theta < 0$
   (c) $\tan\theta < 0$, $\cos\theta > 0$
   (d) $\sin\theta < 0$, $\cos\theta < 0$
   (e) $\tan\theta < 0$, $\sin\theta < 0$.

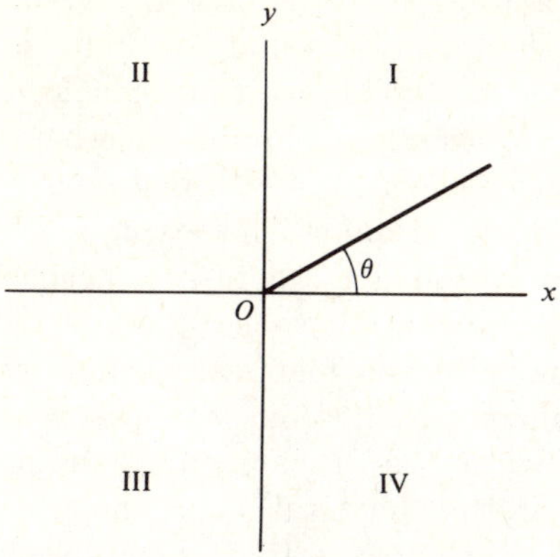

7. An equilateral triangle is divided into two right triangles by drawing a perpendicular from one corner to the opposite side. From this show that $\sin 30^\circ = \frac{1}{2}$, $\sin 60^\circ = \sqrt{3}/2$.
8. Find the components of a vector in the xy-plane 1 unit long such that the x-component is $\frac{3}{4}$ of the y-component, both components being positive.
9. On page 21 the height that a ski trail of slope 15° to the horizontal must have to give it a length of 5 miles was found. How does this change if the trail is to cover a *horizontal* distance of 5 miles instead of 5 miles measured along the slope?

4. VELOCITY AND ACCELERATION

We have learned how to specify the position of a particle and the displacement of a particle from one place to another with the help of vectors. Often we are interested not only in the position of a particle but in its position at a definite time and in how its position is changing with time. Pure geometry is not enough for this because geometry does not involve time. The way the position of a particle depends on the time can be specified by giving its position vector **r** at each value of the time t. We say that **r** is a *function* of t, thereby implying that there is a prescription that enables us to calculate the value of **r** for any value of t. It also is necessary to introduce the idea of the velocity of a particle, the rate at which the particle is changing its position, and the idea of the acceleration of a particle, the rate at which the particle is changing its velocity. We need go no further than this, a fact first clearly understood by Newton that forms the essence of his "second law of motion," discussed in Section 9.

When one quantity $\mathbf{r}$ is a function of another t, so that for any value of the second quantity the value of the first can be found, we indicate this by writing $\mathbf{r}(t)$, which is read as "$\mathbf{r}$ of t." This notation can lead to trouble if it is confused with the use of parentheses in algebra where, for example, $a(b + c)$ means the product of a with the sum of b and c. Usually the context makes it clear which meaning the parentheses have, and we shall try to use the multiplication sign where it is needed to avoid ambiguity, so that we write the above product as $a \times (b + c)$.

A particle cannot just disappear at one point and reappear at another distant point but must follow a continuous path from one place to another. Otherwise it would lose its identity as a particular, unique particle. The mathematical statement of this physical requirement is that $\mathbf{r}$ must depend continuously on t, which means that the position vector of the particle at time $t + \Delta t$ differs from the position vector at time t by less and less as Δt gets smaller and smaller. Here Δt is a time interval, the Greek letter Δ (delta) often being used to indicate a change in the quantity following it. This requirement of continuity is written

$$\mathbf{r}(t + \Delta t) \rightarrow \mathbf{r}(t) \quad \text{as} \quad \Delta t \rightarrow 0 \tag{4.1}$$

The arrow is read "approaches." The continuous path traced out by $\mathbf{r}(t)$ as t varies is called the *trajectory* of the particle.

If at time t the particle being studied is at $\mathbf{r}(t)$, a short time Δt later it will still be near $\mathbf{r}(t)$, and if Δt is very small we would expect the displacement of the particle from $\mathbf{r}(t)$ to be proportional to Δt. Doubling the time interval Δt ought to double the displacement of the particle provided Δt is too short a time for the motion of the particle to change appreciably. This is stated in mathematical language in the following way:

$$\mathbf{r}(t + \Delta t) = \mathbf{r}(t) + \mathbf{v}(t)\,\Delta t \tag{4.2}$$

The left side gives the new position vector of the particle; on the right side are the old position vector $\mathbf{r}(t)$ and the displacement occurring during the short time interval Δt, $\mathbf{v}(t)\,\Delta t$, which is proportional to Δt, as illustrated in Figure 4.1. The vector $\mathbf{v}(t)$ is called the velocity of the particle. Equation 4.2 can be solved to give an explicit expression for $\mathbf{v}(t)$.

$$\begin{aligned} \mathbf{v}(t) &= \frac{\mathbf{r}(t + \Delta t) - \mathbf{r}(t)}{\Delta t} \\ &= \frac{\Delta \mathbf{r}}{\Delta t} \end{aligned} \tag{4.3}$$

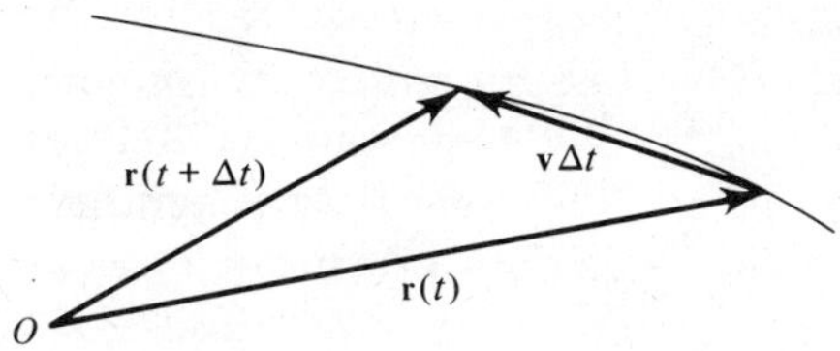

FIGURE 4.1 The change in the position of a moving particle in the short time interval Δt is in the direction of the velocity $\mathbf{v}$ and has the magnitude $v\,\Delta t$.

In Eq. 4.3 the quantities in the numerator and in the denominator both are small but their ratio is generally well-defined and is independent of the size of Δt provided it is small enough. If the particle is at rest, $\mathbf{r}(t + \Delta t) = \mathbf{r}(t)$, and so $\mathbf{v}(t) = 0$.

As an example of the way one can treat small quantities, consider a particle whose x-coordinate is given as a function of the time by

$$x(t) = (1 + t)^2$$

when x is measured in centimeters and t in seconds. Expanding the square by the rules of algebra gives

$$x(t) = 1 + 2t + t^2$$

When $t = 0, x = 1$. When $t = 0.02$, then

$$\begin{aligned} x(0.02) &= 1 + 0.04 + 0.0004 \\ &= 1.0404 \end{aligned}$$

When $t = 0.01$,

$$\begin{aligned} x(0.01) &= 1 + 0.02 + 0.0001 \\ &= 1.0201 \end{aligned}$$

For small values of t, the term t^2 makes only a very small contribution. Compared to the term proportional to t, this contribution becomes smaller as t gets smaller. For sufficiently small t, then, the t^2 term can be neglected and the change in x can be taken as proportional to t. In other words, in this example $\Delta x/t$ hardly depends on t when t is very small compared with unity.

The velocity of a particle is a vector quantity. It has a magnitude and a direction and is obtained by subtracting one vector from another and multiplying the result by a number, $1/\Delta t$, which are operations we have already seen can be applied to vectors to give new vectors. The direction of the velocity at a point on the trajectory is that of the tangent to the trajectory at that point. The length or magnitude of the velocity vector is called the *speed*. Physicists are not very consistent in maintaining the distinction between velocity and speed in everyday speech; the context usually makes the meaning clear. The units of velocity and of speed are those of distance divided by those of time. In this book we usually measure length in centimeters and time in seconds. These units are defined in the next section. The units

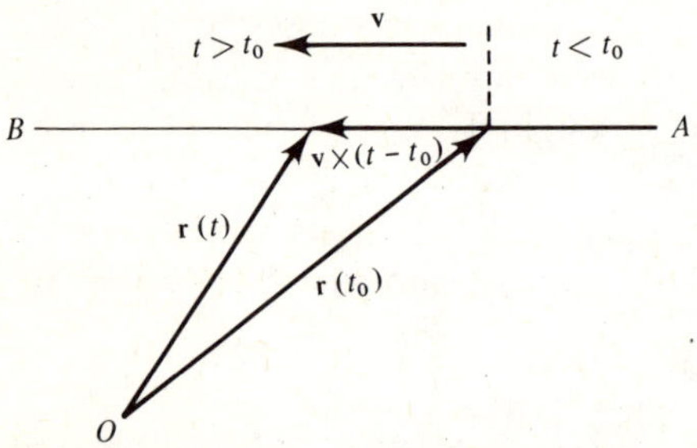

FIGURE 4.2 The change in the position of a uniformly moving particle in the time interval $t - t_0$ is in the direction of the velocity $\mathbf{v}$ and has the magnitude $v \times (t - t_0)$.

of velocity and of speed are then centimeters per second. Other units are often more convenient or are more commonly used. The speed of traffic is measured in miles per hour; the speed of a glacier may be expressed in inches per year.

The simplest motion of a particle is that for which $\mathbf{v}(t)$ is constant in time so that it may be written just as $\mathbf{v}$. In this case the change in position of the particle is proportional to the time interval even when the interval is not short. Equation 4.2 is written in the slightly modified form

$$\mathbf{r}(t) = \mathbf{r}(t_0) + \mathbf{v} \times (t - t_0) \tag{4.4}$$

This says that at $t = t_0$, the particle is at position $\mathbf{r}(t_0)$, and that at time t, its position differs from this by an amount proportional to $t - t_0$. The meaning of this equation is illustrated in Figure 4.2. The particle whose position vector is $\mathbf{r}(t)$ moves uniformly along the line AB and passes the position $\mathbf{r}(t_0)$ at time $t = t_0$. The trajectory of a particle moving with constant velocity is a straight line.

We have spoken of the change of the position vector $\mathbf{r}$ of a particle relative to a set of coordinates as though it were due to a motion of the particle. A change of $\mathbf{r}$ also would result if the particle remained fixed and the coordinate system moved. It is a problem of physics to decide which of the possibilities should be taken seriously in any given case. For many problems it does not matter which is considered to be

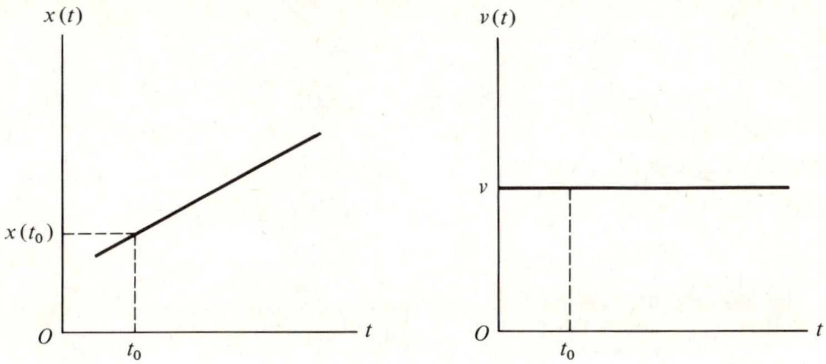

FIGURE 4.3 The distance from the origin of a particle moving uniformly along the x-axis increases linearly with the time (the graph is a straight line), and the speed is constant.

moving as long as the velocity is constant, it being the change in the motion that is of most interest for physics.

In specific problems it is often necessary to consider the components of the vectors entering Eq. 4.4. Because here $\mathbf{v}$ is a constant vector, it is possible to choose the coordinate system in such a way that $\mathbf{v}$ points in the direction of the positive x-axis and therefore has vanishing y- and z-components. The three component equations corresponding to Eq. 4.4 then are

$$x(t) = x(t_0) + v \times (t - t_0) \tag{4.5a}$$

$$y(t) = y(t_0) \tag{4.5b}$$

$$z(t) = z(t_0) \tag{4.5c}$$

Only x changes with time, which is very helpful. The x components of $\mathbf{r}$ and of $\mathbf{v}$ are plotted in Figure 4.3.

Particles in nature do not usually move with constant velocity, so we must study the change in velocity of a particle with time. One would expect the velocity to vary continuously with the time, but the reason is not the same as for the continuous change of position with time. If the position of a particle changed discontinuously, the identity of the particle would be lost. We shall see that it is external influences which cause the velocity of a particle to change, and that a discontinuous change in velocity would imply a catastrophic external influence. It is occasionally useful to use discontinuous changes of velocity as approximations to very rapid changes, especially in studying collisions. For the moment $\mathbf{v}(t)$ is taken to be a continuous function of t.

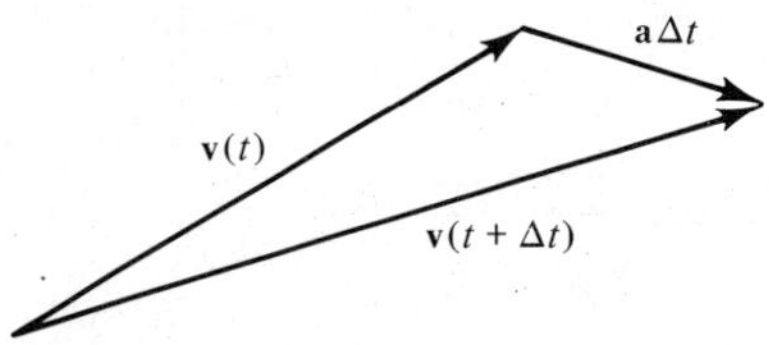

FIGURE 4.4 The change in velocity of a moving particle in the short time interval Δt is in the direction of the acceleration $\mathbf{a}$ and has the magnitude $a\ \Delta t$.

If $\mathbf{v}(t)$ varies continuously with changing t, we may follow the same procedure used in introducing the velocity and write

$$\mathbf{v}(t + \Delta t) = \mathbf{v}(t) + \mathbf{a}(t)\ \Delta t \tag{4.6}$$

as in Figure 4.4. This is exactly of the form of Eq. 4.2, but with different symbols representing different physical quantities. $\mathbf{a}(t)\ \Delta t$ is the change in $\mathbf{v}(t)$ in the short time interval between t and $t + \Delta t$, which for short enough intervals should be proportional to this time interval. When $\mathbf{v}$ is constant, $\mathbf{a}$ vanishes. The vector $\mathbf{a}(t)$ is called the *acceleration* of the particle at the time t. It is change in velocity per unit time and consequently usually will be measured in the units centimeters per second per second or, said another way, centimeters per $(\text{second})^2$. Other units may be used if they are more convenient, but all units must contain a unit of length divided by the product of two units of time. This can be summarized by saying that acceleration has "dimensions" of length per $(\text{time})^2$ or $l/t^2 = lt^{-2}$. From Eq. 4.6 we can get the explicit definition of $\mathbf{a}(t)$. It is

$$\mathbf{a}(t) = \frac{\mathbf{v}(t + \Delta t) - \mathbf{v}(t)}{t}$$

$$= \frac{\Delta \mathbf{v}}{\Delta t} \tag{4.7}$$

Here again the numerator approaches zero at least as fast as Δt does, so the ratio is well-defined as Δt gets arbitrarily small.

It is important to note that while speed is the length of the velocity vector, it is not necessarily the change in the length of the position vector per unit time. In Figure 4.1 it can be seen that the length of $\mathbf{r}(t)$ decreases with time until $\mathbf{r}(t)$ is perpendicular to $\mathbf{v}$ and then increases with time afterward, even though the velocity is constant. In the same way the magnitude of the acceleration vector is not necessarily the rate at which the speed is changing. When a particle moves steadily around in a circle, for example, its speed is constant and the acceleration vector has a constant length, but because neither the velocity vector nor the acceleration vector is constant in direction they are not constant vectors. This is studied in Section 7.

There are several important situations in nature where particles move with constant, or nearly constant, acceleration. If $\mathbf{a}(t)$ does not change with t, we may write it simply as $\mathbf{a}$. Then the change in the velocity is proportional to the time interval even for long intervals so that Eq. 4.6 may be written in the form similar to Eq. 4.4,

$$\mathbf{v}(t) = \mathbf{v}(t_0) + \mathbf{a} \times (t - t_0) \tag{4.8}$$

This says that the velocity at time t differs from that at time t_0 by the amount $\mathbf{a} \times (t - t_0)$, which is proportional to $(t - t_0)$. The x-component of this equation is plotted in Figure 4.5(a).

When a particle moves with constant velocity, its position at time t, $\mathbf{r}(t)$, is related to its position at the time t_0, $\mathbf{r}(t_0)$, by Eq. 4.4. This result cannot be immediately applied to the motion of a particle moving with constant acceleration because in this case the velocity $\mathbf{v}$ depends on the time, and one must determine which value of $\mathbf{v}$ is to be used. It is useful to define the average velocity of the particle between time t_0 and time t by

$$\langle \mathbf{v} \rangle = \frac{\mathbf{r}(t) - \mathbf{r}(t_0)}{t - t_0} \tag{4.9}$$

(Angular brackets are often used to denote averages.) If $t - t_0$ is small,

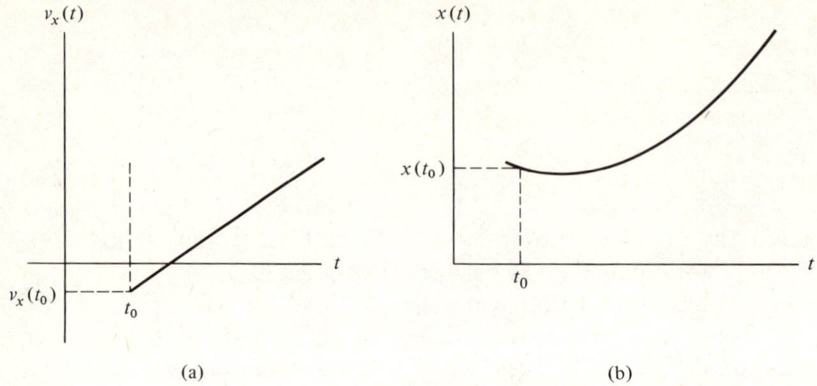

FIGURE 4.5 When the speed of a particle increases with the time, the distance of the particle from the origin is no longer a linear function of the time.

the average velocity is the same as the velocity defined by Eq. 4.3, and if $\mathbf{v}$ is constant, it is its own average. If $\mathbf{v}(t)$ varies in a complicated way, it is difficult to find its average value, but when $\mathbf{a}$ is constant, $\mathbf{v}(t)$ changes uniformly with the time and the average is simply given by

$$\langle \mathbf{v} \rangle = \frac{\mathbf{v}(t) + \mathbf{v}(t_0)}{2} \tag{4.10}$$

This way of averaging is correct only when $\mathbf{v}$ changes uniformly with time, as it does only when the acceleration $\mathbf{a}$ is constant. Using the expression Eq. 4.8 for $\mathbf{v}(t)$ gives the useful result

$$\langle \mathbf{v} \rangle = \mathbf{v}(t_0) + \tfrac{1}{2}\mathbf{a} \times (t - t_0) \tag{4.11}$$

This can now be used in the definition of $\langle \mathbf{v} \rangle$ to give

$$\mathbf{r}(t) = \mathbf{r}(t_0) + \mathbf{v}(t_0) \times (t - t_0) + \tfrac{1}{2}\mathbf{a} \times (t - t_0)^2 \tag{4.12}$$

The x-component of this is plotted in Figure 4.5(b), the value of $v_x(t_0)$ being negative. Because the velocity contains a part proportional to $(t - t_0)$, the position vector contains a part proportional to $(t - t_0)^2$.

The acceleration of a particle in general is not constant. Must we, therefore, go on to study the rate at which the acceleration changes and the rate of change of the rate of change of the acceleration and so on *ad infinitum?* Fortunately, we need go no further than the acceleration because in nature it appears that in a given environment the acceleration of a particle depends on its position and velocity; knowing the position, velocity, and environment of the particle, the acceleration can be found. This is the essence of Newton's laws of motion. Time bombs and alarm clocks obviously do not act in any such simple way. They cannot be considered as particles during a time interval in which they go off because their internal structure instead of their environment then governs their behavior.

Problems

1. A particle moves relative to a set of coordinate axes. At time t_1 it has coordinates (x_1, y_1, z_1), and at time t_2 it has coordinates (x_2, y_2, z_2). Write expressions for the components of its average velocity during the time interval from t_1 to t_2.
2. A particle moves uniformly in a straight line starting from the origin at time $t = 0$. With what speed must it move in order to arrive at the point P with coordinates $x = 200$ cm, $y = 200$ cm, and $z = 500$ cm at time $t = 0.003$ sec?
3. A particle moves uniformly in a straight line lying in the xy-plane. With what speed does it move if $v_x = 5 \times 10^5$ cm/sec and if the angle between the velocity vector and the x-axis is $60°$?
4. The acceleration of gravity at points that are height h above the earth has a magnitude that depends on the height according to Newton's inverse square law (described in Section 10). According to this law, this magnitude can be written as $g(h) = g(0) \times R^2/(R + h)^2$ where R is the radius of the earth, 6.37×10^8 cm, and $g(0)$ is the acceleration of gravity at the earth's surface, 980 cm/sec^2. At a height of 300 km above the surface of the earth, by how much is the value of g reduced below its surface value?

5. UNITS AND NOTATION

The position of a particle relative to a set of coordinate axes is specified by three numbers, the three coordinates of the particle. Each of these numbers is the measure of a distance and has a meaning only when the unit of length has been defined. To know that a certain distance is 10 cubits without knowing what a cubit is, is to know nothing. The selection of a standard of length is not a trivial matter. It must be as precise as any measurement of length that can be made, and there must be good reasons for believing that it does not vary with time but remains the same over decades and centuries.

The unit of length used in scientific work everywhere is the meter, or its submultiple the centimeter. The meter is the everyday unit of length in most Western countries. Great Britain is to be completely on the metric system by 1975, and soon only the United States, Liberia, and some smaller British Commonwealth countries will still be using feet and inches. Until 1960 the meter was defined to be the distance between two lines on a particular platinum alloy bar when the bar is at a temperature of 0°C (the melting point of ice) and is supported in a specified way on two rollers so that the bar is in the same state of flexure each time it is used. This definition is accurate only to one part in 10 million because the rulings are not of zero width and cannot be located more accurately than about one 10-millionth of a meter. For most purposes this accuracy is more than ample, but there are some situations where it is not good enough. For this reason a new definition of the meter was adopted in 1960.

The new definition of the meter is given in terms of the wavelength of the orange light emitted by an isotope of the noble gas krypton when in a certain kind of lamp. The reasons for believing that light is a wave phenomenon, what the wavelength of a wave is, and why atoms of the elements emit light of definite wavelength are discussed to some extent later on. These matters are brought up here only because of the legal definition of the meter. If λ is the wavelength of the orange light emitted by the isotope Kr^{86} in a standard lamp, then

$$1 \text{ m} = 1{,}650{,}763.73\lambda \tag{5.1}$$

This equation is exact because it is the definition of the meter. Anyone anyplace can set up a standard krypton lamp and use the observed wavelength of the orange light as a basis for finding out how long a meter is. If there are scientists on Mars or Venus, lengths measured by them could be communicated to us on earth in meters because there is good evidence that the krypton on Mars or Venus is the same as krypton here.

The unit of length is defined by an atomic standard. Until 1967 the unit of time was defined by astronomical means. The second used to be $\frac{1}{86,400}$ of a mean solar day, the period from midnight to midnight averaged over a year $(86{,}400 = 24 \times 60 \times 60)$. The rotation of the earth is not exactly uniform, so this standard became inadequate. The ephemeris second is now defined by the following relation:

$$\text{The tropical year 1900} = 31{,}556{,}925.9747 \text{ sec} \tag{5.2}$$

This is a most peculiar-sounding definition, because how can one go back to the year 1900 to check a clock? Obviously this cannot literally be done, but enough is known about how the motion of the earth is affected by the moon, the other planets, and other perturbations so that good theoretical formulas connecting the length of any given year with the year 1900 are available if needed. Since 1967 the ephemeris second has been replaced as the fundamental unit of time by one defined by an atomic standard. By action of the General Conference on Weights and Measures, the second is now defined by the relation

1 sec = 9,192,631,770 periods in the vibration of the cesium atom Cs^{133} oscillating in a specified way

This definition was chosen, of course, to make the new definition agree with the old one as accurately as the uncertainty in the old definition allows.

The centimeter is a visible unit of length, and the second is a unit of time that can be sensed directly, so for many purposes these units are of a convenient size. Many lengths encountered in physics are enormously larger or smaller than the centimeter, so the expression of these lengths in units of centimeters requires the use of very large or very small numbers. For example, the distance from the earth to the sun is about 14,900,000,000,000 cm, and the radius of an electron is about 0.000000000000280 cm. Both these numbers are awkward as written. A notation is needed that will make these numbers and others like them easy to write and easy to comprehend. The difficulty with these numbers lies in the large number of zeros they contain. It is time-consuming to count them, and all one learns by counting them is the location of the decimal point. This difficulty is overcome by writing the earth-sun distance, for example, as 1.49×10^{13} cm. The number 1.49 is readily understandable, and the following factor, 10^{13}, tells us that the number is large; the decimal point should be 13 places to the right of where it is written. A positive exponent of the 10 indicates that the decimal point is to be shifted to the right. A negative exponent indicates that it should be shifted to the left. The radius of the electron is therefore written as 2.80×10^{-13} cm. If the decimal point is moved to the left 13 places, 12 zeros will appear in front of the two, as do appear above. This number could equally well be written 0.280×10^{-12} or 28.0×10^{-14}; the essential point is to keep the first factor of an understandable size. The exponential notation is equally effective for large or for small numbers.

The number of digits written in the first factor of a number like 2.80×10^{-13} depends on the accuracy with which the number is known. Writing 2.80 is usually taken to mean that the number is known to lie between 2.795 and 2.805, which also can be written explicitly as 2.80 ± 0.005. If the number is known much more accurately than this, more digits are used, as 2.8037. If it is only slightly better known, not enough to justify more digits, the possible range of values is explicitly written as, for example, 2.800 ± 0.001. The number of digits used to give the known value to full accuracy is called the number of significant figures in the result.

When two numbers given in exponential notation are to be added or subtracted, the exponents must be made equal. Thus

$$(2.80 \times 10^{-13}) + (3.76 \times 10^{-11}) = (0.0280 + 3.76) \times 10^{-11}$$
$$= 3.79 \times 10^{-11}$$

Why 3.79 and not 3.7880? Because 3.76 means 3.760 ± 0.005, so the answer is not accurate beyond the second decimal place. The three-figure number closest to 3.7880 is 3.79, so this is the correct way to write the answer. A more extreme example of this can be given. It is said that there are 80 million motor vehicles in the United States. Suppose you have just demolished your car. Should you then say that there are 79,999,999 motor vehicles in the United States? Obviously not, since the original figure was not accurate within hundreds of thousands or millions to start with.

When two numbers given in exponential notation are to be multiplied, the decimal factors are multiplied and the exponents are added. Thus

$$(2.80 \times 10^{-13}) \times (3.76 \times 10^{-11}) = (2.80 \times 3.76) \times 10^{-13-11}$$
$$= 10.53 \times 10^{-24}$$

Why 10.53? Again because we do not want to introduce spurious accuracy into the answer. 2.80 is not known more accurately than about five parts in 3000 since a ± 0.005 is understood, and 3.76 is not known more accurately than about five parts in 4000. If we used the ordinary product, $2.80 \times 3.76 = 10.5280$, it would seem that the result is accurate to within ± 0.00005, or five parts in 100,000, which is clearly not true. The number is rounded off to 10.53 to keep approximate track of the accuracy of the result.

When two numbers given in exponential notation are to be divided, the decimal parts are divided as usual and the exponent of the divisor is subtracted from that of the dividend

$$\frac{3.76 \times 10^{-11}}{2.80 \times 10^{-13}} = \frac{3.76}{2.80} \times 10^{-11+13} = 1.34 \times 10^{2}$$

Here, too, the number of figures kept in the answer is just enough to keep approximate track of the accuracy of the result.

Some numbers such as π, the number giving the ratio of the circumference to the diameter of a circle, are not determined by measurement but are defined in purely mathematical terms. They can be calculated to any accuracy desired, even though an exact numerical value cannot be given because it would take an infinitely long number to do this. When an experimentally determined number has to be combined with a mathematically defined number, a value of the latter that is at least as accurate as the experimental number must be used so as not to throw away any of the information contained in the experimental number.

Only the units of length and time have been introduced here. Other units will be needed later as we go further in the development of physics. The unit of mass to be introduced is the *gram*, so the entire unit system is called the *centimeter-gram-second* or *cgs system*. The only other common scientific unit system is based on the meter, kilogram, and second, therefore being called the *mks system*.

Problems

1. The American inch is defined by the relation

   1 m = 39.37 in.

   From this derive the conversion factor w connecting miles with kilometers: 1 mile = w km.
2. Explain to yourself or a friend why when two numbers written in exponential notation are multiplied the exponents are added and when one is divided by the other one exponent is subtracted from the other.
3. Evaluate the following, assuming the numbers represent the results of measurements with the usual convention on accuracy, that is, keeping an appropriate number of figures:

   (a) $(1.49 \times 10^{3}) \times (2.30 \times 10^{-7})$

   (b) $(1.44 \times 10^{3}) \times (2.3021638 \times 10^{-7})$

   (c) $(1.44 \times 10^{3}) + (1.44 \times 10^{4})$

   (d) $(1.44 \times 10^{3}) - (2.30 \times 10^{-7})$

6. MOTION WITH CONSTANT ACCELERATION

Motion with constant velocity is extremely simple: The particle moves uniformly along a straight line. While this is a very important kind of motion, it alone does not suffice to describe all situations of interest. External circumstances often cause the velocity of a particle to change with time. The simplest case where the velocity of a particle changes with time is that in which the acceleration is constant, just as the simplest case where the position of a particle changes with time is that in which the velocity is constant. Motion with constant velocity is necessarily a "one-dimensional" motion because it is motion along a straight line. Motion with constant acceleration may be one-dimensional, that is, along a straight line. It will be if the velocity vector is parallel or antiparallel to the acceleration vector to start with. The most general motion with constant acceleration takes place, as we shall see, in the fixed plane containing the acceleration vector and the initial velocity vector. It is therefore a two-dimensional motion. This section treats in detail several cases of motion with constant acceleration.

The commonest example of motion with nearly constant acceleration is the motion of a particle in a small region near the surface of the earth. As is well known, the acceleration produced by gravity is directed toward the center of the earth and is inversely proportional to the square of the distance from the center of the earth. If the motion is over only a small region near the earth's surface, both the change in direction and the change in magnitude of the acceleration may be neglected. When satellite motion is discussed, the variation in magnitude of the acceleration of gravity may not be important but the variation in direction most certainly is. Satellite motion is described in Section 10; here only motion in a limited region is considered. Even with this restriction it is necessary to avoid other complicating features, the principal one of which is friction due to the presence of air. For dense objects that do not move too fast this friction is not very important, but for large bodies like parachutes it is all important. In vacuum this influence is absent.

Another important example of the motion of a particle with constant acceleration that is studied in detail later is that of a charged particle in a uniform electric field.

The motion of particles under the influence of gravity alone has one characteristic that differentiates it from any other kind of motion observed in nature. The acceleration of a particle due to gravity at any point in space is independent of the size and composition of the particle. The acceleration of particles at a given point in space produced by other influences such as electricity or magnetism depends strongly on the kind of particle involved.

In a small region near the surface of the earth, the effect of gravity on particles can be completely described by giving the constant vector **g**, the acceleration of gravity whose magnitude is g. The value of g varies slightly from place to place on the surface of the earth. For our purposes, here it can be taken to have the value

$$g = 980 \text{ cm/sec}^2 \quad \text{or} \quad 32 \text{ ft/sec}^2 \tag{6.1}$$

even though these are not quite consistent with each other, 32 feet being about 975 centimeters. The value given in metric units is the more accurate.

To describe the motion of particles under the influence of gravity in a restricted region, it is, of course, convenient to choose the coordinate system so that **g** lies along a coordinate axis and has, therefore, only one nonzero component. It is customary to draw the y-axis vertically upward, so that in such a coordinate system **g** points along the negative y-axis and $g_y = -g$. This inconvenience can be avoided only by exchanging it for the inconvenience of measuring y-coordinates downward so that lower points have larger y-values, which seems to be even less desirable than a negative value of g_y. A coordinate system is chosen, then, in which the acceleration **g** has the components

$$g_x = 0, \qquad g_y = -g, \qquad g_z = 0 \tag{6.2}$$

This fixes the direction of the y-axis. The direction of the x-axis is still free to be chosen at will as long as it is in a horizontal plane. The simplest choice is one that makes the initial velocity $\mathbf{v}(t_0)$ have at most an x- and a y-component, so that $v_z(t_0)$ is zero and $v_x(t_0)$ is positive if it, too, does not vanish. Having fixed the directions of the x- and y-axes in this way, the direction of the z-axis also is determined.

In the coordinate system just described, the general vector equation for the position vector of a particle moving with constant acceleration given in Eq. 4.11 takes the form

$$x(t) = x(t_0) + v_x(t_0) \times (t - t_0) \tag{6.3}$$

$$y(t) = y(t_0) + v_y(t_0) \times (t - t_0) - \tfrac{1}{2}g \times (t - t_0)^2 \tag{6.4}$$

$$z(t) = z(t_0) \tag{6.5}$$

The negative sign of g in Eq. 6.4 results, as mentioned earlier, from the choice of the positive direction of y as upward. The motion is two-dimensional because the z-coordinate of the particle remains constant. In the special case when $v_x(t_0)$ vanishes, the particle's x-coordinate is also constant and the motion is one-dimensional.

We apply these equations to several special situations of importance. Each situation is described by the same equations, the only differences being in the initial values of position and velocity of the particle, $\mathbf{r}(t_0)$ and $\mathbf{v}(t_0)$.

(1) A particle is dropped from rest at the origin Because the particle is at the origin at time t_0, $\mathbf{r}(t_0) = 0$, and because it is at rest at this time, $\mathbf{v}(t_0) = 0$. The equations then become

$$x(t) = 0$$

$$y(t) = -\tfrac{1}{2}g \times (t - t_0)^2 \tag{6.6}$$

$$z(t) = 0$$

The particle remains on the y-axis and has only negative values of y, an obvious result since a dropped particle does not rise. The distance through which the particle falls increases with the square of the elapsed time. During the first second it falls 490 cm; during the first 2 sec it falls 1960 cm. The speed of the particle increases uniformly. Equation 4.8 yields

$$\mathbf{v}(t) = \mathbf{v}(t_0) + \mathbf{g} \times (t - t_0) \tag{6.7}$$

whose y-component is, with $\mathbf{v}(t_0) = 0$,

$$v_y(t) = -g \times (t - t_0) \tag{6.8}$$

At the end of 1 sec of fall the particle has the speed 980 cm/sec, at the end of 2 sec the speed is 1960 cm/sec, and so on.

For actual objects dropped through air, Eq. 6.8 does not hold for very large values of $(t - t_0)$. There is friction between the object and the air that increases with increasing speed. Eventually the friction is so great that it prevents any further acceleration. This effect is used in an important experiment to measure the charge on an electron, an experiment described in Section 19, and is the effect that makes parachutes work.

(2) A particle is thrown vertically upward from the origin As in (1), because the particle is at the origin at the initial time, $\mathbf{r}(t_0) = 0$. Now,

however, the initial value of $\mathbf{v}$ does not vanish but is given by

$$v_x(t_0) = 0$$

$$v_y(t_0) = v_0, \qquad v_0 > 0 \tag{6.9}$$

$$v_z(t_0) = 0$$

Again the particle will move along the y-axis, the equations describing its motion becoming

$$x(t) = 0$$

$$y(t) = v_0 \times (t - t_0) - \tfrac{1}{2}g \times (t - t_0)^2 \tag{6.10}$$

$$z(t) = 0$$

The y-component equation is plotted in Figure 6.1. In order to understand what is going on here, it is useful to write out the velocity equation, Eq. 6.7, for this case. It is

$$v_x(t) = v_z(t) = 0$$

$$v_y(t) = v_0 - g \times (t - t_0) \tag{6.11}$$

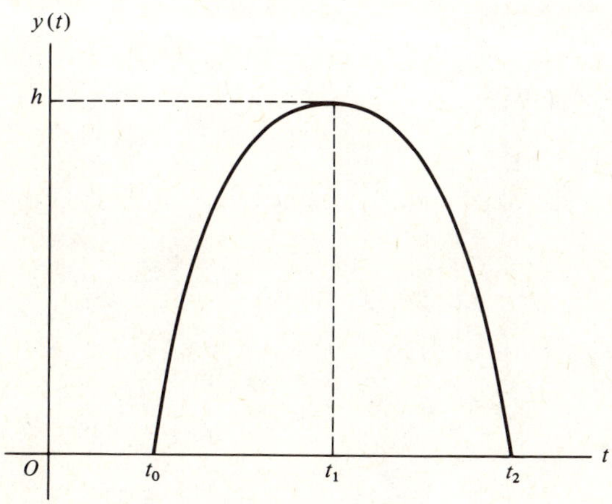

FIGURE 6.1 The vertical motion of a particle moving in the gravitational field near the surface of the earth is symmetrical about the time at which it reaches the highest point of its path. The curve followed is a parabola.

the last of which is plotted in Figure 6.2. This shows that $v_y(t)$ starts out positive and remains positive until it vanishes, which occurs at a time t_1 given by

$$t_1 - t_0 = \frac{v_0}{g} \tag{6.12}$$

At this time the particle has reached its greatest height, as it has stopped going up and has not started to go down. Denoting this greatest value

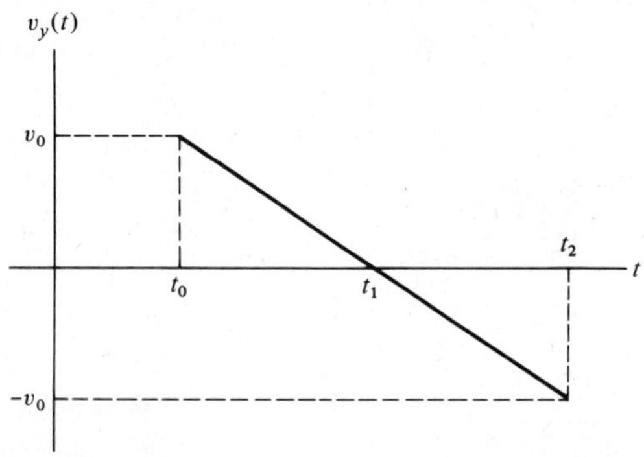

FIGURE 6.2 The y-component of the velocity of a particle moving under the influence of gravity changes uniformly with the time, decreasing and eventually becoming negative.

of $y(t)$ by h, the value of h is obtained by using this time in Eq. 6.10.

$$\begin{aligned} h &= v_0 \times \frac{v_0}{g} - \tfrac{1}{2}g \times \left(\frac{v_0}{g}\right)^2 \\ &= \frac{v_0^2}{2g} \end{aligned} \tag{6.13}$$

The height to which the particle rises increases as the square of the initial upward speed; doubling this speed quadruples the height the particle reaches.

The way $y(t)$ and $v_y(t)$ depend on t has been found. It is possible to find the value of v_y for each value of y. This is done by solving

Eq. 6.11 for $(t - t_0)$ and inserting the result in Eq. 6.10. From Eq. 6.11,

$$t - t_0 = \frac{v_0 - v_y}{g}$$

so that Eq. 6.10 becomes

$$y = v_0 \times \frac{v_0 - v_y}{g} - \tfrac{1}{2}g \times \frac{(v_0 - v_y)^2}{g^2}$$

$$= \frac{v_0^2 - v_y^2}{2g} \tag{6.14}$$

Solving this equation for v_y leads to

$$v_y^2 = v_0^2 - 2gy$$

$$v_y = \pm\sqrt{v_0^2 - 2gy} \tag{6.15}$$

For each value of y there are two values of v_y that are equal in magnitude but opposite in sign. The positive value of v_y occurs when the particle is on its way up, and the negative value occurs when it is on its way down. The particle returns to the origin with its original speed, but going in the opposite direction. Note that y never gets large enough to make the quantity in the square root negative, the maximum value of y just making it vanish. This is necessary because all physical quantities must have real numbers as values, and the square root of a negative number is not real but is what is called an "imaginary" number. We shall not need to consider such numbers.

The time for the particle to reach the top of its motion is given by Eq. 6.12. It is easily seen that it returns to the origin at a time t_2 given by

$$t_2 - t_0 = \frac{2v_0}{g} \tag{6.16}$$

the trip down taking the same time as the trip up. This result is obtained by equating $y(t)$ to zero, which means putting the particle at the origin at time t. From Eq. 6.10 this gives

$$0 = v_0 \times (t - t_0) - \tfrac{1}{2}g \times (t - t_0)^2$$

which is a quadratic equation for $t - t_0$. The two solutions are

$$t - t_0 = 0 \quad \text{or} \quad \frac{2v_0}{g} \tag{6.17}$$

The first solution corresponds to the initial situation; at time $t = t_0$, the particle is at the origin. The second solution corresponds to the return of the particle to the origin.

(3) A particle is thrown obliquely upward from the origin We now consider a particle that at time t_0 is at the origin and has a speed v_0 in a direction other than the vertical. Let the initial velocity $\mathbf{v}(t_0)$ have components

$$\begin{aligned} v_x(t_0) &= v_0 \cos\theta \\ v_y(t_0) &= v_0 \sin\theta \\ v_z(t_0) &= 0 \end{aligned} \tag{6.18}$$

so that θ is the angle between the initial velocity and the x-axis. What is the resulting motion? Again we use Eqs. 6.3, 6.4, and 6.5 with $\mathbf{r}(t_0) = 0$ since the particle is projected from the origin, obtaining

$$\begin{aligned} x(t) &= v_0 \cos\theta \times (t - t_0) \\ y(t) &= v_0 \sin\theta \times (t - t_0) - \tfrac{1}{2}g \times (t - t_0)^2 \\ z(t) &= 0 \end{aligned} \tag{6.19}$$

The x-equation is just that for motion with constant speed in the x-direction. The y-equation is just like Eq. 6.10 except that $v_0 \sin\theta$ replaces v_0, as it should since $v_0 \sin\theta$ is now the initial value of v_y. The trivial form of the z-equation results from our matching the coordinate system to the conditions of the problem.

The only new feature here is the necessity of correlating the motions in the x- and the y-directions. According to Eq. 6.19, the x-coordinate of the particle is proportional to $t - t_0$, so if y is plotted as a function of x instead of t, the resulting curve will be similar to the plot of y against t shown in Figure 6.1, except that the location of the origin is changed, as are the units along the horizontal axis. The result is shown in Figure 6.3. The trajectory of the particle is a parabola, specified by the equation

$$y = -\left(\frac{g}{2v_0^2} \cos^2\theta\right) x^2 + (\tan\theta)x \tag{6.20}$$

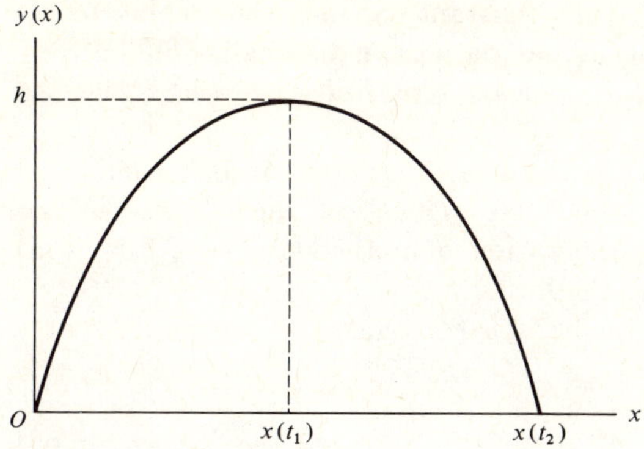

FIGURE 6.3 The graph of the vertical position of a particle moving in the gravitational field of the earth as it depends on the horizontal position is very similar to the graph as it depends on time, because the horizontal motion is uniform in time.

The total time of flight of the particle is the time it takes the particle to reach the highest point of its trajectory and return to its starting level, a time given by Eq. 6.16 modified to allow for the new value of $v_y(t_0)$. It is

$$t_2 - t_0 = \frac{2v_0 \sin \theta}{g} \tag{6.21}$$

During the time interval $t_2 - t_0$ the particle will have traveled a distance in the x-direction given by

$$\begin{aligned} x(t_2) &= v_0 \cos \theta \times (t_2 - t_0) \\ &= 2v_0^2 \frac{\sin \theta \cos \theta}{g} \\ &= \frac{v_0^2 \sin 2\theta}{g} \end{aligned} \tag{6.22}$$

In the last step, the double-angle formula $\sin 2\theta = 2 \sin \theta \cos \theta$ has been used. It will not be used again; it merely gives a simpler form to the quantity in the line above. $x(t_2)$ is the range of the projectile. From the last of Eqs. 6.22 it can be seen that this is a maximum when $2\theta = 90°$ or $\theta = 45°$. This, as all the other results given here, is valid only when the acceleration is truly constant so that friction due to air or other causes is absent. The actual path of a baseball, golf ball, or water from a hose differs from the parabolic path described here because of the presence of friction.

The motion of a particle with constant acceleration will be encountered several times later on in this book, so we collect the results derived here and write them for an acceleration **a** in the positive y-direction; **g** is therefore replaced by —**a** in the formulas.

The horizontal component of velocity is constant.

$$v_x(t) = v_x(t_0)$$

$$x(t) = x(t_0) + v_x \times (t - t_0) \qquad (6.23)$$

The vertical component of velocity varies according to

$$v_y(t) = v_y(t_0) + a \times (t - t_0) \qquad (6.24)$$

The vertical coordinate y is given by

$$y(t) = y(t_0) + v_y(t_0) \times (t - t_0) + \tfrac{1}{2}a \times (t - t_0)^2 \qquad (6.25)$$

The vertical coordinate and vertical velocity are related by

$$v_y^2(t) = v_y^2(t_0) + 2a \times [y(t) - y(t_0)] \qquad (6.26)$$

which comes from Eq. 6.14 with y being replaced by $y(t) - y(t_0)$ because here it is not assumed that $y(t_0)$ is zero.

In the discussion of various motions with constant acceleration, we have considered only times later than the initial time t_0. If the acceleration of the particle is constant for earlier times too, then the formulas given above are also valid for times earlier than t_0. The verbal description of the motion tends to imply that before time $t = t_0$ the physical situation is different from that after this time. Item (1) above is entitled, "A particle is dropped from rest at the origin." The picture conjured up by this is that something is holding the particle at the origin for some time before t_0, and at t_0 this holding process ceases and the particle falls. This is not necessarily the case. Study of the discussion will show that all that matters is the set of initial values of **r** and of **v**. The initial values used in (1) could be achieved as described above but also could be achieved by throwing the particle vertically upward from below the origin at a time earlier than t_0, making the throw so that the particle's trajectory has the origin as its highest point and so that it reaches this point at the time t_0. If the initial values are achieved in this latter way, the formulas describing the motion after time t_0

also describe it before t_0, back to the time that the particle was thrown. In any case, the method of achieving the initial conditions is immaterial if one is interested only in times later than the initial time, and this is usually so.

Problems

1. A stone dropped from rest at the top of a well strikes the water after a time t. Write an expression for the depth of the well, neglecting the time taken for the sound to travel up the well. Evaluate it for $t = 4$ sec. If the well is this deep, find the time it takes sound traveling with speed 3×10^4 cm/sec to get from the surface of the water to the top of the well. Does this mean that one makes an appreciable error if one measures the time of fall by measuring the time between releasing the stone and hearing the stone strike the water?
2. A leaky faucet drips three times a second. At the instant one drop breaks off, how far has the previous drop fallen? If the total fall is 4 ft, how high is the latter drop when the previous drop strikes the bottom? What is the time interval between landings of successive drops?
3. Water from a fountain goes a distance of 750 cm straight up from the nozzle. What is the speed of the water at the nozzle? Neglect air friction.
4. A fire hose can squirt water straight up to a height of 200 ft. What is the farthest it can squirt across a level surface? Neglect air friction.

7. UNIFORM CIRCULAR MOTION

In the previous section we studied the kinds of motion that could result from various initial conditions when the acceleration of the particle was taken to be constant. In this section we reverse the procedure and find what acceleration is needed to produce a specified motion, motion with constant speed in a circle. That this uniform circular motion does imply the existence of an acceleration is clear, because in the absence of acceleration the velocity is constant and motion with constant velocity is uniform straight-line motion. Any motion other than uniform motion in a straight line is accelerated motion.

All motions with constant speed share one characteristic: If the acceleration is not zero, it is perpendicular to the particle velocity. This is easily seen on writing

$$\mathbf{a}(t)\ \Delta t = \mathbf{v}(t + \Delta t) - \mathbf{v}(t) \tag{7.1}$$

which is illustrated in Figure 7.1. The constancy of the speed means that the length of the velocity $\mathbf{v}(t + \Delta t)$ is the same as the length of $\mathbf{v}(t)$, so the triangle in the figure is isosceles. The angles φ at the two ends of $\mathbf{a}\,\Delta t$ are therefore equal. The sum of the interior angles of a triangle is 180°, so

$$2\varphi + \Delta\theta = 180° \tag{7.2}$$

As Δt gets smaller and smaller, the direction of $\mathbf{v}(t + \Delta t)$ approaches the direction of $\mathbf{v}(t)$ and $\Delta\theta$ approaches zero. Therefore, for vanishingly

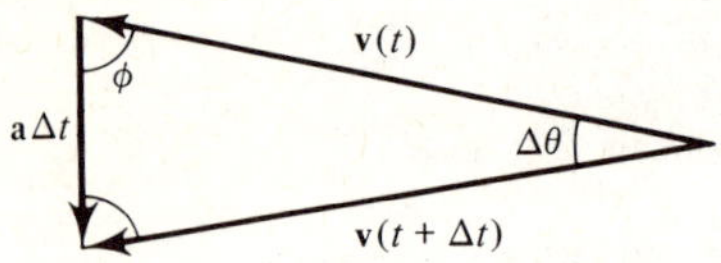

FIGURE 7.1 When the speed of a particle is constant, the velocity triangle is isosceles and the equal angles at the base approach right angles as the time difference Δt becomes smaller.

small values of Δt,

$$2\varphi = 180°$$

$$\varphi = 90° \tag{7.3}$$

This establishes the result. A component of $\mathbf{a}$ parallel to $\mathbf{v}$ will change the speed of the particle.

The acceleration of a particle moving uniformly in a circle is perpendicular to the velocity of the particle. The velocity is tangent to the circle, so the acceleration is along the radius of the circle. The remaining problem is to find the magnitude of the acceleration and in which direction along the radius it points.

Let the circular path of the particle be as shown in Figure 7.2(a), with the origin of $\mathbf{r}(t)$ at the center of the circle so the radius of the circle is r. Part (b) of the figure is similar to Figure 7.1. The angles $\Delta\theta$ in (a) and (b) are equal because a rotation that puts $\mathbf{r}(t)$ on top of $\mathbf{r}(t + \Delta t)$ would also put $\mathbf{v}(t)$ on top of $\mathbf{v}(t + \Delta t)$.

The radian measure of an angle θ is by definition the ratio of the arc length intercepted between the arms of the angle to the radius of

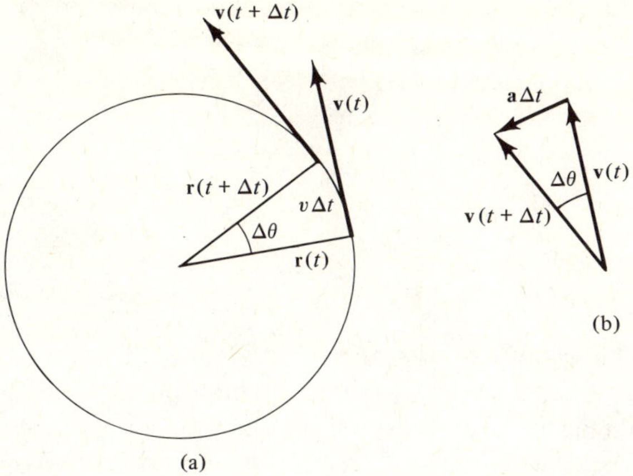

FIGURE 7.2 The angle between the two position vectors in (a) is the same as the angle between the two velocities in (b). For uniform circular motion the magnitudes of r and of v are constant, so two similar isosceles triangles are formed.

the arc. The connection between radian measure and degree measure of angles is established by considering an angle of 180°. Here the arc length between the arms of the angle is the length of a semicircle of radius r, namely πr, and the radius is r, so that

$$\begin{aligned} 180^\circ &= \pi \text{ rad} \\ &= 3.14159\ldots \text{ rad} \end{aligned} \tag{7.4}$$

Radian measure is more useful in physics and geometry than degree measure, but most tables of sines, cosines, and so on are given in degree measure, so one must be able to convert from one to the other.

The length and direction of the acceleration $\mathbf{a}(t)$ are now found by equating the angles $\Delta\theta$ in parts (a) and (b) of Figure 7.2 using radian measure. From (a),

$$\Delta\theta = \frac{v\ \Delta t}{r}$$

and from (b),

$$\Delta\theta = \frac{a\ \Delta t}{v}$$

In the latter equation, $a\ \Delta t$ is not the curved arc length between the arms of the angle but rather the straight-line distance. For small angles this makes a very small difference, and as $\Delta\theta \to 0$ the difference vanishes, so in the limit the equation is correct. We obtain

$$\frac{v\ \Delta t}{r} = \frac{a\ \Delta t}{v}$$

$$a = \frac{v^2}{r} \tag{7.5}$$

The magnitude of **a** is the square of the speed divided by the radius of the circle. The units are readily seen to be the same on both sides of Eq. 7.5 as long as all lengths are measured in the same units. One must not use v in miles per hour and r in feet. The direction of $\mathbf{a}(t)$ is seen in Figure 7.2(b) to be toward the center of the circle, so $\mathbf{a}(t)$ is called a *centripetal* acceleration. Summing up, when a particle moves with constant speed v in a circle of radius r, it is accelerated toward the center of the circle, the magnitude of this acceleration being v^2/r.

In uniform circular motion the vectors $\mathbf{r}(t)$, $\mathbf{v}(t)$, and $\mathbf{a}(t)$ all depend on the time, but all are of constant length and the angles between them are constant. $\mathbf{v}(t)$ is always perpendicular to $\mathbf{r}(t)$, and $\mathbf{a}(t)$ is always antiparallel to $\mathbf{r}(t)$. The three vectors rotate together, making one complete revolution every time the particle makes one circuit of the circle. The time taken for one revolution is called the *period* of the motion and is denoted by τ (the Greek letter tau). The circumference of the circle being $2\pi r$, we see that

$$v\tau = 2\pi r, \qquad \tau = \frac{2\pi r}{v} \tag{7.6}$$

A more useful quantity than the period τ is the *angular velocity* of the particle, denoted by ω (the Greek letter omega). This is the angle through which the radius vector $\mathbf{r}(t)$ turns per unit time. In the time τ this angle is 2π, one complete revolution, so

$$\begin{aligned} \omega\tau &= 2\pi \\ \omega &= 2\pi/\tau \qquad (7.7) \\ &= v/r \end{aligned}$$

The angular velocity is the speed divided by the radius of the circle. In terms of the angular velocity, the centripetal acceleration is

$$a = \omega^2 r \tag{7.8}$$

As an illustration of circular motion we take the motion of the earth around the sun. This motion actually is along an ellipse with the sun at one focus, but it is very nearly along a circle with a radius of 1.49×10^{13} cm. The period of revolution is 1 yr $= 3.16 \times 10^7$ sec (very nearly $\pi \times 10^7$ sec). From this we can find various other quantities.

$$\text{(1) Speed: } v = \frac{2\pi r}{\tau} = \frac{2 \times 3.1416 \times 1.49 \times 10^{13}}{3.16 \times 10^7}$$
$$= 2.97 \times 10^6 \text{ cm/sec}$$
$$= 29.7 \text{ km/sec}$$
$$= 18.4 \text{ miles/sec}$$
$$= 4.8 \times 10^4 \text{ mph}$$

[1 kilometer (km) $= 10^3$ meters $= 10^5$ cm. Also, 1 mile $= 1.61$ km.]

$$\text{(2) Angular velocity: } \omega = \frac{2\pi}{\tau}$$
$$= \frac{2 \times 3.1416}{3.16 \times 10^7}$$
$$= 1.99 \times 10^{-7} \text{ rad/sec}$$

$$\text{(3) Centripetal acceleration: } a = r\omega^2$$
$$= (1.49 \times 10^{13}) \times (1.99 \times 10^{-7})^2$$
$$= 0.59 \text{ cm/sec}^2$$

The acceleration of the earth caused by the sun is, we see, smaller than the acceleration of an object near the surface of the earth caused by the earth, 980 cm/sec^2.

One might think that the acceleration produced by the sun would affect observations on the earth noticeably. For example, the acceleration toward the center of the earth of an object dropped at noon when the sun is overhead might be less than that of an object dropped at midnight when the earth and sun are on the same side of the object and pull it in the same general direction. This is not correct. In a reference frame attached to the sun, the sun alone gives both the earth and the object the acceleration $a = 0.59$ cm/sec^2 toward the sun except for a very small effect due to their different distances from the sun, so they would move along parallel paths. The acceleration of the object *toward the earth* is then produced by the gravitation of the earth alone and is the same on both sides of the earth.

The discussion of uniform circular motion has been carried through without the explicit introduction of coordinate axes and without resolving vectors into components. This is an example of the usefulness of vector methods. By introducing coordinate axes now we shall learn very little new about circular motion, but we find some formulas that are useful in treating another important kind of motion, harmonic motion. Let

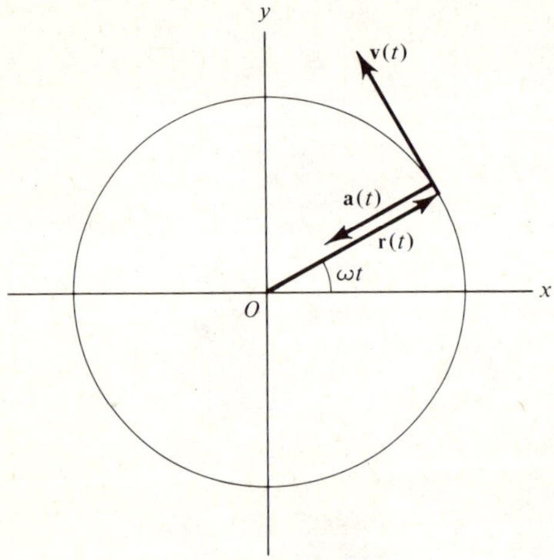

FIGURE 7.3 The acceleration, the velocity, and the position vector of a particle moving uniformly in a circle are related to each other as shown; the velocity is at right angles to the radius vector and the acceleration, while the acceleration is antiparallel to the radius vector.

us, therefore, introduce a set of coordinates with the origin at the center of the circle and with the circle lying in the xy-plane. Further, let the particle be on the x-axis at time $t = 0$. The angle θ between the position vector $\mathbf{r}(t)$ of the particle and the positive x-axis is zero at time $t = 0$ and increases uniformly with the time so that

$$\theta = \omega t \tag{7.9}$$

From Figure 7.3 the components of the various vectors can be read off. All z-components have been arranged to vanish, so they are omitted.

The components of $\mathbf{r}(t)$ are

$$x(t) = r \cos \omega t$$

$$y(t) = r \sin \omega t \tag{7.10}$$

The velocity $\mathbf{v}(t)$, since $v = \omega r$, has components

$$v_x(t) = -\omega r \sin \omega t$$

$$v_y(t) = \omega r \cos \omega t \tag{7.11}$$

in which the signs agree with the figure. Finally the acceleration $\mathbf{a}(t)$ has components

$$\begin{aligned} a_x(t) &= -\omega^2 r \cos \omega t \\ &= -\omega^2 x(t) \\ a_y(t) &= -\omega^2 r \sin \omega t \\ &= -\omega^2 y(t) \end{aligned} \tag{7.12}$$

In all these equations the angle ωt grows indefinitely large as t increases. This does not mean that x or any other vector component increases indefinitely; the functions sine and cosine are periodic, returning to their previous values when the angle increases by 2π rad, corresponding to a complete revolution of the position vector. Therefore

$$\sin(\omega t + 2\pi) = \sin \omega t \tag{7.13}$$

If one must find the sine of a large angle, one may subtract the angle 2π from it over and over again until an angle between zero and 2π rad (360°) is obtained and then determine the sine of this smaller angle.

Equations 7.12 are the results of our introduction of coordinates, which turn out to be useful, as apparent in the next section.

A particle moving uniformly around a circle has an acceleration that is constant in magnitude and always perpendicular to the particle's velocity. Now we ask the kind of question asked in the previous section: What kind of motion can result when the acceleration is constant in magnitude and perpendicular to the velocity? This question is raised here because such an acceleration is produced by a uniform magnetic field acting on a moving charged particle, a situation studied in Section 18 and occurring again later. A possible motion is uniform circular motion, as has just been seen. This is not the most general motion that can result, however. Looking at Figure 7.3, one can imagine that the particle moving around the origin is also moving steadily into or out of the page. A steady motion out of the page involves no acceleration in this direction and involves a new component of velocity only along a direction perpendicular to **a**, which lies in the plane of the page. The acceleration for this modified motion is, therefore, perpendicular to the velocity, and all the conditions specified are met. Looked at in three dimensions, the motion is that illustrated in Figure 7.4. It is along a helix, like the thread of a screw lying along the z-axis. This three-dimensional motion is the most complicated we shall ever need to encounter

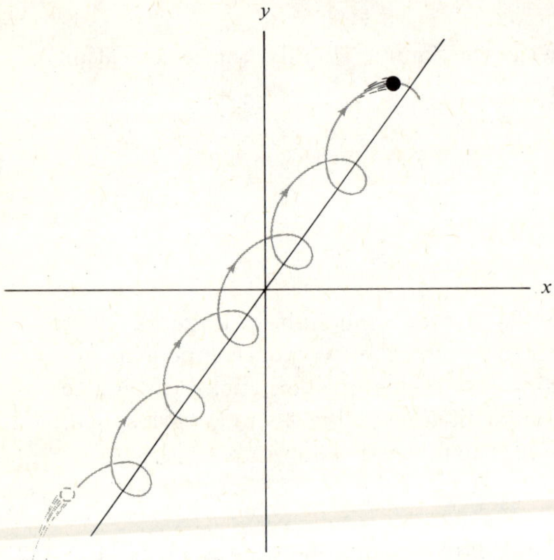

FIGURE 7.4 A particle moving with acceleration constant in magnitude and always perpendicular to the velocity moves along a helix. Here the acceleration is in the xy-plane, so the motion parallel to the z-axis is uniform while the motion parallel to the xy-plane is uniform circular motion.

in our discussion of physics. Tracks of particles moving this way in a magnetic field can be seen later in Figure 27.2.

Problems

1. The earth moves around the sun with a speed of very nearly 3×10^6 cm/sec. The radius of the orbit is 1.5×10^{13} cm. What is the centripetal acceleration of the earth? Compare this with the centripetal acceleration of the moon, due to its motion around the earth, moving with a speed of 1.0×10^5 cm/sec in a circle of radius 3.8×10^{10} cm.

2. A roller coaster track has a radius of curvature of 20 feet, as shown in the figure. How fast can the train be going when it reaches the top and have gravity just able to keep it on the track?

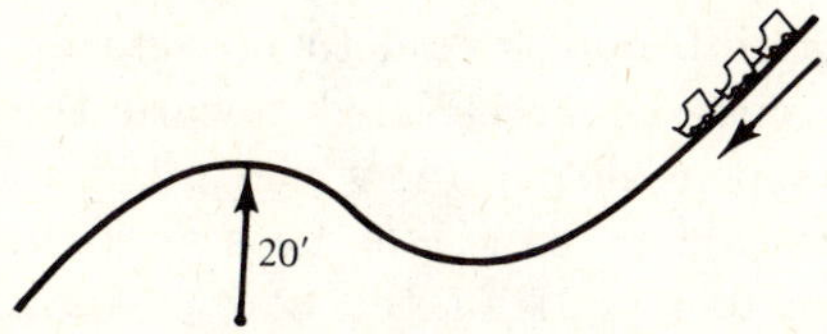

3. How fast must an earth satellite move just above the surface of the earth so that its centripetal acceleration is g? The radius of the earth is 6.4×10^8 cm.
4. A modern ultracentrifuge can spin a container 1250 revolutions per second, producing an acceleration in the container of $500{,}000g$. What is the radius of the circle in which the container moves?
5. How would Equations 7.10 be written if the particle moving in the circle in Figure 7.3 started at the top of the circle at time $t = 0$?
6. How would Equations 7.10 be written if the particle moving in the circle in Figure 7.3 went around clockwise instead of counterclockwise, starting on the x-axis at $t = 0$?

8. HARMONIC MOTION

Periodic motion is one of the most important types of motion encountered in physics. Any motion that repeats itself at regular intervals is periodic. The motion of a planet around the sun or of the moon around the earth, the vibration of a string in a guitar or piano, and the waves emitted by a radio station or by radar all involve periodic, or almost periodic, motion. The simplest of all periodic motions is harmonic motion, the subject of this section. One of the most useful methods of treating other kinds of periodic motion is to decompose them into harmonic motions, but this we do not do here. Because harmonic motions occur frequently in the later parts of the book, it is worthwhile to give a description of it here, where the ideas of motion in general are being developed.

A particle that is at rest and is unaccelerated will remain at rest and is said to be in equilibrium. A place where a particle can be in equilibrium is called a *point of equilibrium,* and a particle at rest at a point of equilibrium will remain there. If a particle is displaced from a point of equilibrium, any one of several things can happen. Imagine that the particle is confined to a curve that goes through the equilibrium point and is moved a short distance x away from the equilibrium point. There are three possible results: (1) The new position may be a point of equilibrium also. In this case the equilibrium is said to be *neutral.* (2) At the new position the particle may be accelerated toward the original equilibrium position. If this happens no matter which way the particle is displaced, the equilibrium is said to be *stable,* as a particle displaced away from it tends to return to it. (3) At the new position the particle may be accelerated away from the original equilibrium position. If this happens for either direction of displacement, the equilibrium is said to be *unstable,* as the particle once displaced away from it ends up going further away.

The three types of equilibrium described above are easily illustrated as in Figure 8.1. A smooth, round ball on a smooth, level table is an example of (1), the neutral equilibrium. The ball will stay anywhere on the table if it is placed there at rest. A smooth, round ball at the bottom of a hemispherical bowl is an example of (2), the stable equilibrium. If the ball is placed at rest anywhere on the bowl except at the bottom, it will be accelerated toward the bottom. A smooth, round ball at the top of an inverted hemispherical bowl exemplifies (3), the unstable equilibrium. If the ball is placed anywhere on the bowl except at the top, it will be accelerated away from the top. In example (1) any motion of the ball will be uniform, straight-line motion. In example (2) any motion will be an oscillatory one, the ball always tending to return to the equilibrium point but not staying there when it arrives because it does not arrive there at rest. The motion in (3) will be a runaway motion. We shall not need to consider this type of motion in any detail.

The situation of interest to us now is (2), when the equilibrium is

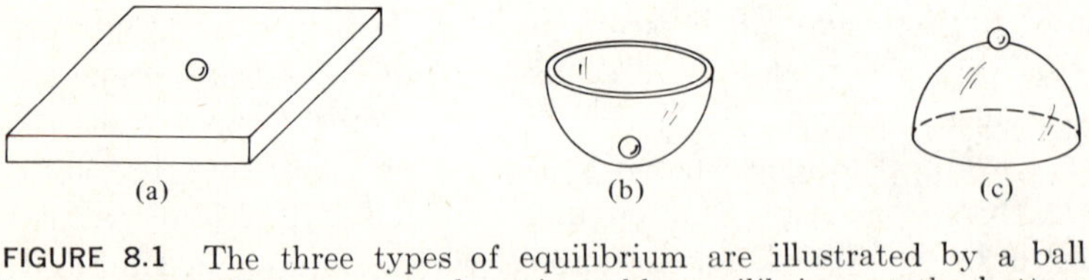

(a) (b) (c)

FIGURE 8.1 The three types of equilibrium are illustrated by a ball in neutral equilibrium on a plane, in stable equilibrium at the bottom of a bowl, and in unstable equilibrium on top of a bowl.

stable. When a particle is displaced not too far away from such an equilibrium point, it is usually a good approximation to say that the magnitude of the resulting acceleration is proportional to the displacement. This can be written as

$$a = -\lambda x \tag{8.1}$$

with λ (the Greek letter lambda) being a positive number. The minus sign describes the fact that the acceleration and the displacement are antiparallel. Because λ is positive, it may be written as the square of another number, ω. If this is done, the above equation becomes

$$a = -\omega^2 x \tag{8.2}$$

Equation 8.2 has been seen before. It is the part of Eq. 7.12 which describes the x-component of the motion of a particle that is moving uniformly in a circle. This means that we can exploit what we know about circular motion to get a description of harmonic motion. Consider the device shown in Figure 8.2. A wheel rotates with uniform angular velocity ω. A distance A from the center is a pin that fits into a vertical slot attached to a rod, which is free to slide parallel to the x-axis. The x-motion of the rod is the same as the x-motion of the pin, but the y-motion of the rod is zero. Because the x-coordinate of the pin, x_P, satisfies the equation

$$a_{Px} = -\omega^2 x_P \tag{8.3}$$

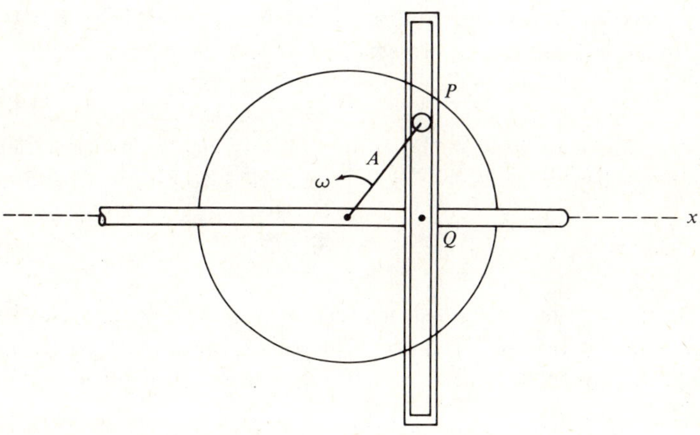

FIGURE 8.2 As the point P is carried around uniformly by the wheel, its projection Q on the x-axis moves with harmonic motion back and forth along the x-axis.

according to Eq. 7.12, the x-coordinate of the point Q that moves with the rod also must satisfy this equation, the x-coordinate of Q being the same as the x-coordinate of P although the y-coordinates of these two points differ. Thus the coordinate x of Q satisfies Eq. 8.2.

We know the motion in the x-direction resulting from Eq. 8.3. According to Eq. 7.10 it is, with r replaced by A,

$$x_P = A \cos \omega t \tag{8.4}$$

Therefore, this also must give the dependence of the x-coordinate of Q on the time, so that

$$x = A \cos \omega t \tag{8.5}$$

is the solution of Eq. 8.2. This solution works for any value of A, since the radius of the circle does not appear in Eq. 8.3 for the circular motion. The conclusion we can draw is that harmonic motion, which is defined to be the motion resulting from the proportionality of acceleration to displacement as written in Eq. 8.1 or Eq. 8.2, can be described by the formula Eq. 8.5 with any value of A, it being agreed to measure the time starting at an instant when $x = A$, because the cosine of zero is unity.

A is called the *amplitude* of the harmonic motion. It is the largest excursion of the oscillating particle from its position of equilibrium. ω is the *angular frequency* of the harmonic oscillator, and the angle ωt is called the *phase* of the motion. The period τ of the oscillator is the time for it to go through one complete cycle of its motion, which is the same as the time required for the reference particle P of Figure 8.2 to go around the circle once, so that as in Section 7

$$\omega\tau = 2\pi \tag{8.6}$$

The frequency of the oscillator is the number of cycles of motion per unit time. Denoting the frequency by ν (the Greek letter nu), we have

$$\nu = \frac{1}{\tau} = \frac{\omega}{2\pi} \tag{8.7}$$

The angular frequency is the most convenient quantity to use in describing an oscillator because its use eliminates factors of 2π from formulas like Eqs. 8.2 and 8.3. In terms of the frequency these become

$$a = -4\pi^2\nu^2 x, \qquad x = A \cos 2\pi\nu t \tag{8.8}$$

which are awkward, hard to remember, and can lead to unnecessary numerical calculation. Therefore we shall use ω rather than ν.

When the assumption Eq. 8.2 is valid, the angular frequency ω of the oscillator is a constant independent of the coordinate x. This means that the frequency of the oscillator does not depend on the amplitude with which it oscillates. If we consider the oscillator as being set in motion by pulling the particle away from its equilibrium position a distance A and then releasing it from rest, it will oscillate with amplitude A and with the same frequency ω for all the values of A for which Eq. 8.2 remains valid. This is an extremely useful characteristic of harmonic oscillators. For this reason all modern clocks are based on some form of harmonic oscillator, be it a pendulum, a balance wheel and spring, a tuning fork, the vibrations of an ammonia molecule, or a maser or laser.

The velocity and the acceleration of the oscillating particle are obtained from the x parts of Eqs. 7.11 and 7.12 relating to circular motion. They are

$$v(t) = -\omega A \sin \omega t$$

$$a(t) = -\omega^2 A \cos \omega t \tag{8.9}$$

The subscript x has been left off, the y-coordinate of the circular motion being irrelevant to harmonic motion. Figure 8.3 shows plots of x, v, and a as functions of t.

As a specific example of harmonic motion let us take a simple pendulum. This consists of a particle suspended from a fixed point by a string

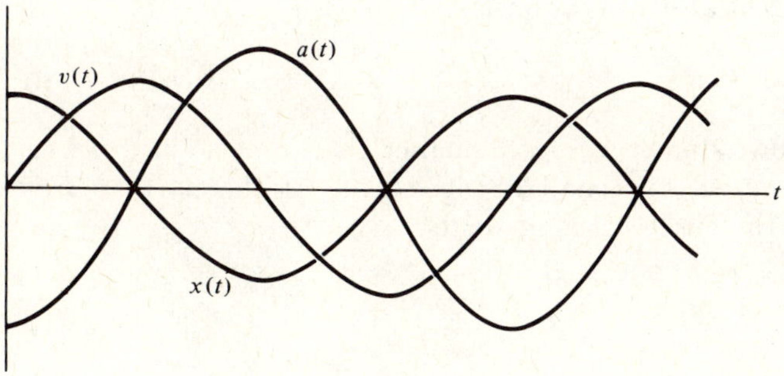

FIGURE 8.3 The displacement, velocity, and acceleration of a particle moving harmonically are plotted against the time. The units are chosen so that ω is greater than unity with the result that the amplitude of the acceleration curve is greater than that of the velocity curve, which in turn is greater than that of the displacement curve.

of length l near the surface of the earth. The situation is seen in Figure 8.4. If it were not for the string, the particle would fall with acceleration **g**. Because the string is there, no acceleration parallel to it is possible but there is no constraint on the motion of the particle in a direction perpendicular to the string. Therefore the particle moves along a circle of radius l with an acceleration that is the component of **g** along the

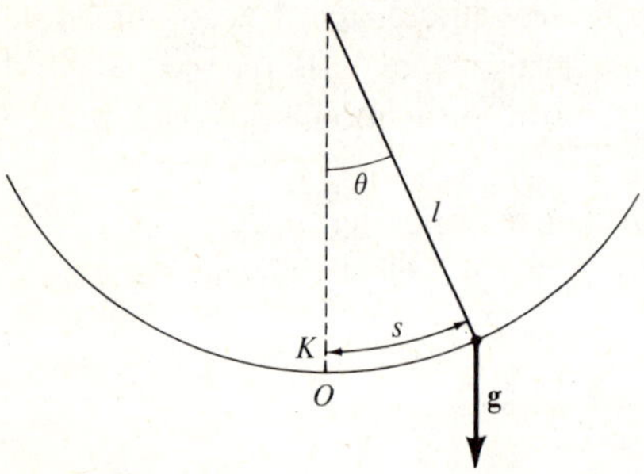

FIGURE 8.4 A simple pendulum provides an example of harmonic motion when the amplitude of the motion is very small compared to the length of the string.

tangent to the circle at the location of the particle. From the figure we see that the tangential acceleration is

$$a = -g \sin \theta \tag{8.10}$$

If θ (as measured in radians) is much smaller than unity, $\sin \theta$ is very nearly the same as θ, so for small displacements of the particle from the lowest point on the circle we may write

$$a = -g\theta = -g\frac{s}{l} \tag{8.11}$$

where s is the distance along the circle from the lowest or equilibrium point. The minus sign indicates that the acceleration is in the direction of diminishing s or θ.

Equation 8.11 is exactly of the form of Eq. 8.2, which defines harmonic motion if we identify ω^2 with g/l. The motion of the particle is, therefore, harmonic motion with an angular frequency of

$$\omega = \sqrt{\frac{g}{l}} \tag{8.12}$$

or a period of

$$\tau = 2\pi\sqrt{\frac{l}{g}} \tag{8.13}$$

according to Eq. 8.6. It is important to note that a pendulum moves with harmonic motion only when the angular range of its motion is small. As s gets larger the acceleration does not increase as rapidly as s and indeed reaches a maximum when s is a quarter of the circumference of the circle or θ is 90°. This is an illustration of why harmonic motion is the typical motion involved in small vibrations of particles around equilibrium even though large motions are usually not harmonic in character.

Problems

1. A particle has an acceleration of 1600 cm/sec^2 when it is released from rest 9 cm away from a position of stable equilibrium. With what angular frequency will it oscillate about that equilibrium position?
2. An oscillator is to have an angular frequency of 3×10^4 rad/sec. What must its acceleration be when the distance from equilibrium is 0.01 cm? What is the particle's maximum speed if the amplitude of oscillation is 0.06 cm?
3. A round pencil is in neutral equilibrium when lying on a level table. With the help of some sketches show that an elliptical or oval pencil has a position of stable and a position of unstable equilibrium when lying on a level table. (The center of the pencil, the lead, tends to get as low as possible.)
4. Two harmonic oscillators are moving so that they have the same maximum speed, but the first has twice the amplitude of the second. What is the ratio of their frequencies?

CHAPTER 2 Physical Laws of Particle Motion

9. THE LAWS OF MOTION

A law of nature is a statement describing a regularity observed in nature. It is an inference drawn from experience; it is not imposed on nature but is inherent in it. It is not absolute but is a summary and generalization of what has been observed. When new kinds of phenomena are observed, it is often found that they do not fit into a description adequate to include everything previously known; the new phenomena may not follow the old laws of nature. This is not a disaster for science, and it does not prove that science is built on false foundations. It does show that at any time our knowledge is incomplete and subject to enlargement. When new phenomena do not fit into the framework of previous ideas this framework must be modified, but in such a way that it still includes what was known before as well as what is new. Thus laws of nature tend to become more generalized and to apply to a wider range of phenomena as our knowledge of nature increases.

In this section some of the most important laws of nature are described, Newton's laws of motion. There are phenomena that do not fit into the Newtonian framework of ideas, especially phenomena in the realm of atomic physics and nuclear physics, so that Newton's laws of motion can no longer be considered to be exact laws of nature. They are absolutely basic to an understanding of physics, however, because they do give an accurate account of the motion of particles that are directly visible to us. It is by direct observation that the concepts of particle and of the motion of a particle are defined, so that the laws governing directly perceived particles must be the starting point for any development of these concepts.

If the position $\mathbf{r}$ of a particle is known for every value of the time t, everything about the motion of the particle is known. For example, the velocity of the particle is known because in the equation defining the velocity,

$$\mathbf{v}(t) = \frac{\mathbf{r}(t + \Delta t) - \mathbf{r}(t)}{\Delta t}$$

all the quantities on the right are known. Knowing the motion of the particle, however, does not tell us how this motion came about; the external circumstances that must be provided in order for the particle to move in the given way are not at all determined by anything that has been said so far. The subject of the motion of a particle becomes part of physics when the relation between motion and external circumstances is introduced.

Newton's three laws of motion provide the means for relating the motion of a particle to its environment. We state these laws as Newton did in his *Principia Mathematica*,* and then we restate them later in this section in a more modern, though equivalent, way.

The First Law of Motion Every body continues in its state of rest or of uniform motion in a right line unless it is compelled to change that state by forces impressed upon it.

According to Newton, then, a particle free of external influences, a "free particle," moves with constant velocity. The law is not an obvious one because it is impossible to free a particle of all external influences. What can be done is to reduce the magnitudes of the external influences to smaller and smaller values and then extrapolate to the case where they vanish. Thus on throwing a ball faster and faster over a horizontal distance of 10 yards, the time gravity has to act on it gets less and

* The quotations are taken from the Andrew Mott translation as revised by F. Cajori, University of California Press, Berkeley, 1946.

less and the more nearly straight the path of the ball is seen to be. Or on rolling a ball along smoother and smoother horizontal surfaces, the more nearly constant the velocity of the ball becomes. It seems reasonable to take a smooth surface as exerting less influence on the motion than a rough one. Even though real particles do not actually move completely freely, Newton's first law is important in telling us that velocity is not the immediate product of external influences but that change in velocity, or acceleration, is such a product.

The Second Law of Motion The change in the motion is proportional to the motive force impressed; and is made in the direction of the right line in which the force is impressed.

Here the relation between force and change in velocity is stated. In mathematical language the force **F** is proportional to the acceleration **a**, or

$$\mathbf{F} \propto \mathbf{a} \tag{9.1}$$

The vector character of this proportion means that Newton's statement on directions is included. This law appears strange because it contains the term "force," which has not been defined, and how can a "law" involve a new, undefined quantity? The answer to this is implicit in Newton's statement that the force is *impressed* on the body. What this means is that the force on a particle is a consequence of the environment of the particle, being something like a push impressed from outside the particle and not arising from any internal structure of the particle. The effect of this force is to accelerate the particle. Newton's genius in framing his second law of motion lay in his identification of the acceleration as the quantity to relate directly to circumstances external to the particle. Most of his predecessors tried to use the velocity for this purpose.

The Third Law of Motion To every action there is always opposed an equal reaction; or, the mutual actions of two bodies upon each other are always equal, and directed to contrary parts.

In this statement "the mutual actions of two bodies upon each other" means the forces exerted on each by the other. If particle A exerts the force $\mathbf{F}_{A \to B}$ on particle B, and if particle B exerts the force $\mathbf{F}_{B \to A}$ on particle A, then Newton says that

$$\mathbf{F}_{A \to B} = -\mathbf{F}_{B \to A} \tag{9.2}$$

This statement is important because it permits us to say that the forces on two different particles are equal in magnitude if these forces are due solely to the interaction of the two particles with each other. No comparison of forces on distinct particles is possible on the basis of the second law alone.

With this the statement of Newton's laws of motion is complete, but it is very convenient to use the third law to introduce the idea of the mass of a particle and then to rewrite the second law as an equality rather than as a proportion.

Consider two bodies, A and B of Figure 9.1, which interact with each other through a light spring that is initially compressed and then pushes

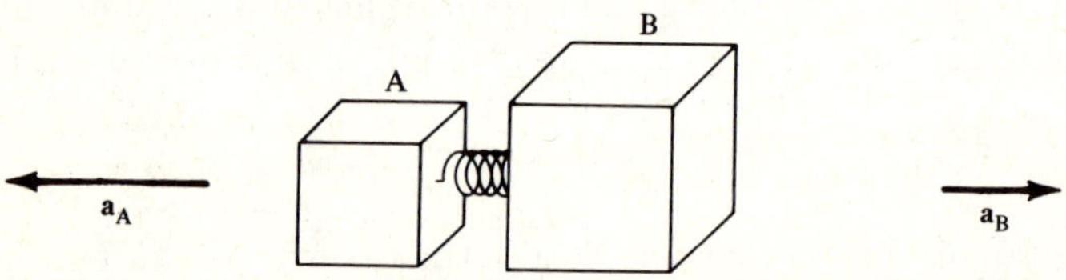

FIGURE 9.1 When two objects exert forces on each other, their accelerations are in opposite directions and the ratio of the magnitudes of these accelerations is constant.

the bodies apart. The accelerations of the two bodies can be measured and indeed always are found to be in opposite directions but not necessarily of equal magnitudes. According to Newton's third law, however, the forces on the two bodies are of equal magnitudes. We say, then, that the particle that has the greater acceleration under these circumstances has the less inertia or the smaller *mass*. The ratio of the masses is defined to be

$$\frac{m_B}{m_A} = \frac{a_A}{a_B} \tag{9.3}$$

so that on cross multiplying we obtain

$$m_B a_B = m_A a_A$$

With the masses introduced in this way, it is now useful to define the force acting on body A by

$$\mathbf{F}_{B \to A} = m_A \mathbf{a}_A \tag{9.4}$$

and similarly for body B. With this definition of force the third law of motion is satisfied automatically. Equation 9.4 is commonly what is meant by the phrase "Newton's second law of motion," and we shall refer to it in this way also. Equation 9.3 determines only the ratio m_B/m_A and not m_B or m_A separately. The mass of a particle is given an absolute value by accepting one particle as a standard with unit mass. The legal standard in the United States is a platinum alloy cylinder whose mass is defined to be 1 kilogram (kg). The common scientific unit of mass is the gram, 1/1000 kg. The mass of any particle in principle can be determined by making it interact with a standard particle of mass 1 g while isolated from all other influences and by measuring the ratio of the accelerations. This method is not that used in practice. The practical way of comparing masses is by weighing, a method whose equivalence to the above method to within one part in 10^{11} has been proved experimentally.

If the mass of a third particle m_C is to be determined, there are two obvious ways to do it. The first is to make C interact with the standard particle A of unit mass and to measure the resulting accelerations as was done in finding m_B. The second way would be to make C interact with B and so determine the ratio m_C/m_B. The ratio m_B/m_A is already known, so that by this means we obtain a value of m_C/m_A as the product of two separately measured numbers:

$$\frac{m_C}{m_A} = \left(\frac{m_C}{m_B}\right) \times \left(\frac{m_B}{m_A}\right)$$

It is found by experiment that these two procedures for measuring m_C always give the same result. If they did not do so, we would have to conclude that mass as defined using Newton's third law is an uninteresting quantity whose value depends on the means used to measure it. The result of experiment is, however, that mass is a useful quantity characteristic of a particle and independent of the means used to measure it.

The force in Newton's second law of motion describes the entire effect produced on the motion of the particle by its environment. Often several external influences are present, each of which can be imagined to act alone. Each separate influence then can be considered as impressing a force on the particle, so that when all are acting at once there are several forces on the particle. The force to be inserted in Newton's second law of motion is the vector sum of these forces. For example, a particle acted on by gravity is suspended by a spring as in Figure 9.2 so that it remains at rest; its acceleration is zero, and the force on it is therefore zero too. If the spring were not there, the particle would accelerate downward with acceleration **g**, so the particle has a downward force of gravity of magnitude mg acting on it. If the earth were not there, the stretched spring would accelerate the particle upward. When both the spring and the earth are present, the particle is subject to both forces, which in this example must add up to zero because the acceleration is zero. The upward force exerted by the spring on the particle is then also of magnitude mg. A spring supporting a mass at rest in a gravitational field of acceleration **g** must be under tension mg. The presence of more than one kind of force on a particle is a common

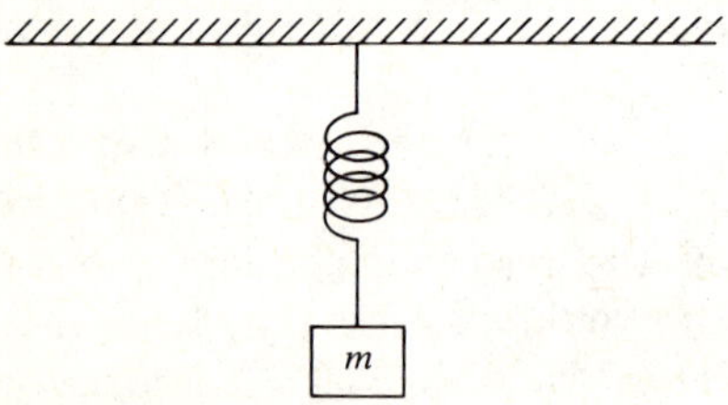

FIGURE 9.2 The total force on a particle in equilibrium is zero, so if two separately identifiable forces are acting on it, they must be equal in magnitude and opposite in direction.

occurrence, and always the vector sum of the various forces is to be used in Newton's second law.

The force on a particle is the product of the mass of the particle and the acceleration of the particle. It is a vector quantity, being the product of a scalar (the mass) with a vector (the acceleration). Its units are determined by Eq. 9.4. A unit force will produce unit acceleration of unit mass. When mass is measured in grams and acceleration in centimeters per (second)^2, the unit of force is the dyne. If m is measured in kilograms and a in meters per (second)^2, the unit of force is the newton. Because 1 kg = 10^3 g and 1 m = 10^2 cm, it is easily seen that one newton is 10^5 dynes. There will be no occasion for us to use British units in connection with Newton's laws, so we do not define the British units of either mass or force.

It is Newton's third law of motion that makes force a more generally useful quantity to talk about than acceleration. When two particles exert forces on one another, the sum of these forces is zero according to the third law, although the sum of the accelerations produced by these two forces in general is not zero because the particles have different masses. Consider a particle made up of two other particles fastened together. The acceleration of the composite particle is determined by its mass and by the vector sum of the forces acting on it. Among these forces are the forces binding the two constituent particles together, but these forces cancel because the binding force on one particle is equal and opposite to the binding force on the other so that together they do not influence the motion of the composite particle as a whole at all. We do not need to know what holds a particle together to be able to study its motion under the influence of forces coming from outside the particle. A ball is cut into two pieces, which then are cemented back together. The behavior of the ball does not depend on whether it is considered to be a single particle or to be two particles held together by cement but depends only on its mass and on the external forces acting on it.

Let us now restate Newton's first and second laws of motion in a way that emphasizes certain features glossed over in the traditional presentation just given. These features play a special role where modifications of Newton's laws are considered, as they must be in atomic physics.

The first law of motion states that a free particle moves with constant velocity. We must recall, however, that the position of a particle is specified by giving its coordinates in some particular coordinate system or reference frame. If the values of these coordinates change with time, it is not clear whether it is because the particle is moving or because the reference frame is moving or because both are moving. It is not even clear what is meant by saying that the reference frame is moving because the motion of a reference frame can be described only as seen from another reference frame. If there are two reference frames moving relative to each other, in which one should we give the coordinates of a particle of interest so as to describe its position, its velocity, and its acceleration? A rider on a merry-go-round sees himself as at rest in the reference frame of the merry-go-round, but a spectator sees him as having both a velocity and an acceleration in the reference frame of the ground. Which is the better reference frame to use, or are they equally good for the purpose of giving a description of what happens in nature? The selection of the most advantageous reference frame to use is a problem of physics. There is a class of reference frames, called *inertial frames,* that are by far the most convenient to use. Newton's first law of motion defines inertial reference frames.

It is always possible to choose the reference frame so that a particular particle is at rest in that frame. One simply has to put the origin of the coordinate system wherever the particle is; the particle is then permanently at the origin. In general it is not possible to keep a second particle permanently at rest in this same reference frame, because if the distance between the particles changes, the value of at least one of the second particle's coordinates must change. Similarly, it is always possible to choose the reference frame so that the velocity of a particular particle is constant in time, but then the velocity of a second particle in general will not be constant. However, if the particles being observed are all remote from one another and from everything else so that each of them can be considered free of any external influences, then it is found that a reference frame exists in which each of the particles has a constant velocity. This is one of the inertial frames mentioned above.

An inertial frame is defined by the distant stars. The forces acting on distant stars are so weak that during the period of any human observation their velocities do not change much. (There are, of course, exceptions to this such as double stars that revolve about a common center with periods typically measured in years. We exclude such systems or regard the double star as a single, composite star system for our purpose here.) A reference frame in which the velocities of the distant stars are constant constitutes an inertial frame. When one frame of reference is an inertial frame, so is another that moves with constant velocity with respect to the first, as a particle unaccelerated in the one system will be unaccelerated in the other, although a particle at rest in one system is moving with respect to the other. A passenger walks steadily down the aisle of an airplane in steady flight. He is unaccelerated both with respect to the airplane and the ground. A sitting passenger is at rest with respect to the airplane but is moving with respect to the ground, although he, too, is unaccelerated with respect to both the airplane and the ground.

A statement that is equivalent to the first law of motion then is the following: There exist inertial reference frames, that is, reference frames in which every free particle moves with constant velocity. The validity of this law on a small enough scale has been amply verified. It is an open question whether it is correct on the cosmological scale where galaxies or clusters of galaxies are the particles and the forces involved are gravitational; this is the realm of Einstein's general theory of relativity, and experimental answers are hard to obtain. From now on we shall always describe the motion of a particle or system of particles by the use of an inertial reference frame. If a noninertial frame were used, even particles remote from all others would be accelerated and the original statement of Newton's first law of motion would not be correct, or else even isolated particles would be subject to impressed forces. This is a complication we choose to avoid.

A statement that is equivalent to the second law of motion can be made as follows: When the external circumstances that can influence the motion of a particle are kept the same, the particle always will move in the same way if it is started from the same initial position and with the same initial velocity. The subsequent motion is independent of how the particle is brought to its initial position and of how it is given its initial velocity. A more mathematical way of saying this is that under given external circumstances the position of a particle at a time t, $\mathbf{r}(t)$, depends only on the initial position $\mathbf{r}(t_0)$ and the initial velocity $\mathbf{v}(t_0)$ of the particle. Consider, for example, a particle suspended from a rigid support by a spring. If the spring is stretched an amount L and the particle is then released from rest, it will oscillate up and down with a gradually decreasing amplitude until it comes to rest. Every time this is done the motion will be the same unless something affecting the particle is changed. If in one trial the air is still and in another a strong wind is blowing past the particle, the two motions will not be the same. The more accurately the motion of the particle is to be observed, the more carefully the external circumstances must be preserved in order to make successive trials appear identical. It is, of course, impossible to reproduce all the circumstances of an experiment; we can hope to come very close to reproducing the relevant circumstances. If this were not possible, experimental science would not exist. We must now see why the statement given above is related to the second law of motion.

If the position $\mathbf{r}$ of a particle is known for each value of the time t, the complete history of the particle's motion is known, being described by the function $\mathbf{r}(t)$. If this function were known for every particle we are ever to be interested in, no general statement about the motion of particles would be needed, as all we could want to know would be known explicitly. Under these circumstances laws of motion would not be of great interest. If in tossing a coin it were known ahead of time what the result of each toss would be, the law of chance that on the average equal numbers of heads and tails should appear would not be a very useful bit of knowledge. The two situations are similar.

Even when we do not know the trajectory of a particular particle, we might be able to make a statement like the following: If the particle is at position **r**, then it has a velocity **v** that is known and that depends only on **r**. This statement does not say that the particle ever is at any particular point **r**; it states that *if* it is at **r**, *then* its velocity must have the value **v** corresponding to that **r**. Such a statement is of a more general character than one giving the position **r** at each time, since it says something about any conceivable motion and not just the actual one. The statement, if true, would be a law of motion. A consequence of this hypothetical law is that the trajectory of a particle would be fixed by giving only the position of the particle at one time, say t_0. To see this, consider the short time interval Δt and the change in the position **r** during this interval. Because the velocity of the particle is known for each position of the particle, the velocity at the time t_0 has the known value $\mathbf{v}(t_0)$ corresponding to $\mathbf{r}(t_0)$. By the definition of the velocity, then,

$$\mathbf{r}(t_0 + \Delta t) = \mathbf{r}(t_0) + \mathbf{v}(t_0)\,\Delta t$$

so that the position of the particle at time $t_0 + \Delta t$ is known. This, however, then determines what the velocity at time $t_0 + \Delta t$ is, so that the next step, finding $\mathbf{r}(t_0 + 2\,\Delta t)$, can be carried out. The process can be repeated as often as desired, and the trajectory can be found over any finite time interval of interest. In simple situations the process can be carried out analytically with the result that a formula for **r** as a function of t can be given. When this is possible, the limit $\Delta t \to 0$ necessary for perfect accuracy actually is approached and the result is exact. In more complicated situations the process must be done numerically using a computer. Then Δt must be chosen small enough so that the error made at each step through having used a time increment that is finite does not accumulate to an intolerable degree. The essential point is that if the velocity is known for each possible position of the particle the trajectory of the particle is determined as soon as the position at one time is known. Our hypothetical law of motion giving **v** as a function of **r** does not work in nature because there we know that we can give a particle any velocity we want no matter where it is.

We now show that if the acceleration of a particle is determined by the position and velocity of the particle, then the motion of the particle is fixed once its initial position and initial velocity are known. The argument is very similar to the one given above when **v** was specified for each **r**, but now one more step is required. Let the particle be at

$\mathbf{r}(t_0)$ moving with velocity $\mathbf{v}(t_0)$ at the initial time t_0. A short time Δt later the particle is at a position given by

$$\mathbf{r}(t_0 + \Delta t) = \mathbf{r}(t_0) + \mathbf{v}(t_0)\,\Delta t \tag{9.1}$$

just as before. Because the acceleration is known for every $\mathbf{r}$ and $\mathbf{v}$, it is known at the initial time t_0, being $\mathbf{a}(t_0)$. The velocity at time $t_0 + \Delta t$ is then

$$\mathbf{v}(t_0 + \Delta t) = \mathbf{v}(t_0) + \mathbf{a}(t_0)\,\Delta t \tag{9.2}$$

Consequently, both the position and velocity of the particle are known at time $t_0 + \Delta t$, so that the acceleration at this time also is known. The information needed to make another step to find the position and velocity at time $t_0 + 2\,\Delta t$ is at hand. The process can be repeated over and over until all the trajectory of interest has been traced out. Here, as above, in simple situations the stepwise procedure can be carried out analytically so that the limit $\Delta t \to 0$ can be approached and exact results obtained. In complicated situations the calculation must be done numerically using a value of Δt small enough to give the required accuracy.

The path-tracing method just described can be illustrated in the example of constant acceleration $\mathbf{a}$. Let us start a particle from rest, $\mathbf{v} = 0$, at the origin, $\mathbf{r} = 0$, at time t_0, and let us suppose the external circumstances are such that wherever the particle is it has an acceleration $\mathbf{g}$ downward in the negative y-direction. This is then the example of Section 6, "A particle is dropped from rest at the origin." To carry out the process we make a table of the quantities t, y, and v_y. The first row of the table contains the initial values of these quantities. The second row contains the new time, later than the previous one by Δt; the new position obtained from the previous one by adding Δt times the previous velocity to it; the new velocity obtained from the previous one by adding Δt times the previous known acceleration to it. Thus the table can be written down row after row (Table 9.1).

The values of y appearing in the table are plotted against the time in Figure 9.3, being shown as dots. Also plotted on the same graph is the value of y as given by Eq. 6.6, this being the result of our earlier discussion of this kind of motion. The dots do not fall exactly on the curve. The reason for this is that here we used the velocity at the beginning of each time interval while in Section 6 we used the average velocity in each time interval. This makes a difference that becomes

TABLE 9.1
The Free Fall of a Particle from Rest, Followed in Discrete Steps in Time

t	y	v_y
t_0	0	0
$t_0 + \Delta t$	0	$-g\,\Delta t$
$t_0 + 2\,\Delta t$	$-g(\Delta t)^2$	$-2g\,\Delta t$
$t_0 + 3\,\Delta t$	$-g(\Delta t)^2 - 2g(\Delta t)^2 = -3g(\Delta t)^2$	$-3g\,\Delta t$
$t_0 + 4\,\Delta t$	$-3y(\Delta t)^2 - 3g(\Delta t)^2 = -6g(\Delta t)^2$	$-4g\,\Delta t$

smaller as the length of the time interval becomes smaller. When the calculation of a particle's path is actually done in this way using a computer, the time interval Δt must be chosen small enough to give the desired accuracy. The method is exact only in the limit as Δt gets smaller and the number of steps gets larger.

Newton's second law of motion states that the acceleration of a particle is the quantity determined by external circumstances. This amounts to saying that if the position and velocity of the particle are known, the particle's relation to its environment is known and its acceleration is

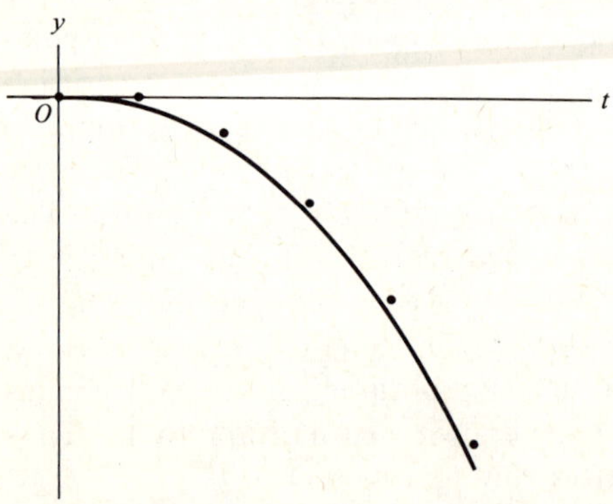

FIGURE 9.3 The position of a particle dropped from rest at the origin at time $t = 0$ is plotted as obtained from Table 9.1 and from Eq. 6.6. The dots representing the table values do not fall exactly on the curve because the table is calculated using the velocity at the beginning of each time interval rather than the mean velocity during the interval.

determined. But this is just what we have used above in determining the path of a particle knowing its initial position and velocity. This shows, therefore, that the traditional statement of Newton's second law is equivalent to our statement that the trajectory is determined, under given external circumstances, by the initial position and initial velocity of the particle.

Our modified statement of Newton's second law of motion is a useful one in practice, as is Newton's original statement. Many experiments in physics are done using beams of particles produced at a certain point moving with a certain velocity. It is now clear that what happens after the beam is produced is independent of the machinery used to produce the beam; the experimenter is free to use whatever equipment he can as long as it causes the beam particles to appear at the right place with the right velocity. Another example is provided by the launching of a space vehicle. Small errors in the vehicle's course during the initial rocket firing are not vital because later on course corrections can be made that give the vehicle the right velocity at some point on its desired trajectory, and from then on it will move according to plan.

Problems

1. A particle moves along the x-axis of a coordinate system in such a way that at any value of x, $v_x = 0.1x$. Follow the motion of this particle for five 1-sec steps starting from (a) $x = 1$ and (b) $x = -2$.
2. A particle moves along the x-axis of a coordinate system in such a way that at any value of x, $a = 0.2x$. Follow the motion for five 1-sec steps starting from (a) $x = 5$, $v_x = 10$ and (b) $x = 10$, $v_x = -2$.
3. What kind of law would replace Newton's second law of motion if the future motion of a particle required the knowledge of the initial position, velocity, and acceleration of the particle for its specification? Would Newton's definition of force still be useful under these circumstances?
4. A reference frame is rigidly attached to the earth. Imagine a free particle, not even affected by gravity, that is at rest at the earth's equator at one moment in this coordinate system. Sketch its path for the next 24 hr as seen from this system. Neglect the motion of the earth around the sun, and consider only the earth's rotation about its axis.
5. If the free particle of Problem 4 has a mass of 2000 g, what force would have to be applied to it to keep it at rest in the earth-fixed reference frame? What fraction of normal gravity (remember our particle is abnormal in that it is not supposed to feel gravity) is this force?

10. SOME APPLICATIONS OF NEWTON'S LAWS OF MOTION

The meaning and utility of Newton's laws of motion become apparent only when some examples of their application are studied. In this section physical situations that result in the motions described in the preceding chapter are given.

Sometimes the agency that exerts a force on a particle is clearly visible, as when the force is exerted by pushing with a rod or pulling with a string. Other times the agency is invisible, as when the force is that of gravity on a particle near the earth or is that of one electrically charged particle on another a considerable distance away. When the force on a particle is not exerted by a visible mechanism, it is often convenient to introduce the idea of a *field of force.* If at any point in a region of space it is known what force would be exerted on a particle located there, we say there is a force field in that region. The kinds of force for which this idea is most useful are gravitational and electric forces. With a little more complication, magnetic forces also can be described in terms of a magnetic field, and this is done later on. The electric and magnetic field first introduced in this way appear to have a physical reality of their own that shows up when wave motion is considered.

An important example of motion is that of a *free particle,* a particle on which no external force acts. In an inertial coordinate system a free particle moves with constant velocity, its acceleration being zero. This example is trivial but is important because often a particle moves as a free particle before some event takes place or after some event has taken place, such as a collision between two billiard balls on a table.

The simplest nontrivial motion of a particle is that which takes place in a *uniform field of force,* that is, in a field of force that everywhere exerts the same force on a particle. This motion is one in which the acceleration is constant. From Newton's second law of motion, $\mathbf{F} = m\mathbf{a}$, it follows that

$$\mathbf{a} = \frac{\mathbf{F}}{m}$$

and when $\mathbf{F}$ is constant, so is $\mathbf{a}$. The trajectory for such motions was given in Section 6 and is

$$\mathbf{r}(t) = \mathbf{r}(t_0) + \mathbf{v}(t_0) \times (t - t_0) + \frac{1}{2}\frac{\mathbf{F}}{m} \times (t - t_0)^2 \tag{10.1}$$

This vector equation can be expressed in terms of components most simply if we chose a coordinate system in which **F** lies along one axis and in which $\mathbf{v}(t_0)$ lies in a plane containing **F** and another axis. Then **F** has only one nonzero component and $\mathbf{v}(t_0)$ has at most two nonzero components. The equations for the components of $\mathbf{r}(t)$ then are those given in Section 6 with **a** replaced by $\mathbf{F}/m$.

There are two especially important sets of circumstances when a particle is in a uniform field of force. The first is when the particle moves under the influence of gravity alone in a limited region of space, and the second is when a charged particle moves in a uniform electric field. This latter case is taken up in Section 16; here we consider gravity, both alone and in conjunction with strings.

When gravity alone acts on a particle, it gives it an acceleration that is quite independent of the characteristics of the particle—an acceleration denoted by **g** that has the magnitude of about 980 cm/sec^2 near the surface of the earth. This means that the force on the particle must be

$$\mathbf{F} = m\mathbf{a} = m\mathbf{g} \tag{10.2}$$

When gravity alone acts, the ideas of mass and force are superfluous, the acceleration being given directly. When gravity and some other kind of force act simultaneously on a particle, the ideas of mass and force regain their importance. An example of this was given in Figure 9.2 where a mass m is shown suspended by a spring in a gravitational field. The tension in the spring depends on the mass of the suspended particle, being of magnitude mg if the particle is in equilibrium.

Another example of particles in a uniform gravitational field is furnished by Atwood's machine, illustrated in Figure 10.1. This consists of two particles of masses m_1 and m_2 that are connected by a string of fixed length and negligible mass, this string running over a pulley also of negligible mass, as shown. Here each particle is subject to two forces, the force of gravity downward and the force of the string, the tension T, acting upward. By assuming that the string and the pulley have negligible mass, we are assuming that the force exerted by the string on m_1 is equal to the force exerted by the string on m_2, no force being required to accelerate the string or the pulley. Newton's second law may now be applied to the two particles separately, which yields

$$\begin{aligned} m_1g - T &= m_1a_1 \\ m_2g - T &= m_2a_2 \end{aligned} \tag{10.3}$$

if the accelerations are taken as positive in the downward direction.

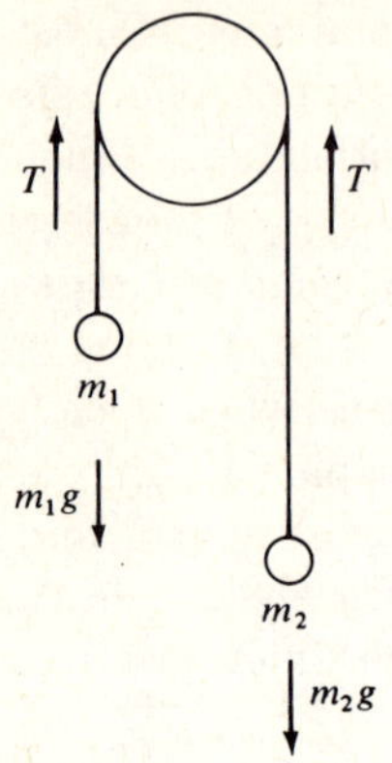

FIGURE 10.1 In Atwood's machine two particles are acted on by the same upward force, the tension T in the string, but by different downward forces, so they are not in equilibrium.

The condition that the string has a fixed length requires particle 2 to go up if particle 1 goes down, so that

$$a_2 = -a_1 \tag{10.4}$$

Using this and subtracting the second of Eqs. 10.3 from the first to eliminate the tension T from the equations, we obtain

$$(m_1 - m_2)g = (m_1 + m_2)a_1$$

$$a_1 = \frac{m_1 - m_2}{m_1 + m_2} g \tag{10.5}$$

The result given by Eq. 10.5 would be difficult to obtain without the use of Newton's second law. That the result makes sense is fairly obvious. When the two masses are equal, $a_1 = 0$, and if the system starts out at rest it remains at rest, the two equal masses balancing each other. If m_1 is larger than m_2, it accelerates downward; if m_1 is less than m_2, it accelerates upward. If $m_2 = 0$, m_1 is attached to nothing and accelerates downward with acceleration g. This device is essentially a balance and provides a way of measuring m_1. What one has to do is to adjust m_2, using particles of known mass, until the acceleration a_1 is zero when the balance is balanced. Then $m_1 = m_2$.

Instead of a circular pulley, two equal arms are used as in Figure 10.2. The apparatus is adjusted so that the acceleration of the pans is zero when they are empty. Then the mass to be determined is placed in the left pan and standard masses in the right pan. Any acceleration is due to the added masses, and if they are equal, the acceleration is zero. The effect of massive pans and arms is to reduce the sensitivity of the balance by reducing the acceleration produced by unequal masses in the pans.

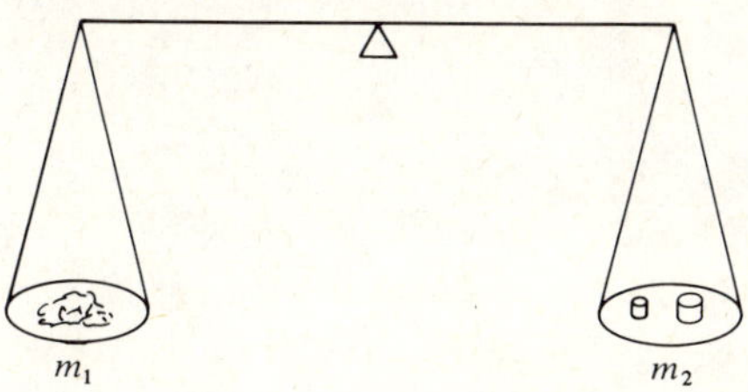

FIGURE 10.2 A common balance is a form of Atwood's machine, the pulley being replaced by the beam. The mass m_2 is adjusted to produce equilibrium.

The next example is *circular motion.* A particle moving uniformly in a circle was described in Section 7, the principal result being that a particle moving with a constant speed v in a circle of a radius r is accelerated toward the center of the circle with an acceleration of magnitude v^2/r. According to Newton's second law of motion, there is then a force acting on the particle, this force also being directed toward the center of the circle and having the magnitude

$$F = \frac{mv^2}{r} \tag{10.6}$$

Such a force is called a *centripetal force.*

In discussing the effect of gravity on particles in a limited region near the surface of the earth, the concept of a uniform field of force was introduced. We now introduce the concept of a central field of force, a field in which a particle experiences a force that is always directed toward a *center of force* located at a fixed point O and in which the magnitude of the force depends only on the distance r of the particle from the force center. Let the magnitude of the force at distance r

be called $F(r)$. When this central force is an attractive one, pulling the particle in toward O, a particle can move uniformly in a circular orbit of radius r about the center provided the particle's speed is such that

$$F(r) = \frac{mv^2}{r} \tag{10.7}$$

The central force must just provide the needed centripetal force. Orbits other than circular orbits are also possible, but more mathematics is needed for their description and they are not discussed here. At any

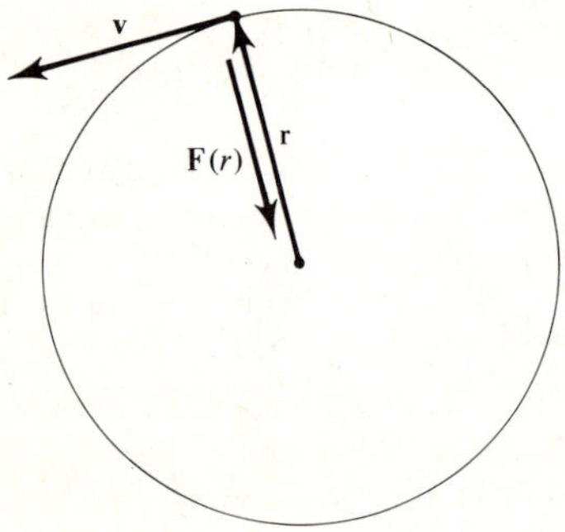

FIGURE 10.3 A force directed toward the center must be exerted on a particle that is moving uniformly in a circle to provide the needed centripetal acceleration.

point on the circular orbit shown in Figure 10.3, the force $F(r)$ is just that needed to keep the particle moving in the circle. If the circle's radius were different, the particle would have to move with a different speed, the difference depending on how the force varies with the distance r.

Central fields of force are important in physics. Newton proposed the first one when he stated his law of universal gravitation. This law states that every particle attracts every other particle with a force proportional to the product of the particle masses and inversely proportional to the square of the distance between them. Thus

$$F(r) = \frac{Gm_1m_2}{r^2} \tag{10.8}$$

The constant of proportionality G is the gravitational constant, to be distinguished from the acceleration of gravity at the surface of the earth g. Usually the law is applied to systems in which one of the masses is much larger than the other, so that the acceleration of the larger mass is very small and the massive body can be considered to be fixed in position. This is very accurate for the sun and earth or for the earth and an artificial satellite. The force of gravity on the lighter body is then a central one with the larger body as the center of force. The force between two electrically charged particles is also an "inverse square law" force given by Coulomb's law which is taken up in Section 15. If the particles involved in Newton's law of gravitation are large but round, the distance r is to be measured between their centers. It is a fortunate circumstance that for an inverse square law of force, the force on a particle due to a massive body is the same whether the mass of the body is concentrated at a point or spread symmetrically throughout a sphere, as long as the particle is outside the body. For any other force law the force would depend on how the mass is distributed. As a consequence, the inverse square law of gravitation with r measured from the center of the earth holds right down to the surface of the earth but not below the surface, except for every small effects due to the not quite spherical distribution of the earth's mass.

For a particle of mass m to move in a circular orbit about the earth, whose mass is M_E, according to Eqs. 10.7 and 10.8 its speed must satisfy the condition

$$\frac{mv^2}{r} = \frac{GmM_E}{r^2} \tag{10.9}$$

The mass m cancels out, and there remains

$$v^2 = \frac{GM_E}{r} \tag{10.10}$$

The product of the gravitational constant G and the mass of the earth M_E is easily found, because the force of gravity on a particle of mass m at the surface of the earth, a distance $R_E = 6.37 \times 10^8$ cm from the center, is mg with $g = 980$ cm/sec.^{2} Thus

$$mg = \frac{GmM_E}{R_E^2}$$

or

$$GM_E = R_E^2 g = 3.98 \times 10^{20} \text{ cm}^3/\text{sec}^2 \tag{10.11}$$

The speed necessary for a particle, an earth satellite, to move in an orbit a height h above the earth now can be found. The radius of the orbit is $R_E + h$, and so according to Eq. 10.10

$$v^2 = \frac{R_E^2 g}{(R_E + h)} \tag{10.12}$$

For $h = 1000$ km, which is 621 miles, this works out to

$$v^2 = \frac{3.98 \times 10^{20}}{6.37 \times 10^8 + 1 \times 10^8}$$
$$= 5.40 \times 10^{11} \text{ cm}^2/\text{sec}^2$$

so that

$$v = 7.35 \times 10^5 \text{ cm/sec}$$
$$= 1.64 \times 10^4 \text{ mph}$$

Earth satellites in general do not follow circular orbits but elliptical ones. The general motion of a particle in an inverse square law force field is in an elliptical orbit with the force center at one focus of the ellipse, a law stated by Kepler for the planets before Newton proposed his law of gravitation and known as Kepler's first law. The circular orbits considered here are special cases of elliptical orbits.

The moon is the original earth satellite. Knowing its period of revolution, its distance from the earth can be found. If its period is τ and its orbit radius is r, then its speed is $2\pi r/\tau$. This can be inserted in Eq. 10.10 to yield

$$\left(\frac{2\pi r}{\tau}\right)^2 = \frac{GM_E}{r}$$

or

$$\frac{r^3}{\tau^2} = \frac{GM_E}{(2\pi)^2} = \frac{R_E^2 g}{(2\pi)^2} \tag{10.13}$$

with the help of Eq. 10.11. For $\tau = 27\frac{1}{3}$ days $= 2.36 \times 10^6$ sec, we obtain

$$r^3 = 3.98 \times 10^{20} \times \frac{(2.36 \times 10^6)^2}{(2\pi)^2}$$
$$= 5.63 \times 10^{31} \text{ cm}^3$$

$$r = 3.83 \times 10^{10} \text{ cm}$$
$$= 2.38 \times 10^5 \text{ miles}$$

This is the distance also known from other measurements. Equation 10.13 embodies Kepler's third law of planetary motion, that the square of the period of a planet's motion about the sun is proportional to the cube of the radius of its orbit. (Applied to the motion of planets around the sun, M_E must be replaced by the mass of the sun.) Kepler's second law involves angular momentum and is discussed in Section 12.

It is possible to measure the gravitational attraction between two bodies whose masses can be measured in the laboratory. This was done by Cavendish in 1785 and has been repeated with greater accuracy since. The technique, due to Coulomb, is described briefly in Section 15 in connection with Coulomb's law. When this is done all the quantities except the gravitational constant in Newton's law are known, so G can be solved for. When G is known, the mass of the earth can be determined. The results are

$$G = 6.673 \times 10^{-8} \text{ dyne cm}^2/\text{gm}^2$$

$$M_E = \frac{R_E^2 g}{G} = 5.98 \times 10^{27} \text{ gm} \tag{10.14}$$

This value for the mass of the earth shows that the average density of the earth is considerably greater than the density near the surface. The volume of a sphere of radius r is $4\pi r^3/3$, so the volume of the earth is

$$\begin{aligned} V_E &= \frac{4\pi}{3} \times (6.37 \times 10^8)^3 \text{ cm}^3 \\ &= 1.08 \times 10^{27} \text{ cm}^3 \end{aligned}$$

The average density $\langle\rho\rangle$ (the Greek letter rho) is then

$$\begin{aligned} \langle\rho\rangle &= \frac{M_E}{V_E} \\ &= 5.53 \text{ gm/cm}^3 \end{aligned}$$

The density near the surface has an average value of only about 2.7 gm/cm³. Any model of the structure of the earth must account for this difference.

It is a remarkable fact that the force of gravity on a particle is exactly proportional to the mass of that particle, so that the acceleration of the particle does not depend on its mass. In the statement of Newton's law of gravity,

$$ma = G\frac{mM}{r^2}$$

the mass m inserted on the right is often called the *gravitational mass* of the particle, and the mass inserted on the left is called the *inertial mass* of the particle. One could conceive of a law of gravitation in which the gravitational mass is not equal to the inertial mass, so the two would not cancel, but experiments have been done that show them to be equal to within 1 part in 10^{11}. Since all particles in a gravitational field have the same acceleration, it is an attractive idea to attribute this acceleration to the properties of space and time themselves. This is a subtle idea that is difficult to describe accurately and briefly, so we shall not pursue this path. It leads to the theory of general relativity.

The last motion analyzed here is harmonic motion, which was discussed in Section 7. In harmonic motion the acceleration of the particle executing the motion is proportional to its distance from a position of equilibrium. We consider the motion to take place along the x-axis of a coordinate system. Let the origin be at the equilibrium point so that the acceleration is proportional to x. According to Newton's second law, then, the force is also proportional to x and we may write

$$F = ma = -kx \tag{10.15}$$

The statement that the force resulting from a displacement from an equilibrium position is proportional to the size of the displacement is known as Hooke's law. A frequently occurring system to which this

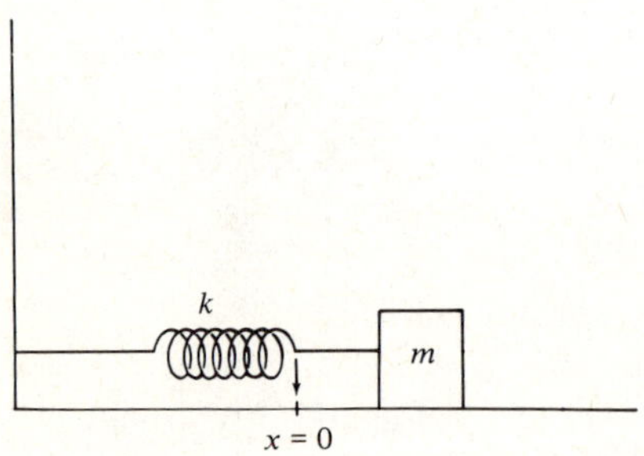

FIGURE 10.4 A mass m on a smooth horizontal table is attached to a vertical wall by a spring of constant k. If started, it will move back and forth with harmonic motion.

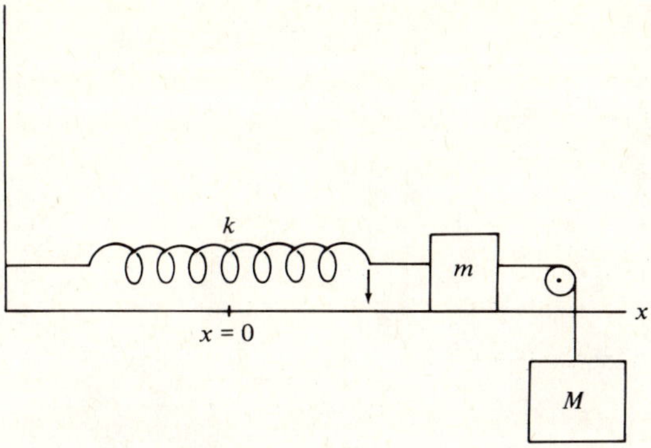

FIGURE 10.5 The spring constant k of a spring can be determined without reference to harmonic motion by using the arrangement shown to exert known forces on the spring and observing the amount it is stretched.

law applies consists of a particle attached to a fixed support by a spring, as shown in Figure 10.4. The displacement of the particle is chosen to be horizontal so that gravity will not affect the motion and friction is assumed to be negligible. Equation 10.15 can be justified experimentally in a manner independent of harmonic motion. The force exerted by the spring at any extension can be measured in a static fashion by attaching a string to the end of the spring, running it over a pulley, and suspending known masses from it as in Figure 10.5. If the system is in equilibrium with the pointer at $x = 0$ when the suspended mass is zero, then when $M > 0$, it is found that

$$Mg = kx \tag{10.16}$$

as plotted in Figure 10.6. In this equation everything except k is known so that k can be calculated.

When the force on a particle is that given by Hooke's law, Newton's second law becomes

$$ma = -kx$$

$$a = -\frac{k}{m}x \tag{10.17}$$

Equation 10.17 shows what the acceleration of the particle is at each

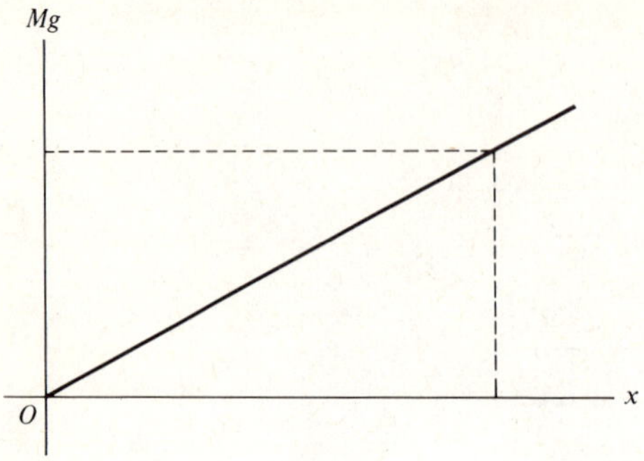

FIGURE 10.6 The amount a spring is stretched by hanging mass M from it is proportional to Mg. The slope of the graph gives the spring constant k.

position x; in this case the acceleration does not depend on the velocity. We know that this determines the motion uniquely once the initial position and velocity are given. Equation 10.17 is just like Eq. 7.1, namely,

$$a = -\omega^2 x$$

provided that we take

$$\omega^2 = \frac{k}{m}, \qquad \omega = \sqrt{\frac{k}{m}} \tag{10.18}$$

The motion resulting from Eq. 7.1 is known to be harmonic motion with a period of $\tau = 2\pi/\omega$. One can conclude, therefore, that the motion of the mass m on the spring of constant k is harmonic motion with a period τ given by

$$\tau = 2\pi\sqrt{\frac{m}{k}} \tag{10.19}$$

This period is independent of the amplitude of the motion.

The period τ, the mass m, and the spring constant k are all separately measureable, so that the relation of Eq. 10.19 can be verified by experiment. It always is valid. What this checks is that the force exerted by the spring at a given extension x is the same when the spring is moving as when it is stationary. If the mass of the spring is not negligible compared with the mass attached to it, some force is required to accelerate the spring, and an effective mass of the spring must be added to the particle mass in Eqs. 10.17 and 10.18. This effective mass is not the entire mass of the spring because the fixed end of the spring is not accelerated at all and parts of the spring near the fixed end are only slightly accelerated.

In the examples of this section, the properties of the particles and the force laws were assumed to be known and the motion was inferred from Newton's laws. Often the situation is reversed, the motion of the particle being observed and some property of the particle or of the force acting on the particle being inferred from this motion. It is in this way that the properties of many atomic and subatomic particles are established.

Problems

1. A mass of 25 kg hangs from a rope running over a pulley. What mass must be fastened to the other end of the rope so that the 25-kg mass will descend with an acceleration of $0.1g$?
2. Two boys of equal mass hold onto a rope running over a pulley, one at each end. One starts to climb the rope, the other just holds on. What happens if the rope is massless and the pulley is frictionless?
3. A 3-g particle is projected horizontally into a uniform field of force directed downward, the initial speed being 40 cm/sec. If the particle moves a horizontal distance of 240 cm while dropping 60 cm, what is the acceleration of the particle? What is the strength of the force field in dynes?
4. A particle of mass 0.9×10^{-24} g is projected horizontally with a speed of 1.5×10^{9} cm/sec into a uniform field of force directed downward and of strength 3.6×10^{-8} dynes. What is the value of its vertical component of velocity after it has traveled 15 cm in the horizontal direction? What angle does its velocity make with the horizontal at this point?
5. A nylon monofilament fishing line will support 3500 g. How fast can a 25-g particle be whirled in a horizontal circle at the end of 120 cm of this line and just not cause it to break?
6. A particle of mass m is whirled with constant angular velocity ω in a vertical circle of radius r at the end of a rod. At what value of ω will the tension in the rod be zero when the particle is at the top? What happens if the angular velocity is less than this?

7. If in the situation in Problem 6 the tension in the rod when the particle is at the top is zero, what is it at the bottom?
8. A spring stretches 8 cm when a 1000-g mass is hung from it. What is the spring constant in dynes per centimeter? What mass must be hung from this spring to give an oscillator with an angular frequency of 5 rad/sec? What is the period of this oscillator?
9. By what factor must the mass of an oscillating particle be changed in order to double the frequency of its oscillation? By what factor must the spring constant be changed to do this?
10. Two identical springs, each with spring constant k, are connected end to end as shown. What is the spring constant of the combination?

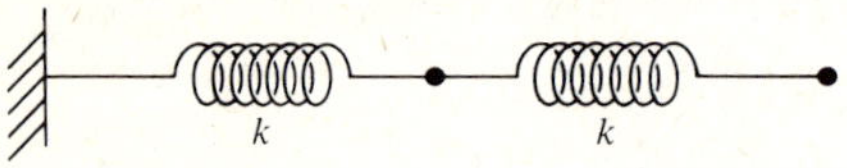

11. Two identical springs, each with spring constant k, are connected side by side as shown. What is the spring constant of the combination?

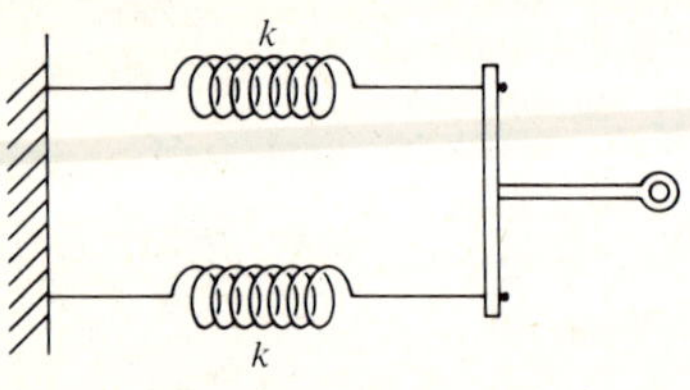

11. MOMENTUM

The motion of a particle is, as we have said before, completely specified if its position is known as a function of the time. When this is known, the velocity and the acceleration of the particle can be evaluated from their definitions. When it came to discussing the physical causes of a particular motion we found that it is very convenient to talk about force instead of acceleration, the two being related by the second law of motion, $\mathbf{F} = m\mathbf{a}$. For almost the same reasons it is often much more convenient to replace the velocity $\mathbf{v}$ of the particle by the momentum

$\mathbf{p}$ defined by

$$\mathbf{p} = m\mathbf{v} \tag{11.1}$$

This quantity was used by Newton under the descriptive name "quantity of motion." In the definition of $\mathbf{p}$, the mass m is that defined earlier with the help of the third law of motion. Later we shall see that m itself depends on the speed of the particle, but this is important only when the particle speed approaches the speed of light c, which has the enormous value of 3×10^{10} cm/sec = 186,000 miles per sec. Let us ignore this here; we return to it in Section 22.

Newton's first law of motion states that the momentum of a free particle is constant, because the velocity of such a particle is constant and multiplying a constant vector by a constant scalar yields another constant vector.

Newton's second law of motion can be stated neatly using the idea of momentum. Let us calculate the change in momentum caused by a change in velocity of a particle, the mass being fixed.

$$\begin{aligned} \Delta\mathbf{p} &= m\mathbf{v}(t + \Delta t) - m\mathbf{v}(t) \\ &= m[\mathbf{v}(t) + \mathbf{a}(t)\,\Delta t] - m\mathbf{v}(t) \\ &= m\mathbf{a}(t)\,\Delta t \\ &= \mathbf{F}\,\Delta t \end{aligned} \tag{11.2}$$

In the last step, the second law in the form of Eq. 9.4 has been used. Solving for the force $\mathbf{F}$ then yields

$$\mathbf{F} = \frac{\Delta\mathbf{p}}{\Delta t} \tag{11.3}$$

and so force is the rate of change of momentum with time. This turns out to be very useful.

Newton's third law of motion states that when two bodies exert forces on one another, these forces are equal in magnitude and opposite in direction. If the mutual interaction is the only external influence acting on either of the two particles, then the third law of motion implies that the sum of the momenta of the two particles is constant in time. Let the particles be called A and B. Then under the influence of the force $\mathbf{F}_{B\to A}$, in the time interval Δt, the momentum of A will change by the amount

$$\Delta\mathbf{p}_A = \mathbf{F}_{B\to A}\,\Delta t$$

and under the influence of the force $\mathbf{F}_{A\to B}$, during the same time interval, the momentum of B will change by the amount

$$\begin{aligned}\Delta\mathbf{p}_B &= \mathbf{F}_{A\to B}\,\Delta t\\ &= -\mathbf{F}_{B\to A}\,\Delta t\\ &= -\,\Delta\mathbf{p}_A\end{aligned} \tag{11.4}$$

Any change in $\mathbf{p}_B$ is balanced by an opposite change in $\mathbf{p}_A$, so the sum

$$\mathbf{p} = \mathbf{p}_A + \mathbf{p}_B$$

is constant. When two particles exert forces on each other but are otherwise free, their total momentum is constant in time, or the momentum of the system of two particles is conserved.

It is not hard to see that this result also holds when more than two particles are involved. Each particle exerts a force on each other particle, and according to the third law of motion these forces are equal and opposite in pairs and therefore produce equal and opposite momentum changes in any time interval. The total momentum of the system, that is, the sum of the momenta of the individual particles, is conserved whenever the only forces acting are forces between particles included within the system. This does not mean that the momentum of any one particle of the system is conserved; only the total momentum is conserved.

As a first example of momentum conservation consider two particles that are confined to a line, as two ideal railroad cars on a track or two beads on a wire, where the particles can move frictionlessly along the line. If one is initially at rest and the other is approaching with speed v as shown in Figure 11.1, the momentum of the first particle is zero and the other has a component mv along the line. Before they

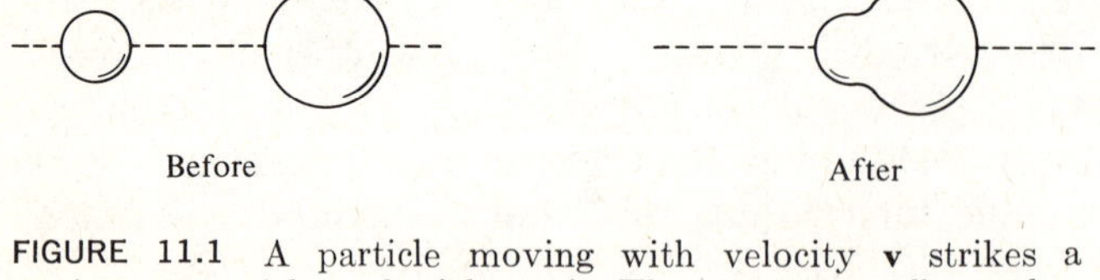

FIGURE 11.1 A particle moving with velocity $\mathbf{v}$ strikes a stationary particle and sticks to it. The two move off together with final velocity $\mathbf{V}$. Momentum is conserved in the collision.

collide, then, the total momentum **p** has a component along the line

$$p_{\text{before}} = mv \tag{11.5}$$

Suppose that after the collision the two particles stick together and move with the common speed V, so that the final total momentum has a component along the line

$$p_{\text{after}} = (m + M)V \tag{11.6}$$

In such a collision the particles act only on each other, at least as far as motion along the line is concerned, so that the component of momentum along the line is conserved and the two p's in the equations above are equal. Therefore

$$mv = (m + M)V \tag{11.7}$$

which can be solved to give an expression for the final speed V:

$$V = \frac{mv}{m + M} \tag{11.8}$$

In this example the final speed is determined by the conservation of momentum alone, but this is because the two particles have a common velocity after the collision. If they bounced apart afterward, there would be two unknown velocity components and only one equation to determine them, so momentum conservation alone would not suffice to fix the motion after the collision. As we shall see when the idea of energy is introduced, one often can say something about the energy of the system before and after, and this added information may be enough to determine completely the motion afterward. This is treated in detail in later sections.

As another example, consider two particles of equal mass attached to each other by a light string. A possible motion of the particles is motion in a circle with the string length as diameter, as shown in Figure 11.2. The momentum of either particle separately is changing with time, but one's momentum change is the opposite of the other's, and the total momentum is constant. In the case illustrated in the figure, the total momentum is zero. This can be looked at from another inertial reference frame, however, in which the center of the circle moves to the right with velocity **V**. As far as Newton's laws are concerned, one inertial reference frame is as good as any other, so the total momentum of the two particles as seen from the new frame also will be conserved, Newton's third law holding here, too. The paths of the particles are now much more complicated looking, being illustrated in Figure 11.3.

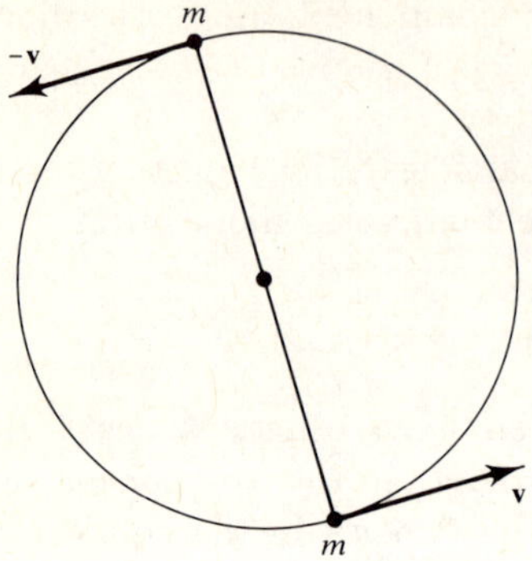

FIGURE 11.2 Two particles of equal mass attached to each other by a string whirl around in circles, the velocity of one particle always being opposite to that of the other.

At each point of the trajectory the velocity in Figure 11.3 is obtained from the velocity at the corresponding point in Figure 11.2 by adding the velocity **V** to it. At the top of one of the loops the particle speed is $v - V$, and at the bottom it is $v + V$. When the particle crosses the line midway between the top and bottom, in Figure 11.2 it would be moving straight up (or down) with speed v. In Figure 11.3, then, its vertical component of velocity is v and its horizontal component

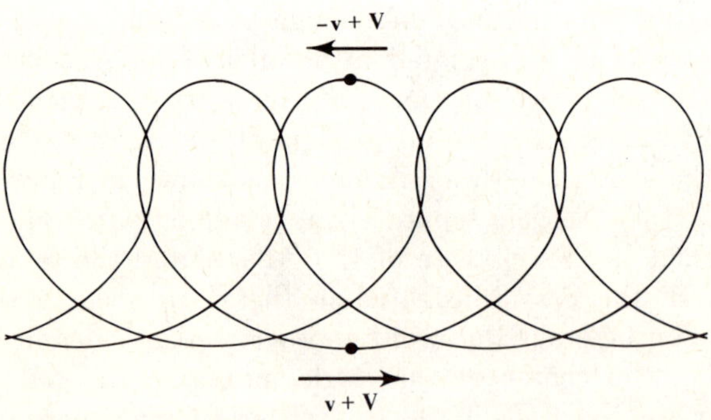

FIGURE 11.3 The system of Figure 11.2 as seen from a reference frame moving to the left with velocity $-\mathbf{V}$ in the reference frame of Figure 11.2. All velocities in the previous figure must be modified by adding **V** to them to obtain the velocities seen in this figure.

is V. Because the total momentum of the system is constant, it can be evaluated at any time during the motion, and the most convenient time is when one of the particles is at the top and the other is at the bottom. From the figure one sees that at such a time

$$\begin{aligned} \mathbf{p} &= \mathbf{p}_1 + \mathbf{p}_2 \\ &= m(-\mathbf{v} + \mathbf{V}) + m(\mathbf{v} + \mathbf{V}) \\ &= 2m\mathbf{V} \end{aligned} \tag{11.9}$$

even though at this time neither particle has momentum $m\mathbf{V}$.

Another example is the golf ball of Section 9, which we considered either as one particle or as two particles cemented together. The total momentum of the ball is the sum of the momenta of the two halves and is constant if the ball as a whole is a free particle. If the ball is spinning in flight, the two halves move like the two particles just described and neither half has a constant momentum, but the change in momentum of one half exactly compensates for the change in that of the other half.

If more than two particles are present, each particle can exert a force on each of the other particles, but according to Newton's third law, these forces will always be equal and opposite in pairs just as when only two particles are present, so that the sum of all the forces exerted by the particles on each other is zero. The total momentum of the system is, therefore, conserved unless external forces are acting upon it.

The conservation of momentum is most important in the study of collisions. During a two-particle collision the colliding particles usually exert forces on each other that are much larger than the external forces acting on the particles. Collisions being abrupt events, these collision forces last for only a short time. If this is so, it is a good approximation to neglect the change in the total momentum produced by the external forces during the collision and to say that the total momentum just before the collision is equal to that just after the collision. An example is provided by the situation shown in Figure 11.4, which is very like the example of Figure 11.1. A block of wood lies on a real (not frictionless) table, and a bullet is fired into it. In the short time that it takes the bullet to come to rest inside the block the force of friction between the bullet and the block is much greater than the force of friction between the block and the table, so that it is a good approximation to write

$$\mathbf{p}_{\text{after}} = \mathbf{p}_{\text{before}}$$

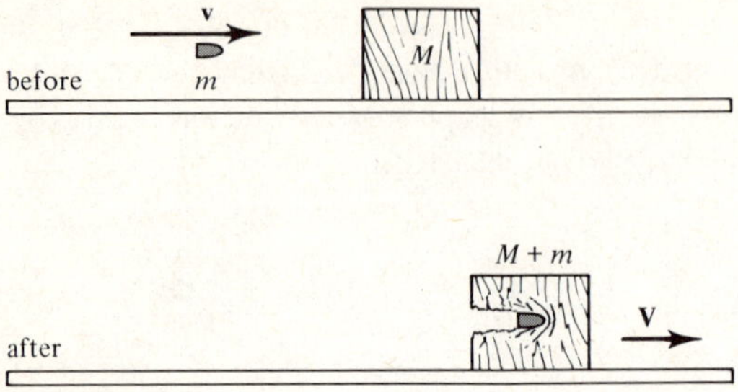

FIGURE 11.4 When a bullet lodges in a block of wood that is free to slide on a table, momentum is conserved during the collision.

where $\mathbf{p}_{\text{before}}$ is the momentum $m\mathbf{v}$ of the bullet and $\mathbf{p}_{\text{after}} = (M + m)\mathbf{V}$ is the momentum of the bullet and the block together right after the bullet has come to rest in the block. If the block is clamped to the table, the force of friction between the block and the table is vastly increased, and neglecting it would be grossly inaccurate.

Much of what we know about atomic and subatomic particles is learned through the study of collisions. It is seen in later chapters how the principle of the conservation of momentum is used in analyzing collisions between such particles.

Problems

1. Two boys standing on smooth ice push each other so that they go apart at 6 feet/sec. If the boys weigh 110 and 90 lb, respectively, what is the speed of each?
2. A 100-kg football player runs at a speed of 30 km/hr. What is his momentum? If he is stopped in $\frac{1}{2}$ sec, what is the average force exerted on him? How does his average acceleration compare with the acceleration of gravity?
3. If the earth should enter a cloud of stationary dust and if all the dust striking the surface of the earth should stick, explain how the momentum of the earth and the speed of the earth would be affected.

12. ANGULAR MOMENTUM

Linear momentum, as the momentum discussed in the last section is often called, is a concept that is readily understandable. The consequences of the conservation of linear momentum seldom appear surprising or paradoxical but are consistent with "common sense." Angular momentum, the topic of this section, is a very difficult concept to grasp and leads to the description of such things as the behavior of tops and gyroscopes, which do seem surprising at first acquaintance.

To introduce the subject, consider a particle of mass m that can move on a smooth, frictionless table. The particle is attached to the end of a string that passes through a small hole in the table and is held there, as shown in Figure 12.1 As the particle moves, the tension in the string

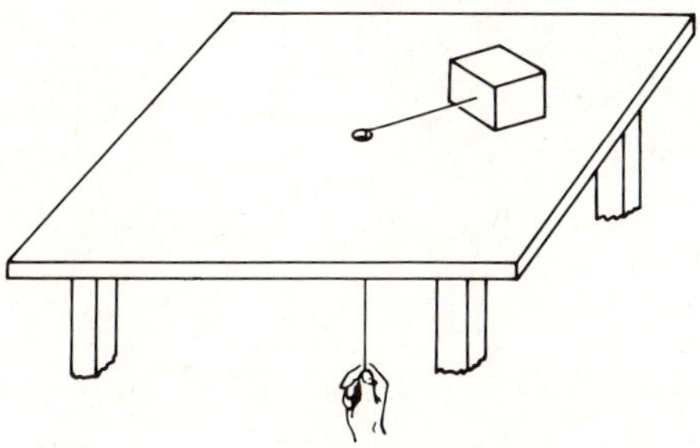

FIGURE 12.1 A particle moves on a smooth table at the end of a string whose length can be varied at will. Its energy and linear momentum are not conserved, but its angular momentum about the hole is conserved.

can be varied at will. Is it possible to define a quantity depending on the mass of the particle and on its motion in such a way that this quantity is conserved? Linear momentum is not conserved, because the string exerts a force on the particle. Energy has not been defined as yet, but energy depending only on the position and velocity of the particle is not conserved. A property of the force on this particle is that it is always directed toward a fixed point, the hole in the table; this makes possible the definition of a conserved quantity, the *angular momentum* about that fixed point. All central forces as defined in Section 10 share this property, so in all central force fields the angular momentum about the center of force is conserved.

A portion of the trajectory of the particle is shown in Figure 12.2. The hole through which the string passes is at O, and the particle is at P and moving with momentum $\mathbf{p}$. The angular momentum l of the

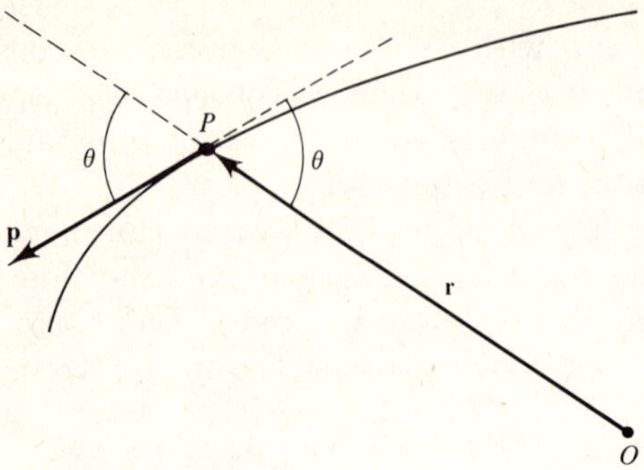

FIGURE 12.2 The momentum **p**, the radius vector **r**, and the angle θ between them are combined in the definition $l = rp \sin \theta$ of the angular momentum about the origin O.

particle about O is defined by

$$l = rp \sin \theta \tag{12.1}$$

where θ is the angle between the position vector **r** and the momentum **p** or velocity **v** of the particle. If the particle moves in a circle about O, θ is a right angle, $\sin \theta = 1$, and the $l = rp$.

In the definition of l, the sine can be combined either with **r** or with **p**, giving the component of **r** in a direction perpendicular to **p** or giving the component of **p** in a direction perpendicular to **r**. Thus writing

$$l = (r \sin \theta) \times p \tag{12.2}$$

it is the product of the component of **r** in a direction perpendicular to **p** and the magnitude of **p**. This is illustrated in Figure 12.3(a). The

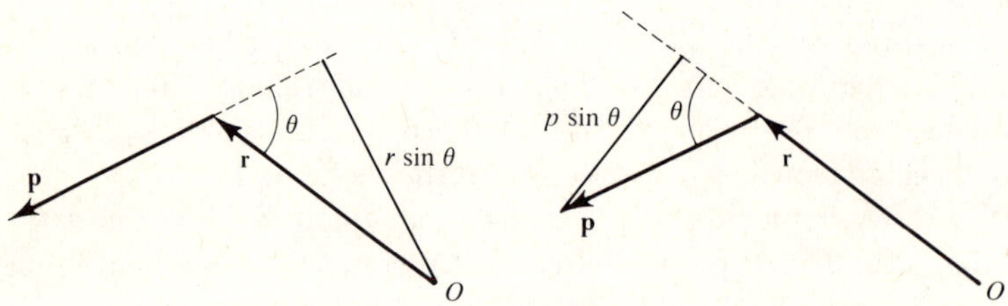

FIGURE 12.3 The factor $\sin \theta$ in the definition of the angular momentum l can be combined with **r** as in (a) to give the component of **r** perpendicular to **p** or with **p** as in (b) to give the component of **p** perpendicular to **r**.

other combination is

$$l = r \times (p \sin \theta) \tag{12.3}$$

which makes l the product of $\mathbf{r}$ and the component of $\mathbf{p}$ in a direction perpendicular to $\mathbf{r}$, as illustrated in part (b) of the figure. The two ways of writing l are, of course, equivalent to each other. If the particle is moving directly toward or directly away from O, $\sin \theta$ is zero and the angular momentum about O vanishes.

We now want to show that the angular momentum of the particle about O is constant no matter what force is exerted by the string, assuming only that the force is in the direction of the string, toward O. (If a string could push, forces directed away from O also could be included.) There are only two things that can change l, a change in $\mathbf{r}$ or a change in $\mathbf{p}$, since l depends only on these two vectors. During the short time interval Δt, the change in the vector $\mathbf{r}$ is given, as always, by

$$\Delta \mathbf{r} = \mathbf{v}\, \Delta t = \frac{\mathbf{p}}{m} \Delta t$$

which is the direction of $\mathbf{p}$. This means that the change in $\mathbf{r}$ does not involve any change in the component of $\mathbf{r}$ perpendicular to $\mathbf{p}$ and so does not change the angular momentum which, according to Eq. 12.2 and Figure 12.3(a), depends on $\mathbf{r}$ only through its component perpendicular to $\mathbf{p}$. This is true for any force, central or otherwise, or no force at all. During this same time interval, $\mathbf{p}$ changes by an amount

$$\Delta \mathbf{p} = \mathbf{F}\, \Delta t$$

When $\mathbf{F}$ is a central force and is, therefore, parallel or antiparallel to $\mathbf{r}$, then $\Delta \mathbf{p}$ also is either parallel or antiparallel to $\mathbf{r}$ and the component of $\mathbf{p}$ perpendicular to $\mathbf{r}$ is unchanged. According to Eq. 12.3 and Figure 12.3(b), however, only a change in the component of $\mathbf{p}$ perpendicular to $\mathbf{r}$ can change the angular momentum, so the change in $\mathbf{p}$ caused by this $\mathbf{F}$ does not change l. This last argument breaks down if $\mathbf{F}$ is not always parallel or antiparallel to $\mathbf{r}$, that is, if the force is not a central one with O as its center. Whenever $\mathbf{F}$ is central, the angular momentum about the force center is constant in time.

The conservation of angular momentum may be applied to the motion of a particle in the gravitational field of another, very massive particle, for example, the motion of an earth satellite about the earth. The orbit is generally elliptical with the earth at one focus of the ellipse, as shown in Figure 12.4. At perigee, when the satellite is closest to the earth, and at apogee, when it is farthest from the earth, the velocity of the satellite is perpendicular to the position vector from the center of the earth. Because the angular momentum l has the same value at all points of the orbit,

$$l = r_p m v_p = r_a m v_a \tag{12.4}$$

The speed at perigee therefore is larger than the speed at apogee in the inverse ratio of the two distances from the earth, which depends

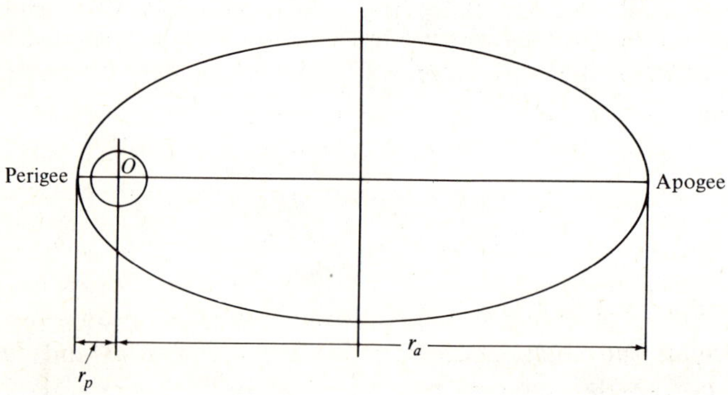

FIGURE 12.4 The velocity of an earth satellite moving in an ellipse with the earth at one focus is perpendicular to the radius vector from the earth at perigee, closest approach, and at apogee, farthest recession.

on the eccentricity of the orbit. For a circular orbit, $r_p = r_a$ and the speed is constant.

Kepler stated the law of conservation of angular momentum for the motion of the planets about the sun in geometrical language in his second law of planetary motion. The first law states that the planets move in elliptical orbits with the sun at one focus, and the third law relates the periods of rotation to the size of the orbits, as was stated in Section 10. Kepler's statement of his second law is that the position vector from the sun to any planet sweeps over equal areas in equal times. That this amounts to the conservation of angular momentum follows from the definition of angular momentum in the form Eq. 12.3, as can be seen with the help of Figure 12.5. Here the position vectors at two times Δt apart are drawn. The area of the triangle enclosed between

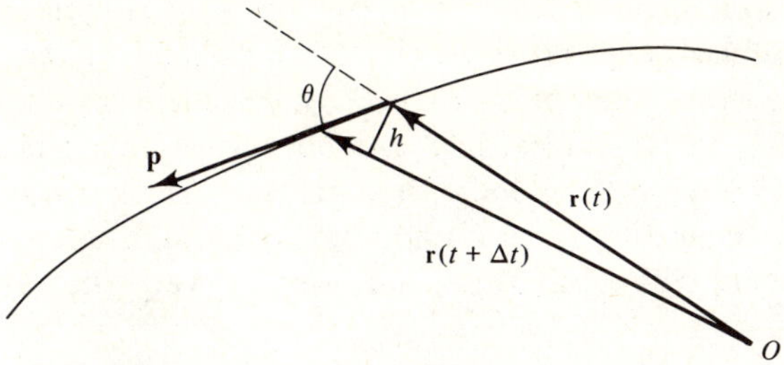

FIGURE 12.5 The angular momentum of a particle about O is proportional to the area swept out per unit time by the position vector from O to the particle. This "areal velocity" is constant when the angular momentum is conserved.

them, which is one-half the product of the base and the altitude of th triangle, is $\frac{1}{2}rh$. The altitude h is related to the velocity $\mathbf{v} = \mathbf{p}/m$ by

$$\begin{aligned} h &= v\,\Delta t \sin\theta \\ &= \frac{p \sin\theta}{m}\,\Delta t \end{aligned}$$

Thus the area ΔA swept over in the time Δt is

$$\begin{aligned} \Delta A &= \frac{1}{2}\frac{rp \sin\theta}{m}\,\Delta t \\ &= \frac{l}{2m}\,\Delta t \end{aligned}$$

and $\Delta A/\Delta t$ is constant if l is constant.

The earth is not, in fact, quite spherical, so that the force of gravity exerted by the earth on a satellite is not exactly a central force, although it very nearly is. Accurate observations of the orbits of artificial satellites show that their angular momentum is not exactly conserved, and the way in which it fails to be conserved can be interpreted to give the deviation of the shape of the earth from a sphere. It has long been known that the earth bulges at the equator, the equatorial radius being 6.378×10^3 km and the polar radius being 6.357×10^3 km. It was discovered by satellite observations that the earth has a shape slightly like a pear, the northern hemisphere being the narrow one.

The behavior of comets can be understood on the basis of angular momentum conservation. Comets that return to the neighborhood of the sun periodically are moving about the sun in extremely eccentric orbits, so that r_a is much greater than r_p. When they are far away from the sun they are moving slowly and take a long time to come back near the sun, but when they are near they are moving fast and soon leave the neighborhood. They are visible, therefore, for only a very small fraction of the time.

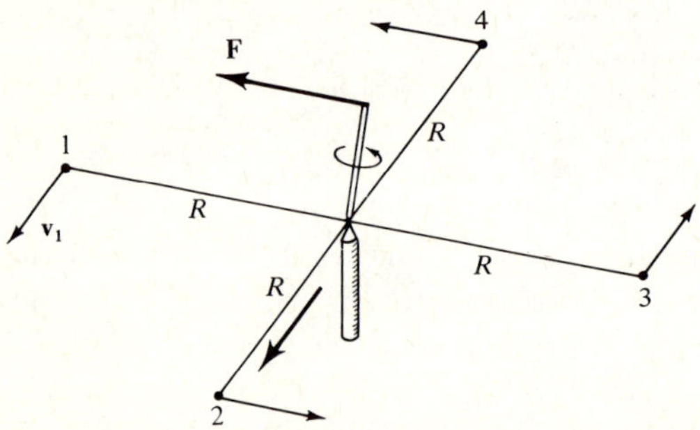

FIGURE 12.6 Four particles are held together at the corners of a square by a pair of light rods and pivoted at the center. If they are rotating about a handle perpendicular to the rods, they will continue to do so with constant speed when no forces are applied. When the force **F** is applied to the end of the handle as shown, the plane of motion tips about the line 1-3, the position 2 going down and the position 4 coming up.

Perhaps the most important application of the concept of angular momentum is in atomic and nuclear physics. The angular momentum of the electrons in atoms is one of the principal things that determine atomic structure. This is briefly described in Section 39.

Let us now consider a more complicated situation. Suppose that four identical particles are fastened together with light rods, the midpoints of which are fixed in position in such a way that the rods are free to turn in any direction, as shown in Figure 12.6. A handle is provided at right angles to the rods, and the whole thing is set spinning about the handle. It is balanced so that it is not affected by gravity, and in the absence of friction continues to rotate indefinitely. One now asks:

"What happens if a force **F** is applied to the handle, as shown in the figure, for a time interval so short that during it the particles do not move far from their indicated positions?" To keep the center from moving off the pivot, a force opposite to **F** must be exerted by the pivot, but this will not tend to cause any motion about the pivot point. The particle near 1 acquires a downward momentum, which it retains after **F** stops acting, so that subsequently it will move into a position below 2. The particle near 3 acquires an upward momentum while **F** is acting, so that subsequently it will move to a position above 4. The particles near 2 and 4 while **F** acts are not given any new momentum by this particular force and subsequently move to 3 and 1, respectively. The effect of the force, then, is to cause the particle at 2 to be lowered and the particle at 4 to be raised, so the plane of the four particles is tilted but not in the direction one might have expected. Although the force **F** is directed toward 1, the handle moves toward 2. If the direction of rotation of the system is reversed, the same force **F** will produce a tilt in the opposite direction, the handle moving toward 4. This so-called *gyroscopic behavior* is rather surprising and is difficult to analyze quantitatively in terms of forces and linear momenta because the force **F** applied to the handle does not act directly on the particles of the system but only indirectly through the rods. We shall now look at this behavior in terms of angular momentum, which is much simpler once one has accepted the idea of a *vector* angular momentum.

Equation 12.1 defines the magnitude of the angular momentum, but it is not at all obvious how to associate a single direction with the type of motion being discussed. The linear momentum vector points in the direction of motion of the particle having that momentum, but a particle moving about a fixed point may be going first one way and then another. There is, however, a unique axis that is determined by such motion, an axis perpendicular to the two vectors that enter the definition of angular momentum, **r** and **p**, and therefore perpendicular to the plane in which the particle moves. In the system of Figure 12.5 this is the direction of the handle, although a choice must be made whether to draw the angular momentum upward or downward along the line of the handle. The universally accepted rule for choosing this direction is the right-hand rule, which can be stated in many equivalent ways. One way is the following: If the fingers of the right hand are extended so as to point in the direction of the vector **r** and can be bent so as to point in the direction of **p**, then the right thumb points in the direction of l. Another way is like that given in Section 2, where a right-handed coordinate system was defined. If one looks at the vectors

r and **p** from such a position that a clockwise rotation of **r** through an angle of less than 180° makes it parallel to **p**, then one is looking in the direction of l. The justification for this seemingly arbitrary definition of the angular momentum vector is that it is useful as soon will be seen.

The change of momentum per unit time is force, a familar concept. The change of angular momentum per unit time has a very similar

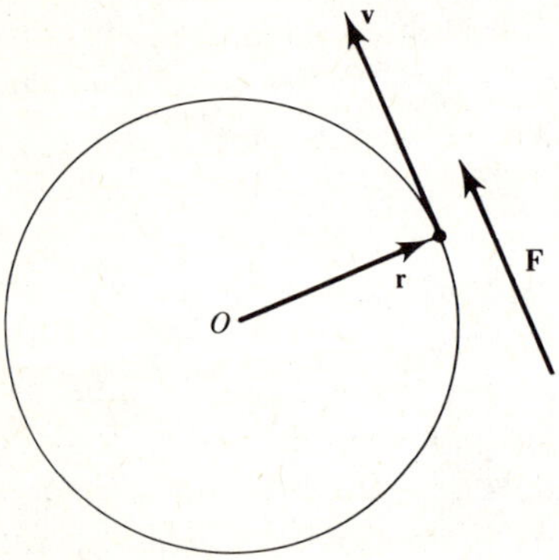

FIGURE 12.7 The change in angular momentum of a particle produced by a force parallel to the velocity is proportional to the product of the force and the perpendicular distance from the point about which the angular momentum is defined.

meaning. This is most easily seen in an example. If a particle of mass m is moving in a circle about a point O, as shown in Figure 12.7, it has angular momentum rmv directed out of the page according to the right-hand rule. Now let a force **F** act on the particle in the direction shown for a short time Δt. The change in the angular momentum produced is, since r does not change during this time interval,

$$\begin{aligned}\Delta l &= \Delta(rmv)\\ &= r\,\Delta(mv)\\ &= rF\,\Delta t\end{aligned} \tag{12.5}$$

The vector **l** is not changed in direction, only in length. Thus suggests that the turning effect of a force on a particle as measured by rF should be represented by a vector perpendicular to both **r** and **F** just as **l** is perpendicular to both **r** and **p**. This vector is called the *torque* and is often denoted by **M**. The torque is the change in angular momentum per unit time. The direction of the torque is also defined by a right-hand rule: If the fingers of the right hand are extended so as to point in the direction of the vector **r** and can be bent so as to point in the direction of **F**, the right thumb points in the direction of the torque.

Now consider the system of Figure 12.4 in terms of angular momentum and torque. With the particles turning about the handle as shown, the

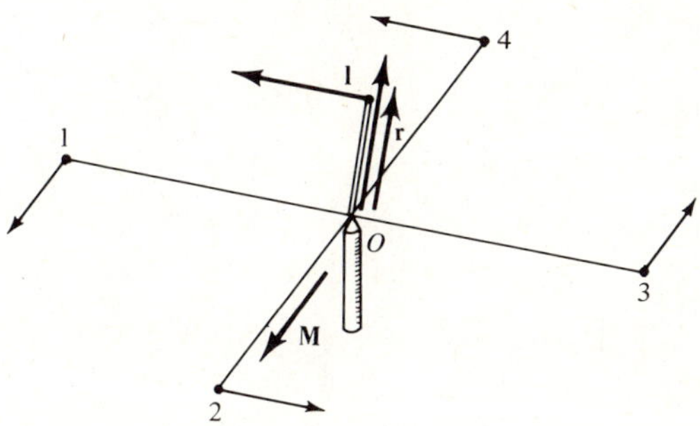

FIGURE 12.8 The system of Figure 12.6 is analyzed according to angular momentum. The angular momentum vector **l** is along the handle, and the torque **M** produced by the force applied to the handle is pointing toward position 2, causing the angular momentum vector to tilt in this direction.

angular momentum vector is up along the handle, as in Figure 12.8, having the magnitude 4 Rmv. The torque due to **F** about the pivot O is **M**, drawn according to our definition perpendicular to both **r** and **F**, pointing in the direction of position 2, and having a magnitude of rF. The torque due to the reaction at the pivot is zero. The change in the angular momentum vector $\Delta\mathbf{l}$, produced by this torque in the time Δt, points in the direction of **M**, and the effect therefore is to tip **l** toward 2, which is just the conclusion we reached before. Once one accepts the notion of a vector angular momentum and a vector torque in the directions assigned, the analysis of such effects is greatly simplified, but the notions are far from obvious ones. The use of angular momentum and torque avoids the necessity for analyzing the magnitudes and directions of forces transmitted from one point of an extended system to another, as the force **F** is transmitted to the particles at 1 and at 3 but not to those at 2 and at 4 by the structure of the handle, rods, and pivot in Figure 12.6. Gyroscopic effects occur in an important way only in the motion of extended systems because only in them can forces acting at one point be transmitted to another point.

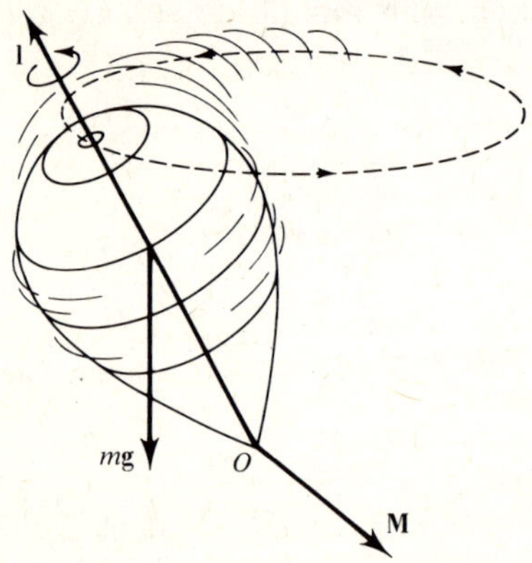

FIGURE 12.9 The force of gravity acting on the top produces a torque about the point of the top in a direction perpendicular to this force and to the radius vector from the point to the center of the top. This torque causes the angular momentum vector **l** to change its direction, and the top precesses about the vertical direction.

It is now easy to understand why a spinning top does not fall down. If the axis of the spinning top is not vertical, the force of gravity tends to pull the axis toward the ground, as shown in Figure 12.9. This force produces a torque about the point of the top that is on the ground, the torque vector being horizontal. The contact force of the ground on the point produces no torque about the point but does prevent the top as a whole from falling with acceleration **g**. The angular momentum vector does not tend to fall over, but rather to move sideways so that the top precesses about a vertical axis through the point.

It is impossible to build a box that contains particles with a large total linear momentum and that will stay in one place. The particles eventually will hit the side of the box and will transfer momentum to it. The presence of linear momentum is therefore always rather apparent. With angular momentum things are different. One can build a box and put a wheel inside that is rotating fast about an axis whose ends are fastened to the box. From the outside the box looks completely inert, and if one moves it without turning it, its behavior is normal, but if one tries to turn it in one direction, it may move in another because of the large angular momentum contained in the box. Figure

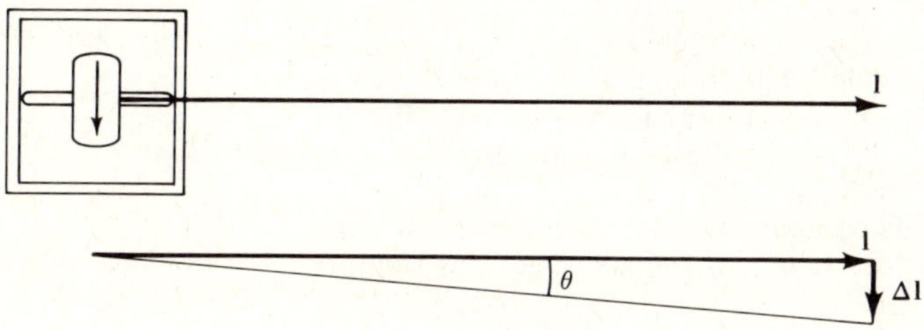

FIGURE 12.10 A box containing a spinning flywheel has a large angular momentum associated with it. To turn this angular momentum vector through even a small angle involves considerable change in angular momentum and therefore considerable turning effort.

12.10 gives an X-ray view of the box with its (invisible) angular momentum vector. To turn the box through the angle θ shown requires a change of angular momentum

$$\Delta l = l\theta \tag{12.6}$$

The bigger l is, the bigger Δl corresponding to a given θ is and the more surprising the reaction of the box to attempts to turn it will be. Only rotations of the box about an axis parallel to $\mathbf{l}$ do not produce such reactions, such rotations not causing any change in the direction of $\mathbf{l}$.

The particles of which matter is made, electrons and atomic nuclei, act a little like the box described above. Even when they are at rest they have an angular momentum, called spin, which affects their behavior. This will come up later when the properties of such particles are discussed.

Problems

1. Two particles of masses m_1 and m_2 have opposite momenta, **p** and —**p**, and pass each other at a distance b as shown. Find the angular momentum of the two particles about

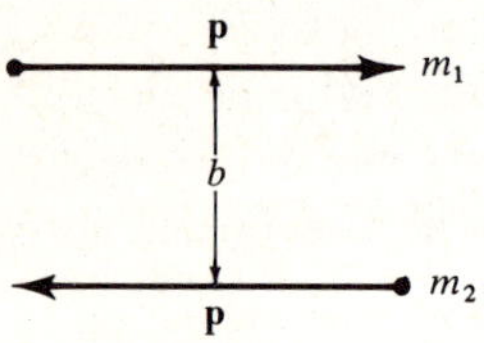

   (a) the position of 1 when it is closest to 2, (b) the position of 2 when it is closest to 1, and (c) the point midway between these two. Can you explain why all the answers are the same? Would they be the same if the two momenta were not opposite to each other?
2. Does the spin angular momentum of the earth point from south to north or from north to south? Show your solution in a sketch.
3. If an enormous rocket were launched from the north pole horizontally along the circle of longitude 0°, in what direction would the earth's polar axis turn? (Along what circle of longitude would the north pole move?)
4. The perhelion distance of the earth (the closest approach of the earth to the sun) is 1.47×10^{13} cm, and the aphelion distance is 1.52×10^{13} cm. If the speed at aphelion is 3×10^6 cm/sec, what is its speed at perhelion? Does the answer depend on the fact that an inverse square force law is acting or only on the fact that angular momentum is conserved?

13. ENERGY

Energy is one of the most fundamental concepts in physics. It appears in a number of forms, and energy in one form can be transformed into energy of another form, although this transformation cannot always be complete. The great importance of the energy concept in physics and in all science and technology comes from the possibility of defining energy in such a way that it is conserved. In any system that is isolated so that it does not interact with anything outside itself, the total energy including all forms energy can have is constant and does not change with time. When a system is not isolated but does interact with its surroundings, the energy of that system need not be conserved, but any change in its energy must be at the expense of the energy of the surroundings. If the system's energy increases, the surrounding's energy must decrease by the same amount, and one is able to identify the means by which this energy exchange takes place.

Two forms of energy are of especial importance in the study of the motion of particles. One of these is associated with motion and inertia and is called *kinetic energy*. The second is associated with the relative positions of particles and is called *potential energy*. Together these are called *mechanical energy*. If one imagines that in the end everything is made of particles, then ultimately these forms of energy should be the only ones. It is not useful, however, to be so restrictive in the definition of energy. In physics we must often deal with large-scale objects such as electronic devices or golf balls, and these do not always behave like particles without structure but can behave in ways that depend strongly on their construction. In a large object there can be a form of energy called *heat* that can be resolved into the mechanical energy of many, many particles, but it is useful to be able to discuss the energy of the large object without bothering about how it is made of particles, so we keep heat as a useful idea. The energy associated with sound can be similarly resolved into mechanical energy, but it is often useful not to have to do this but rather to retain sound energy as a distinct form of energy.

In this section we are concerned only with mechanical energy. To make this restriction we have to give up considering certain kinds of force, in particular the force of friction. The action of friction forces converts mechanical energy into heat, as in the brakes of an automobile. This is a very serious omission for many purposes, but for the purposes of this book it is not so serious and we will be able to say something about frictional forces later on.

Energy is defined to be a scalar quantity. Even kinetic energy, the energy due to motion, does not depend on the direction of that motion but only on the speed. It is defined so that the total energy of two independent systems is the sum of the energies of the systems considered separately. If it were not defined in this way, it would be difficult to say what is meant by "energy conservation." We shall see that this is not a trivial point.

Kinetic energy is energy of motion, associated with the inertia of the moving particle. It is not associated with how the particle came to be moving but only with its instantaneous motion. The simplest expression that can be written down that has the desired properties is

$$K = \tfrac{1}{2}m\mathbf{v}^2 = \tfrac{1}{2}m(v_x^2 + v_y^2 + v_z^2) \tag{13.1}$$

We proceed to examine why this is a good definition of kinetic energy.

First, K is a scalar. It depends only on the mass m, measuring the inertia, and on the square of the speed or velocity, a quantity independent of the direction of $\mathbf{v}$. Second, it has a very useful additivity that can be seen by considering two identical particles of mass m moving side by side with the common velocity $\mathbf{v}$, as in Figure 13.1. According to the definition, each particle has kinetic energy $\frac{1}{2}m\mathbf{v}^2$. One may, however, want to consider these two particles as constituting a single particle

FIGURE 13.1 The kinetic energy of two particles moving side by side with the same velocity is the sum of the individual kinetic energies by definition.

of mass $2m$ moving with velocity $\mathbf{v}$. The total kinetic energy is the same no matter which way the two particles are thought of, because

$$\begin{aligned} K_{\text{tot}} &= \tfrac{1}{2}m\mathbf{v}^2 + \tfrac{1}{2}m\mathbf{v}^2 \\ &= \tfrac{1}{2}(2m)\mathbf{v}^2 \end{aligned} \tag{13.2}$$

The first line corresponds to the two particles considered separately; the second line, to the composite particle picture. If kinetic energy had been defined as $\frac{1}{2}m^2\mathbf{v}^2$, for example, this additivity would be absent, as the first picture gives a total "kinetic energy" of

$$\tfrac{1}{2}m^2\mathbf{v}^2 + \tfrac{1}{2}m^2\mathbf{v}^2 = m^2\mathbf{v}^2$$

while the second gives

$$\tfrac{1}{2}(2m)^2\mathbf{v}^2 = \tfrac{1}{2}(4m^2)\mathbf{v}^2 = 2m^2\mathbf{v}^2$$

which is twice as big.

The numerical factor $\frac{1}{2}$ is chosen purely because it makes formulas we encounter later simple.

The kinetic energy of a particle is closely connected with its momentum $\mathbf{p}$. From the definition of K and of $\mathbf{p}$ we see that

$$K = \frac{1}{2}\frac{m^2\mathbf{v}^2}{m} = \frac{1}{2m}\,\mathbf{p}^2 \tag{13.3}$$

When the particle is free, $\mathbf{p}$ is constant and so too is the kinetic energy K.

The definition of kinetic energy determines the units and dimensions of energy. The dimensions are $m(l/t)^2 = ml^2t^{-2}$. When the mass is measured in grams and the velocity in centimeters per second, the resulting unit of energy is the *erg*. When the mass is measured in kilograms and the velocity in meters per second, the energy unit is the *joule*. These two units are of very different sizes, one joule equaling 10^7 ergs. Another energy unit, the electron volt, is much used in physics. One erg is equal to 0.625×10^{12} electron volts. How this unit arises is shown in Section 16.

To see that the definition Eq. 13.1 of the kinetic energy of a particle can be a useful one, take the case of a particle moving in a uniform field of force, say in the gravitational field near the surface of the earth. This motion was studied in Section 6, and according to Eq. 6.14 the vertical motion of a particle under these circumstances satisfies the equation

$$y = \frac{v_{0y}^2 - v_y^2}{2g}$$

where v_{0y} is the vertical speed of the particle at the initial point and y is the height above the initial point, the acceleration $\mathbf{g}$ being in the negative y-direction. The equation can be rearranged to read

$$\tfrac{1}{2}mv_y^2 + mgy = \tfrac{1}{2}mv_{0y}^2 \tag{13.4}$$

For motion in this force field, the horizontal components of the velocity are constant, so that to the left-hand side of Eq. 13.4 one may add $\frac{1}{2}m(v_x^2 + v_z^2)$ and to the right-hand side one may add the equal quantity $\frac{1}{2}m(v_{0x}^2 + v_{0z}^2)$. As a result, since $\mathbf{v}^2 = v_x^2 + v_y^2 + v_z^2$ and similarly for $\mathbf{v}_0^2$ we see that

$$\tfrac{1}{2}m\mathbf{v}^2 + mgy = \tfrac{1}{2}m\mathbf{v}_0^2 \tag{13.5}$$

This equation is of the form of a conservation law. It says that the kinetic energy plus something depending only on the position but not on the velocity of the particle has a constant value, that of the initial kinetic energy. For this system, then, the potential energy V can be identified as

$$V = mgy \tag{13.6}$$

The potential energy vanishes at the initial position, the entire energy $E = K + V$ being equal to the initial kinetic energy. K and V are plotted as functions of y in Figure 13.2. Note that a constant can always be

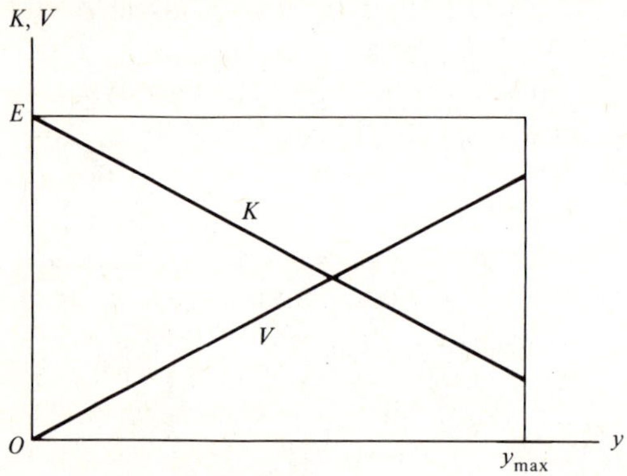

FIGURE 13.2 The sum of the kinetic and potential energies of a particle moving in a uniform gravitational field is constant. At the maximum height when $y = y_{max}$, the kinetic energy is just that associated with the horizontal motion of the particle. If the particle is moving straight up and down, the kinetic energy at y_{max} is zero.

added to V, the only effect being to add the same constant to E but not changing the fact that E is conserved. V, therefore, can be made to vanish at any convenient point. The kinetic energy K is chosen to vanish when the particle is at rest, so no constant can be added to it.

The definition of the kinetic energy of a particle, Eq. 13.2, has led to a potential energy in a gravitational field which, combined with the kinetic energy, gives a conserved total energy. Instead of a freely falling body, consider an ideal car moving frictionlessly along a roller coaster track, as shown in Figure 13.3. There are now two forces on the particle, the force of gravity and the force provided by the track. Is there a

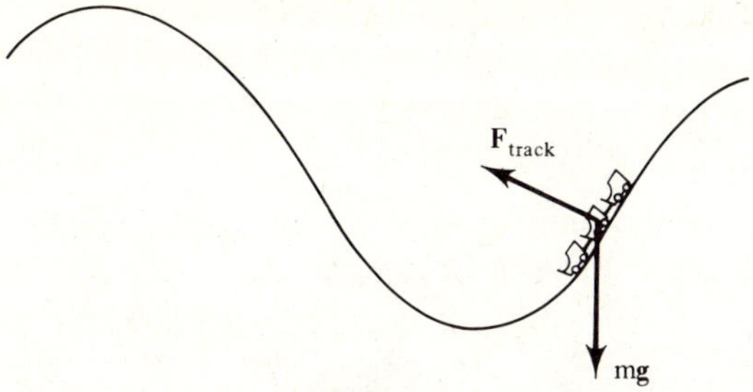

FIGURE 13.3 The forces on a roller coaster car are those of gravity acting down and the force of the track acting perpendicular to the track in the absence of friction. The latter does not change the speed of the car and therefore does not change its kinetic energy.

potential energy associated with the force exerted by the track such that

$$\begin{aligned} E &= K + V_{\text{grav}} + V_{\text{track}} \\ &= K + mgy + V_{\text{track}} \end{aligned}$$

is conserved? The force exerted on the car by the track (in the absence of friction) is just that needed to prevent the train's leaving the track and has no component parallel to the track. At the beginning of Section 7 it was shown that such a force does not change the value of $\mathbf{v}^2$, though it does change the direction of $\mathbf{v}$. If this force alone were acting, the kinetic energy would be constant and energy conservation would be satisfied by choosing the potential energy V_{track} to be zero everywhere. The effect of the track on the expression for energy conservation is to leave it as it is, so Eq. 13.4 describes the conservation of energy for a roller coaster as well as for a freely falling body. This has as a consequence that during any motion with total energy E, the speed of the roller coaster depends only on its height and not on whether it is going uphill, downhill, or on the level at that height. On different runs with different values of E, the speed at any height varies from run to run. Obviously a real roller coaster suffers from friction and eventually comes to rest.

As another example, let us try to construct a potential energy for a harmonic oscillator in such a way that total mechanical energy $E = K + V$ is conserved. According to Eqs. 8.3 and 8.8, the displacement and the velocity of the oscillator are

$$x(t) = A \cos \omega t$$
$$v(t) = -\omega A \sin \omega t \tag{13.7}$$

where A is the amplitude of oscillation and ω is the angular frequency of the oscillator. The kinetic energy of the oscillator is then

$$\begin{aligned} K &= \tfrac{1}{2}mv^2 \\ &= \tfrac{1}{2}m\omega^2 A^2 \sin^2 \omega t \end{aligned} \tag{13.8}$$

Remembering the identity $\sin^2 \theta + \cos^2 \theta = 1$, which is true for any angle θ, one sees right away what potential energy must be added to K to get a constant. It is

$$\begin{aligned} V &= \tfrac{1}{2}m\omega^2 A^2 \cos^2 \omega t \\ &= \tfrac{1}{2}m\omega^2 x^2 \end{aligned} \tag{13.9}$$

This is usually written as

$$V = \tfrac{1}{2}kx^2 \tag{13.10}$$

where k is the spring constant defined in Eq. 10.15 and the relation $m\omega^2 = k$ given in Eq. 10.18 is used. V is plotted as a function of x in Figure 13.4. With this potential energy V,

$$\begin{aligned} E &= K + V \\ &= \tfrac{1}{2}m\omega^2 A^2 \end{aligned} \tag{13.11}$$

is the constant total energy of the oscillator.

The existence of a potential energy that leads to a conserved total mechanical energy $E = K + V$ has been shown in two cases. The method used to find it is not a good one, however, because in both cases the known motion of the particle had to be used to discover what the potential energy is. A conservation law is most useful when it can be applied to a system in which the forces acting are known but whose detailed motion has not necessarily been found. In order to make use of the conservation of energy, one must be able to construct the potential energy directly from the knowledge of the force acting on the particle without having to find the complete motion of the particle first.

The connection between force and potential energy can be derived by generalizing the example treated first in this discussion, that of a particle moving in a uniform field of force. Over any sufficiently short

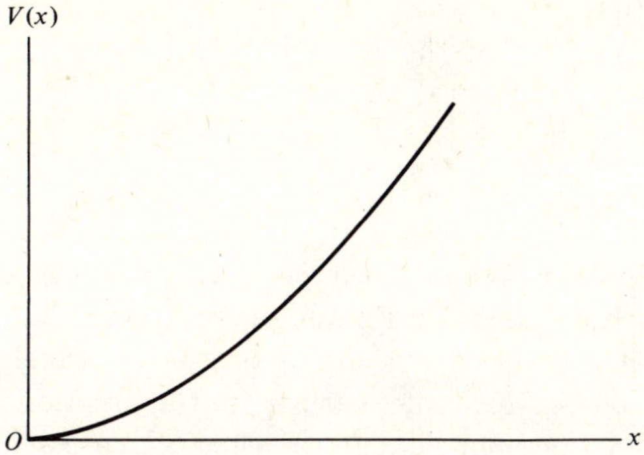

FIGURE 13.4 The potential energy of a harmonically bound particle increases as the square of its displacement from equilibrium. This makes the force proportional to the displacement so that Hooke's law is valid.

time interval the force acting on a particle may be treated as constant, both in magnitude and direction. If the positive x-axis of a coordinate system is chosen to lie in the direction of this force, Eq. 13.5 may be rewritten in the form

$$\tfrac{1}{2}m(\mathbf{v} + \Delta\mathbf{v})^2 - F\,\Delta x = \tfrac{1}{2}m\mathbf{v}^2(t) \qquad (13.12)$$

with $-F$ replacing mg, which is the same as

$$K(\mathbf{v} + \Delta\mathbf{v}) = K(\mathbf{v}) + F\,\Delta x \qquad (13.13)$$

This relates the change in the kinetic energy to the product of the force F and the component of the displacement of the particle in the direction of the force, here chosen to be the x-direction.

If there is a conserved mechanical energy $E = K + V$, then K can be replaced by $E - V(\mathbf{r})$ and the above equation becomes

$$E - V(\mathbf{r} + \Delta\mathbf{r}) = E - V(\mathbf{r}) + F\,\Delta x$$

or

$$V(\mathbf{r} + \Delta\mathbf{r}) = V(\mathbf{r}) - F\,\Delta x \qquad (13.14)$$

The quantity $F\,\Delta x$ appearing in these equations is called the *work* done by the force on the particle in the displacement from $\mathbf{r}$ to $\mathbf{r} + \Delta\mathbf{r}$.

It is the product of the magnitude of the force $\mathbf{F}$ and the component of the displacement $\Delta\mathbf{r}$ in the direction of the force $\mathbf{F}$, which here has been chosen to be the x-direction. In general the work ΔW done by a force $\mathbf{F}$ in a displacement $\Delta\mathbf{r}$ is written

$$\Delta W = F\,\Delta r \cos\theta \tag{13.15}$$

where θ is the angle between $\mathbf{F}$ and $\Delta\mathbf{r}$. When the force is at right angles to the displacement, no work is done and the kinetic energy, according to Eq. 13.13, is not changed. This is the old result of Section 7, which was used again in the roller coaster example earlier in this section; an acceleration perpendicular to $\mathbf{v}$ does not change the speed. Expressed in terms of work, Eq. 13.14 is written

$$V(\mathbf{r} + \Delta\mathbf{r}) = V(\mathbf{r}) - \Delta W \tag{13.16}$$

This equation provides a connection between the difference in the potential energy at two nearby points and the force acting. When $\mathbf{F}$ is known, this equation can be solved for $V(r)$, but the general methods are too mathematical for discussion here. A few examples will be given.

Nothing would be learned by considering the uniform force field; this force is the one that led us to Eq. 13.16, and to rederive the potential mgy with use of this relation would be to go around in a complete circle. The harmonic oscillator of Section 10 is worth doing. First let us recognize that $\Delta r \cos\theta$ in Eq. 13.15 is the same as Δx, since the oscillator in Section 10 is moving along the x-axis. For the oscillator, the force is also along the x-axis and has the value

$$F = -kx$$

This force is variable in magnitude, being twice as large when the displacement is twice as great. The variation of F with x is shown in Figure 13.5. The potential energy when the particle is at $x + \Delta x$ is related to the potential energy when it is at x by the equation

$$V(x + \Delta x) = V(x) - F\,\Delta x$$

like Eq. 13.16 that now assumes the specific form

$$V(x + \Delta x) = V(x) + kx\,\Delta x$$

The change in V is just the area of the narrow strip of height $-F = kx$ and width Δx shown in the figure. Starting at the origin, the sum of all the changes in V on going out to the value x is the sum of the

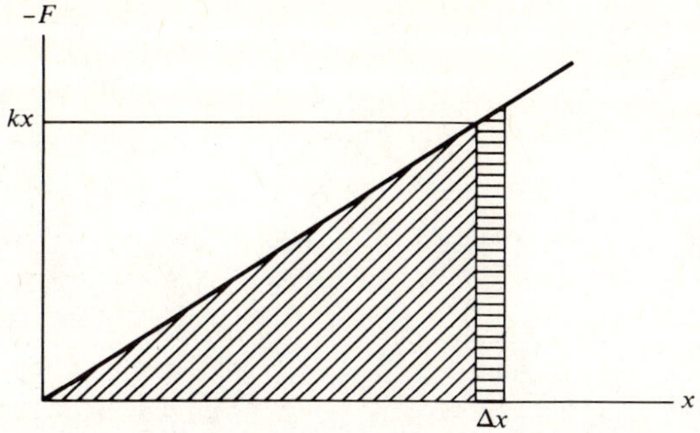

FIGURE 13.5 The work done in moving a harmonically bound particle from the equilibrium point to the point x is the sum of all the strips into which the shaded triangle can be divided, namely, the area of the triangle itself.

areas of all the little vertical strips into which the shaded triangle of Figure 13.5 can be divided, that is, the area of the triangle itself. This is one-half of the base times the height of the triangle, or

$$V(x) = -\tfrac{1}{2}F(x)x$$
$$= \tfrac{1}{2}kx^2$$

which is the same as Eq. 13.10, but here it has been derived without using the formulas for x and v as functions of the time as was done before.

Another example of a potential energy that is very important is that associated with Newton's law of gravitation. The general motion resulting from this force has not been found here, so the conservation of energy in this case will lead to some new results. According to Newton. there is a force of attraction between any two particles whose magnitude is

$$F(r) = \frac{Gm_1m_2}{r^2} \tag{13.17}$$

The potential energy associated with this force is

$$V(r) = \frac{-Gm_1m_2}{r} \tag{13.18}$$

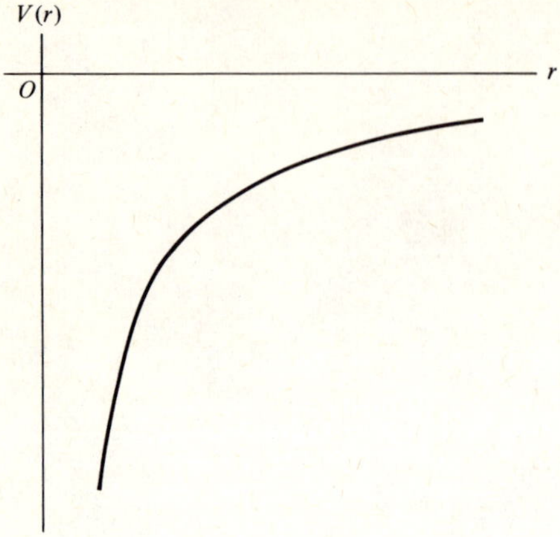

FIGURE 13.6 The potential energy associated with Newton's law of gravity is negative and becomes infinitely large when the separation r becomes zero. This has never been known to happen in nature because no point masses are known, and at very short distances other forces also act.

which is plotted in Figure 13.6. To see that this is consistent with Eq. 13.16, we evaluate

$$V(r + \Delta r) - V(r) = -Gm_1m_2\left(\frac{1}{r + \Delta r} - \frac{1}{r}\right)$$

Putting the two fractions over a common denominator gives

$$V(r + \Delta r) - V(r) = Gm_1m_2 \frac{\Delta r}{r(r + \Delta r)}$$

In the denominator the small quantity Δr can be neglected relative to r, but it cannot be neglected in the numerator where it is all there is. Then

$$\begin{aligned} \Delta V &= \frac{Gm_1m_2}{r^2}\,\Delta r \\ &= -F_r\,\Delta r \end{aligned}$$

which verifies Eq. 13.16. The sign is correct, the $F(r)$ in Eq. 13.17 being an attractive force tending to reduce r and therefore being in the negative

r-direction. The r-component of the force of gravitation is given by

$$F_r = \frac{-Gm_1m_2}{r^2}$$

The potential energy $V(r)$ of Eq. 13.18 is chosen to vanish when $r = \infty$, that is, when the two particles are infinitely far apart. It is then necessarily negative for finite r. To see this consider a particle at rest far from the earth. It has no kinetic energy and almost no potential energy, so its total mechanical energy is negligibly small. It is attracted to the earth, moving toward it with increasing speed and increasing kinetic energy. Its total energy being conserved during the fall, its potential energy must decrease correspondingly, and because it starts from almost zero it must become negative. The energy conservation equation is

$$E = \tfrac{1}{2}mv^2 - \frac{GmM_E}{r} \qquad (13.19)$$

Therefore, if a particle starts from rest far from the earth, its total mechanical energy E is essentially zero, so the speed with which it will hit the surface of the earth where $r = R_E$ is given by putting $E = 0$ and $r = R_E$ in Eq. 13.19. This gives the result

$$v^2 = \frac{2GM_E}{R_E}$$

Using the values given in Section 10, one obtains

$$\begin{aligned} v &= 1.12 \times 10^6 \text{ cm/sec} \\ &= 2.5 \times 10^4 \text{ mph} \end{aligned}$$

The argument can be turned around. If a particle is projected upward from the surface of the earth with this speed, it will just be able to get away from the earth. If its speed is smaller, it will not be able to get beyond a certain finite distance and then will return. This critical speed is called the *escape speed*.

It is interesting to find the total mechanical energy E of a particle moving in a circular orbit in a gravitational field. The kinetic energy can be found from the condition for circular motion, that the force of gravitation be just the centripetal force needed to give the circular motion, or

$$\frac{GmM}{r^2} = \frac{mv^2}{r}$$

From this,

$$K = \tfrac{1}{2}mv^2 = \frac{1}{2}\frac{GmM}{r}$$

$$= -\tfrac{1}{2}V(r) \tag{13.20}$$

Therefore,

$$E = K + V = \tfrac{1}{2}V(r)$$

$$= -\frac{1}{2}\frac{GmM}{r} \tag{13.21}$$

The total mechanical energy is negative and is just one-half of the potential energy. The meaning of this negative energy is simply that energy must be provided to separate the particles, the kinetic energy of the motion not being big enough to do this. In this system both energy and angular momentum are conserved.

A graph of the potential energy of a particle as a function of its position is very helpful in visualizing the kinds of motion that can take place. Because the kinetic energy of a particle is always positive, the total mechanical energy of a particle can never be less than the potential energy. If on a potential energy graph one draws a horizontal line at a height E corresponding to the total energy, as in Figure 13.7 where the potential energy is the gravitational potential energy of Figure 13.6, one sees that the range of the coordinate r over which the motion can extend may be limited. If the energy is E_1 as shown, in this case negative, the particle cannot get further from the attracting center than the distance r_1, the distance at which $V(r) = E_1$, as beyond this distance energy could be conserved only by having a negative kinetic energy present. If the energy is E_2, greater than zero, the figure shows that no matter how large r may be, $V(r)$ is always less than E_2 so that there is no upper limit on r. The least value of E for which r is not limited is clearly $E = 0$. If R_E is the radius of the earth, E_0 in the figure is the total energy of the particle when it is at rest on the surface of the earth, a negative quantity. To enable the particle to leave the earth completely, it must be given a kinetic energy of at least $-E_0$. The speed corresponding to this kinetic energy is the escape speed.

Not all forces are related to a potential energy through Eq. 13.16. In fact, only forces that do not depend on the velocity of the particle can be related to a potential energy in this way, because if $\mathbf{F}$ in Eq. 13.15 changes with $\mathbf{v}$, so must V, and it has been assumed that the

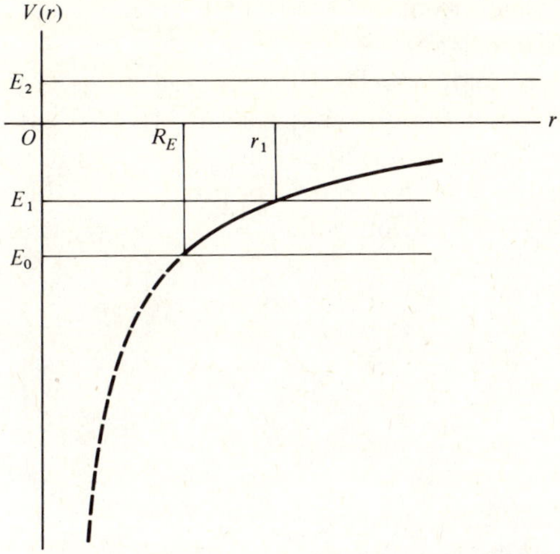

FIGURE 13.7 For a particle moving in the gravitational field of the earth, not necessarily close to the surface, the Newtonian potential energy is correct only when the particle is above the earth's surface. If it has energy E_1, it cannot get further from the center of the earth than r_1, while if it has energy E_2, it can escape from the earth.

potential energy is energy of *position*. One important example of a force not connected to a potential energy is the force of friction. It always acts in opposition to motion and is therefore always antiparallel to $\mathbf{v}$. The work done by such a force is always negative, decreasing the energy of the particle it acts on. This energy is converted into heat, conserving total energy but not mechanical energy alone. Another example we shall meet is the force on a charged particle due to a magnetic field. This is always perpendicular to the velocity of the particle and so does no work on the particle, leaving its kinetic energy unchanged although changing the direction of motion of the particle. Forces related to a potential by Eq. 13.16 are called *conservative* forces.

The principle of conservation of mechanical energy leads to very important results when it can be combined with the principle of conservation of linear momentum. This can be done, for example, when two particles interact with each other through conservative forces and are otherwise free. The conservative character of the forces means that mechanical energy is conserved, and the freedom from external forces means that total linear momentum is conserved.

Suppose a projectile of mass m_1 and velocity $\mathbf{v}_1$ strikes a target of mass m_2 which is at rest, the collision forces being conservative. What is the maximum energy that the target can acquire as a result of the collision? It is intuitively plausible and not hard to prove that this maximum energy transfer occurs in a head-on collision so that $\mathbf{v}_2'$, the

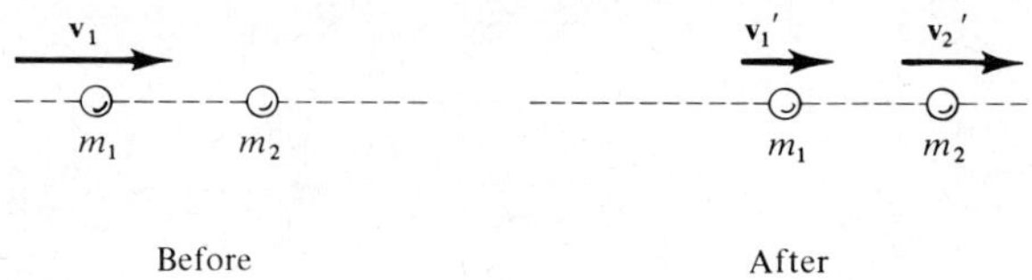

FIGURE 13.8 In a collision between two particles that exert conservative forces on one another, both energy and momentum are conserved. A head-on collision is shown in which neither particle has any momentum transverse to the original velocity direction after the collision.

velocity of the target after the collision, is parallel to $\mathbf{v}_1$, as shown in Figure 13.8. In this case the conservation of energy requires that

$$\tfrac{1}{2}m_1\mathbf{v}_1^2 = \tfrac{1}{2}m_1\mathbf{v}_1'^2 + \tfrac{1}{2}m_2\mathbf{v}_2'^2 \tag{13.22}$$

because the potential energy is zero when the particles are separated from each other. Conservation of momentum requires that

$$m_1\mathbf{v}_1 = m_1\mathbf{v}_1' + m_2\mathbf{v}_2' \tag{13.23}$$

This equation is a vector equation, but for head-on collisions only the components along the direction of motion matter, the rest vanishing. To see what these equations imply rewrite them as

$$v_1^2 - v_1'^2 = \frac{m_2}{m_1} v_2'^2$$

$$v_1 - v_1' = \frac{m_2}{m_1} v_2'$$

Recalling that $a^2 - b^2 = (a - b)(a + b)$, one sees that the first equation can be divided by the second to obtain

$$v_1 + v_1' = v_2'$$

There are now two linear equations v_1' and v_2'. Adding the two yields

$$2v_1 = \left(\frac{m_2}{m_1} + 1\right) v_2'$$

or

$$v_2' = \frac{2m_1}{m_1 + m_2} v_1$$

This expresses the recoil speed of the target in terms of the initial projectile speed. Finally, then, the kinetic energy of the target particle after the collision is

$$\begin{aligned} \tfrac{1}{2}m_2 v_2'^2 &= \tfrac{1}{2}m_2 \left(\frac{2m_1}{m_1 + m_2}\right)^2 v_1^2 \\ &= \frac{4m_1 m_2}{(m_1 + m_2)^2} \times \tfrac{1}{2}m_1 v_1^2 \qquad (13.24) \end{aligned}$$

Let us consider three special cases of this result:

(1) m_1 is much greater than m_2. In the denominator m_2 may be neglected in comparison with m_1, and therefore

$$K_2' = 4\frac{m_2}{m_1} K_1$$

Here, then K_2' is very small compared with K_1 because of the small factor m_2/m_1. A heavy projectile cannot transfer a large fraction of its energy to a light target initially at rest.

(2) m_2 is much greater than m_1. Because the formula Eq. 13.24 is symmetric in m_1 and m_2, we expect the result to be similar to that in (1), and it is if m_1 in the denominator is neglected. Then we get

$$K_2' = 4\frac{m_1}{m_2} K_1$$

which is small because of the small factor m_1/m_2. A light projectile cannot transfer a large fraction of its energy to a heavy target.

(3) $m_1 = m_2$. The denominator in Eq. 13.24 becomes $(2m_1)^2 = 4m_1^2$, so that here

$$K_2' = K_1$$

A projectile can transfer its entire kinetic energy to a stationary target of equal mass. This will happen only in a head-on collision.

These three simple cases occur frequently in atomic and nuclear physics. Case (1) might be a nucleus hitting an electron; case (2), an electron hitting a nucleus; and case (3), one hydrogen atom striking another. We shall encounter these kinds of collisions later.

Problems

1. The potential energy graph describing the situation of a bouncing ball is shown in the figure. The vertical line represents the contact force exerted by the ground on the ball.
   (a) If the ball is dropped from a height y_0, show how to locate the total energy line on this graph.
   (b) From the graph find the heights where the kinetic energy of the bouncing ball is a maximum and a minimum.

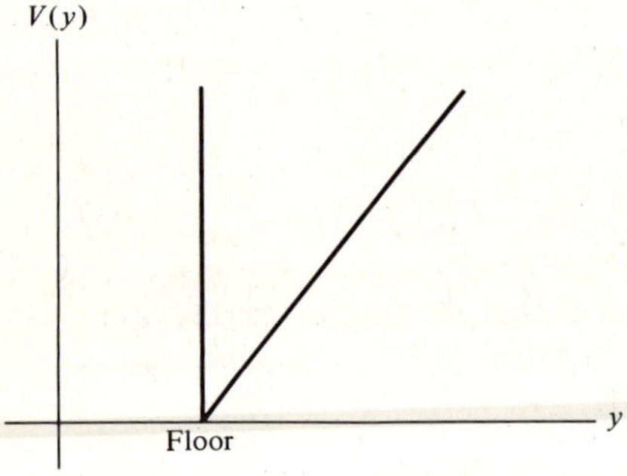

2. A particle is whirling at the end of a horizontal string as indicated, one end of the string being held fixed. If the midpoint of the string strikes an immovable obstacle, explain why each of the quantities below does or does not

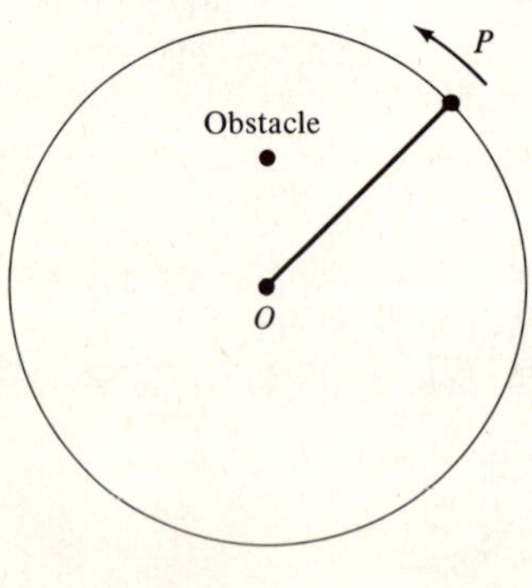

immediately change its value as a consequence: (a) the linear momentum of the particle, (b) the angular momentum of the particle about the obstacle, (c) the angular momentum of the particle about the end of the string, and (d) the kinetic energy of the particle. Which angular momentum is conserved before the collision and which afterward?

3. An upward force of 2000 dynes is applied to a 1-g object initially at rest in the gravitational field of the earth. Find the work done by this force during the first 25 cm of the object's subsequent motion. How much of this work appears as kinetic energy? How much appears as potential energy?
4. An upward force of 500 dynes is applied to a 1-g object initially at rest in the gravitational field of the earth. Find the work done by this force during the first 25 cm of the object's subsequent motion. Find the change in the object's kinetic and potential energies. Is the result self-consistent?
5. A particle of mass 3 g is suspended by a spring of constant 4000 dynes/cm in the earth's gravitational field. If the relaxed spring is at height $y = 0$, sketch the potential energy of the spring as a function of y, the gravitational potential energy of the particle as a function of y, and the total potential energy of the particle as a function of y. Locate the height at which the particle hangs in equilibrium on your sketch. What is the value of y here?

14. CENTER OF MASS

A visible particle is not a mathematical point but is an extended body that has been represented by a point particle for the purpose of simplifying its description. This raises the question of just what point in the extended body should be taken as the position of the representative point particle. There is a best point to choose, the center of mass of the extended body. The center of mass of an extended system is also a useful idea even when the system is by no means equivalent to a single macroscopic particle.

To start with, consider a system of two particles that can be thought of as point particles, and label them 1 and 2. The total momentum $\mathbf{p}$ of the system is the sum of the momenta of the two constituent particles.

$$\begin{aligned}\mathbf{p} &= \mathbf{p}_1 + \mathbf{p}_2 \\ &= m_1\mathbf{v}_1 + m_2\mathbf{v}_2 \qquad (14.1)\end{aligned}$$

The mass of the system is the sum of the masses of the two particles,

$$m = m_1 + m_2 \tag{14.2}$$

It is then reasonable to define the velocity $\mathbf{v}$ of the two-particle system so that the momentum of the system is its mass times its velocity. Therefore

$$\begin{aligned}\mathbf{p} &= m_1\mathbf{v}_1 + m_2\mathbf{v}_2 \\ &= (m_1 + m_2)\mathbf{v}\end{aligned}$$

and solving for $\mathbf{v}$ one finds that

$$\mathbf{v} = \frac{m_1\mathbf{v}_1 + m_2\mathbf{v}_2}{m_1 + m_2} \tag{14.3}$$

The velocity $\mathbf{v}$ is called the velocity of the center of mass of the pair of particles.

The form of Eq. 14.3 suggests the most convenient definition for the position of the center of mass of the two particles. Denoting the position vector of this position by $\mathbf{r}$, then the choice

$$\mathbf{r} = \frac{m_1\mathbf{r}_1 + m_2\mathbf{r}_2}{m_1 + m_2} \tag{14.4}$$

leads directly to Eq. 14.3. The center of mass of a pair of particles lies on the line joining them and nearer to the more massive one. This is most easily seen if a coordinate system is chosen in which one particle, say 1, is at the origin so that $\mathbf{r}_1 = 0$. Then, as shown in Figure 14.1, $\mathbf{r}$ lies along $\mathbf{r}_2$. The center of mass being shown nearer m_2 than m_1, here m_2 is greater than m_1.

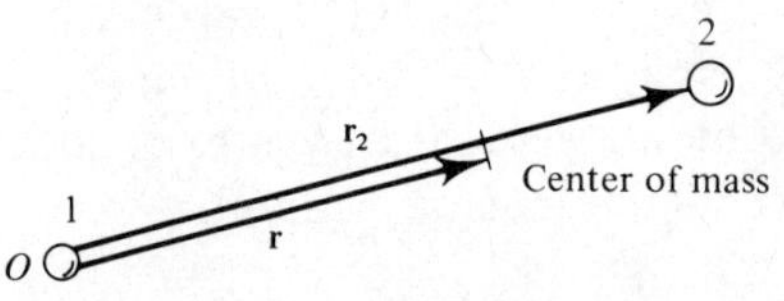

FIGURE 14.1 The center of mass of two particles lies on the line segment between them, nearer to the larger mass. It is most easily found by choosing coordinates so that one particle is at the origin. Here m_2 is the larger mass.

The definition of the center of mass is easily extended to any number of particles. It is given by

$$\mathbf{r} = \frac{m_1\mathbf{r}_1 + m_2\mathbf{r}_2 + m_3\mathbf{r}_3 + \cdots}{m_1 + m_2 + m_3 + \cdots}$$

$$= \frac{\Sigma m_i \mathbf{r}_i}{\Sigma m_i} \tag{14.5}$$

with all particles in the system included in the sum. The symbol Σ (the Greek capital sigma) is often used to indicate a sum of the quantities immediately following. It is a convenient shorthand notation. According to its definition, the position vector of the center of mass of a system of particles is the weighted average of the position vectors of the constituent particles.

If in a two-particle system the two masses m_1 and m_2 are equal, the velocity of the center of mass is the average of the two velocities. If m_2 is very much larger than m_1, then the terms in Eq. 14.3 containing m_1 may be neglected so that to a good approximation $\mathbf{v} = \mathbf{v}_2$. The velocity of the center of mass of a system consisting of a very light particle and a very heavy particle is effectively the velocity of the heavy particle.

When the two particles do not interact with anything except each other, their total momentum $\mathbf{p}$ is conserved, and so is the velocity of their center of mass $\mathbf{v}$. The kinetic energy associated with the center of mass motion then is also constant, being given by

$$K_{\text{cm}} = \tfrac{1}{2}m\mathbf{v}^2 \tag{14.6}$$

This has important consequences for collisions. Suppose that a spring of constant k is attached to the particle of mass m_2, which can be compressed by a plunger and latched if it is compressed by the amount a or more, as shown in Figure 14.2. If particle 2, the target, is initially at rest, with what minimum speed must particle 1, the bombarding particle, be projected at it in order to cause the spring to be latched in the compressed position? This very artificial sounding example is actually a good model for collisions of a kind to be studied later. Latching the spring requires the potential energy of the spring to be increased from zero to $\frac{1}{2}ka^2$, and this energy must come from the bombarding particle. Not all of the bombarding particle's kinetic energy is available to compress the spring, because when particle 1 strikes the spring and starts to compress it, the spring will exert a force on the target particle and accelerate it, giving it some kinetic energy; this energy must also come from the bombarding particle and is not available to compress

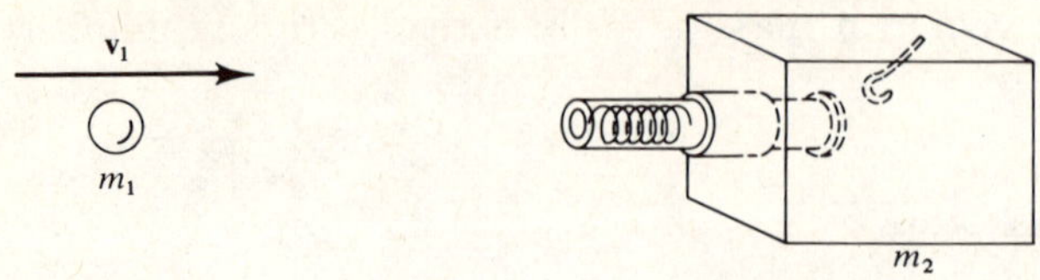

FIGURE 14.2 A ball of mass m_1 is thrown at a box fitted with a spring-loaded piston that can latch in the compressed position. The ball must have a certain threshold energy before it can cause the piston to latch. This threshold energy depends on the masses m_1 and m_2 and on the potential energy of the compressed spring.

the spring. The whole business seems complicated, but the idea of the center of mass energy makes it simple.

The kinetic energy of the center of mass is conserved, so that the energy available to produce effects like the compression of a spring is the total energy minus the kinetic energy of the center of mass, which is called the *relative kinetic energy.* Here the total energy is the initial kinetic energy of particle 1,

$$E = \tfrac{1}{2}m_1\mathbf{v}_1^2$$

The velocity of the center of mass is, from Eq. 14.3 with $\mathbf{v}_2 = 0$,

$$\mathbf{v} = \frac{m_1}{m_1 + m_2}\,\mathbf{v}_1$$

so the kinetic energy of the center of mass is

$$\begin{aligned} K_{\text{cm}} &= \tfrac{1}{2}(m_1 + m_2)\mathbf{v}^2 \\ &= \frac{1}{2}\frac{m_1^2}{m_1 + m_2}\,\mathbf{v}_1^2 \end{aligned} \tag{14.7}$$

If $m_2 = 0$, this is E, the entire energy. If $m_2 = \infty$, the center of mass kinetic energy is zero because the velocity of the center of mass is zero. The relative kinetic energy is now

$$\begin{aligned} K_{\text{rel}} &= E - K_{\text{cm}} \\ &= \frac{1}{2}\frac{m_1 m_2}{m_1 + m_2}\,\mathbf{v}_1^2 \end{aligned} \tag{14.8}$$

This must be equal to or greater than $\tfrac{1}{2}ka^2$ in order to be able to latch the spring during the collision. The smallest value of E that will latch the spring is called the *threshold energy* for the reaction.

The fraction of the kinetic energy of the bombarding particle that is available to produce a reaction during the collision depends on the masses of both the bombarding and the target particles,

$$\frac{K_{\text{rel}}}{K_1} = \frac{m_2}{m_1 + m_2} \tag{14.9}$$

The heavier the target, the larger is the fraction of the bombarding particle's energy available for a reaction. When a reaction between two particles of different masses is to be produced, it is generally better to use the heavier one as the stationary target.

The inertial reference frame in which the target is initially at rest is called the *laboratory reference frame.* An inertial frame moving relative to the laboratory system with the velocity of the center of mass, in which the momentum of the center of mass is zero, is called the *center of mass reference frame.* Often it is easier to describe the motion of a system in the center of mass frame than in any other. An example is furnished by Figures 11.1 and 11.2.

When two particles interact with each other and with nothing else, the momentum of their center of mass is constant, and an inertial reference frame, the center of mass frame, exists in which that momentum is zero. The center of mass of the system is then at a fixed point on the line joining the particles. If the forces exerted by the particles are forces of attraction or repulsion, they act along this line and hence act always toward or away from the center of mass of the two particles, which is a fixed point in the center of mass reference frame. In Section 10 a central force was defined as a force that always acts toward or away from a fixed point, so the force of attraction or repulsion between two particles is a central force with the center of mass as the force center. This is true whatever the particle masses. The conservation of angular momentum in a central force field applies to the angular momentum of the two interacting particles about their center of mass If one particle is much heavier than the other, the center of mass is essentially at the position of the heavier particle, which was the situation discussed in Section 10.

As an illustration consider the system of two particles of equal mass connected by a light string shown in Figure 11.1. Here the center of mass is stationary, being at the center of the circle, and if the length of the string is a, the total angular momentum about the center of mass is $mv \times a/2 + mv \times a/2 = mva$. It is conserved. If one now looks at Figure 11.2, which shows the same particles but viewed from a reference frame traveling to the left with speed V so that the center of mass is seen to be moving to the right in that frame with speed V, one can see that the angular momentum about the center of mass is still conserved and still has the value of mva, which it had in the center of mass frame. When one particle is at the top and the other at the bottom, the angular momentum about the center of mass is

$$m \times (v - V) \times \frac{a}{2} + m \times (v + V) \times \frac{a}{2} = mva$$

The same holds for other positions. This is an example of a general result that we will not derive: The angular momentum of a system of particles about its center of mass does not depend on the motion of the center of mass. This makes the center of mass the logical point about which to evaluate the angular momentum of almost any system.

Problems

1. Why must the center of mass of two particles always lie between them and not on an extension of the line joining them?
2. Three particles lie on the x-axis of a coordinate system, as shown.

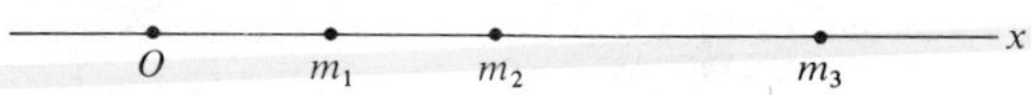

   (a) Write down the expression for the x-coordinate of the center of mass of these three particles.
   (b) Show that the center of mass of the three can be found by considering m_1 and m_2 as a single particle of mass $m_{12} = m_1 + m_2$ located at the center of mass of these two particles. (This is true in general. The center of mass of two uniform spheres, for example, can be found by treating each sphere as a point particle whose mass is that of the sphere and whose location is the center of the sphere.)
3. A 500-g billiard ball is cracked, so that if more than 10^4 ergs of work is done on it, it will break. What is the minimum kinetic energy of a 100-g ball striking the cracked ball at rest that can cause it to break? If the cracked ball is thrown against the 100-g ball, what is the minimum kinetic energy it must have to break? Explain why these two energies are different.

CHAPTER 3
Particles Moving in Electric and Magnetic Fields

15. ELECTRIC FORCES

Newton's laws of motion give a complete picture of the motion of a particle subject to known forces, at least when the particle is large enough to be directly observable and when it is not moving too fast. In the world of directly observable particles there are many kinds of force that act to change the motion of particles, such as the force of gravity, the contact force that prevents one object from penetrating another, the force exerted by springs and strings, and frictional forces between objects sliding over each other. These are all real forces, and on the level of macroscopic objects, they are distinct kinds of forces, all having applications in the operation of devices ranging from pendulum clocks to jet airplanes and rockets.

From the microscopic view, things are a lot simpler. Most of the forces listed above involve forces acting between objects in contact with each other, and ultimately such forces must be related to forces between atoms. An atom of iron cannot penetrate another atom of iron easily, so a piece of iron cannot penetrate another piece of iron easily either. It is now known that an atom of iron or of any other chemical element consists of a nucleus that carries a positive electric charge surrounded by a cloud of electrons each of which carries a negative electric charge. The forces that hold an atom together are the attractive forces between oppositely charged particles and hence are electric forces. The forces that hold atoms together in solids are also electric forces, though here magnetic forces sometimes also play a role. Except for gravity all the forces listed above are thus electric or magnetic in their nature. This makes the study of electric and magnetic forces of great interest.

Of course one should ask how it is known that the forces within and between atoms are electric and magnetic. This comes about through studying these forces as they act on macroscopic particles. The forces between charged particles obey rather simple laws, so that by applying the principles of Newtonian mechanics using these forces, the behavior of visible charged particles can be described theoretically. This same theory then can be applied to the motion of particles that are too small to be observed directly but whose motion can be inferred from observations that can be made. The theory works provided each kind of microscopic particle is assigned its own fixed values of charge and mass. This is the reason for ascribing these values of charge and mass to these particles. Eventually one finds that these particles do not obey all the laws of Newtonian mechanics under all circumstances, and then the laws of mechanics must be modified. In this way quantum mechanics was discovered. This long chain of combined experimental and theoretical development could not have taken place without starting at the beginning: the study of the electric and magnetic forces between macroscopic particles. We start at this beginning, and in the course of the book we shall continue quite far down this chain.

The existence of nongravitational, noncontact forces is shown by the following experiment: A light ball such as a table tennis ball painted with aluminum paint is suspended from a support by a string so that it is free to swing. A rubber rod has no visible effect on the ball when brought near it until the two touch, when the contact force between them acts and prevents the rod from penetrating the ball. Now if the rubber rod is rubbed briskly with a piece of fur before bringing it near the ball, the behavior is radically different. When the rod is a short distance from the ball, the ball is attracted to the rod, the attraction getting stronger the closer the rod is. When the ball touches the rod, the attraction is changed to repulsion, and the ball is repelled by the rod. This repulsion gradually disappears, and it disappears completely

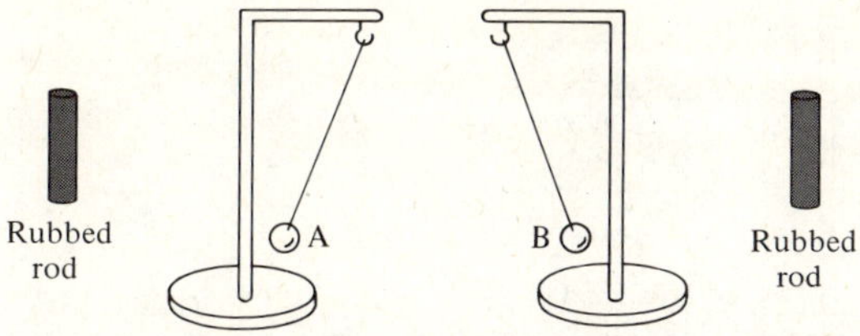

FIGURE 15.1 Two charged balls repel one another when they have been charged in the same way and, therefore, have charges of the same kind.

if the ball and the rod are wiped by the experimenter's hand. This behavior clearly demonstrates the existence of forces other than those met with before.

As a first step in studying these forces, let us have two balls hung from separate supports. If the two balls are treated in the same way, they should acquire the same characteristics. The rubber rod of the first experiment is rubbed on the fur and then touched to ball A. It is rubbed again and touched to ball B. If the two balls are now brought near each other by moving their supports, they are observed to repel one another, as shown in Figure 15.1. One says that the process of touching the ball with the rubbed rod has *charged* it and that the repulsion is caused by the presence of charge.

The existence of two different ways of charging an object can be shown by repeating the above experiment with one variation. After ball A has been charged, instead of charging ball B by touching it with the rubbed rod, it is touched with the rubbed fur. The two balls are now seen to attract each other, as in Figure 15.2. There are, therefore, at least two ways in which a particle can be charged, one the result of touching it with the rubbed rod and the other of touching it with the rubbed fur. If both balls are charged by touching them with the fur, repulsion again results.

If the process of charging a ball is thought of as putting an amount of something called "electric charge" on the ball, there must be two kinds of charge. One kind is supplied by the rubbed rod and the other by the rubbed fur. No third kind of charge has ever been observed. The experiments described above then show that *like charges repel one another* while *unlike charges attract one another*. The force of attraction or repulsion due to the charges on two charged particles is called an *electric force*.

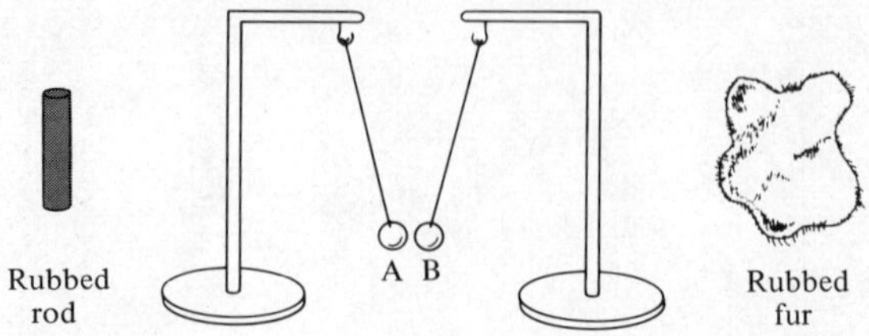

FIGURE 15.2 Two charged balls attract one another when they have been charged differently, one with the rubbed rod and the other with the rubbed fur.

If charge is free to move around in a body, it will tend to distribute itself as thinly as possible because each little bit of the charge repels every other little bit of charge of the same kind. If a quantity of charge stays concentrated on part of an object, it is evidently not free to move around in the material of which the object is made. Materials that do not allow charges to move freely on them are called *insulators,* while those on which charges can move freely are called *conductors.* Whether a material is considered to be an insulator or a conductor depends on how long one is willing to wait to have the charge move. For the kinds of experiments described here, where times of a few tenths of a second are short, the human body and the ground are good conductors while rubber and glass are good insulators. This means that in a few tenths of a second a charge placed on one part of the earth distributes itself over the entire earth as in Figure 15.3. When times very short compared with these are involved, a few millionths or billionths of a second perhaps, the human body and the earth are rather poor conductors. There is an enormous variation in the times it takes charges to distribute themselves over different materials. A charge placed at one point on

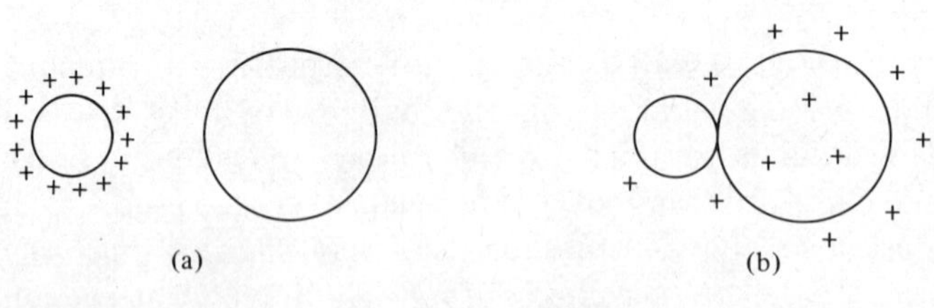

FIGURE 15.3 When a charged conductor comes in contact with an uncharged one, the charge distributes itself as widely as possible over both conductors.

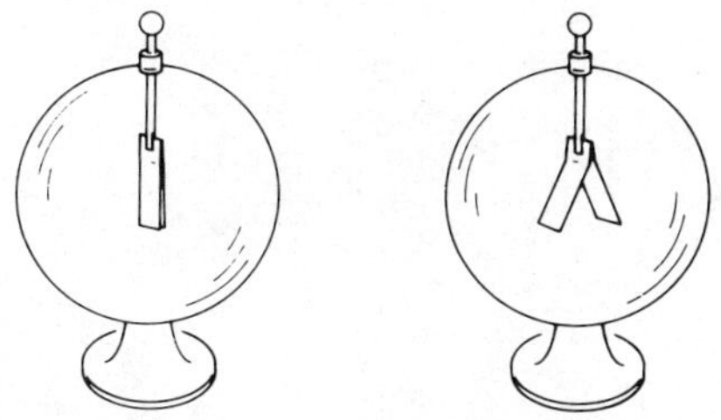

FIGURE 15.4 The leaves of an uncharged electroscope hang side by side as in (a), while those of a charged electroscope repel each other and hang at an angle as in (b).

a copper object will distribute itself in a very small fraction of a second, while it would remain essentially where it was put on a glass object for years. If a charge can get to the ground, it will spread out so far that for all practical purposes it is lost. This is why touching the rod and the ball with the hand destroys all electric effects in the experiment described above: The charge all goes to the ground and is lost.

The mutual repulsion of like charges makes a very simple detector of charge possible, the gold leaf electroscope. It consists of a conducting rod mounted on an insulating stand from which a pair of gold leaves is suspended, as shown in Figure 15.4. The end of the rod is accessible, the gold leaves being enclosed in a glass case to protect them from disturbance by moving air. When the electroscope is uncharged, the leaves hang down side by side as in (a), but when it is charged with either kind of charge, the leaves become charged also and repel one another, separating until the repulsive effect is balanced by gravity, as in (b). This instrument is not adapted to quantitative measurement but shows qualitative effects very clearly.

The general correctness of the picture of charges attracting and repelling each other can be verified in some simple experiments with the electroscope. The process of charging by induction is a good example. Starting with an uncharged electroscope and a rubbed rubber rod, the sequence illustrated in Figure 15.5 is gone through. (a) The rod is brought near the top of the electroscope. To facilitate the discussion, let us call the charge on the rod "negative," the other kind of charge being then "positive" charge. The negative charge on the rod repels

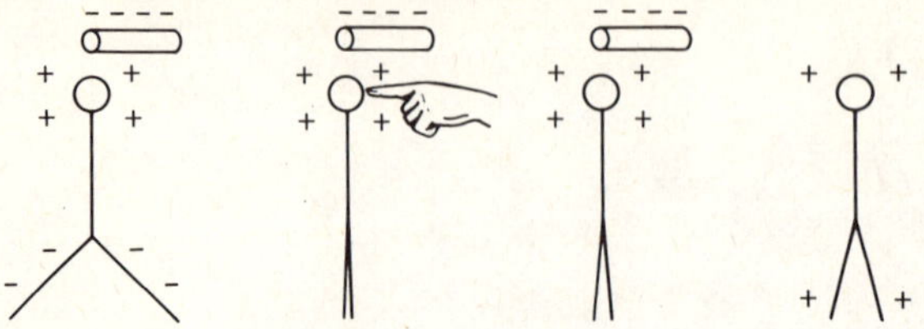

(a) (b) (c) (d)

FIGURE 15.5 An electroscope is charged by induction in four steps: (a) A charged rod is brought near the top of the electroscope. (b) The charge on the leaves is allowed to escape to the ground while that on the ball is held by the rod's attraction. (c) The ground path is removed, leaving the situation unchanged. (d) The rod is removed, allowing the charge formerly held on the ball to spread over the whole electroscope.

negative charge to the leaves of the electroscope and attracts positive charge to the top. (b) The experimenter touches the electroscope with his finger, the rod still being nearby. The attraction of the negative charges on the rod holds the positive charge on top of the electroscope, but the negative charge can now get further away and does. It goes to the ground, and its departure causes the leaves to collapse. (c) The experimenter now removes his finger, the rod still being present. The electroscope leaves remain collapsed. (d) The rod is removed, so the positive charge on the electroscope distributes itself as widely as possible, some going to the leaves and causing them to diverge. The electroscope has been charged positively without touching it with a positively charged body; the process is called "charging by induction." The charge on the rod is not affected by the process at all, so the same negative charge can be used to induce any number of positive charges. No one induced positive charge can be larger than the inducing charge.

The reader should verify that the process just described can be accounted for equally well by considering positive charge or negative charge to move. If negative charge moves away from a neutral or uncharged body, it leaves positive charge behind, an uncharged body being one in which there are equal amounts of positive and negative charge. It requires much more subtle experiments than this to decide what kind actually does move in any given case. Some of these experiments are described in Section 43.

We must now become more quantitative and ask how the forces between charges depend on the size of the charges and on the distance between them. First we must be able to decide when two charges are equal. One says that the charges on two particles, 1 and 2, are equal if the force exerted on a third charged particle by particle 1 is the same as that exerted by particle 2 when at the same distance from the third particle. The comparison must be made at a fixed separation to prevent any variation of force with distance influencing the result. Two charges are equal in magnitude but opposite in sign if they produce forces of equal magnitude but of opposite direction when placed at the same distance from a third charged particle. It would be hard to imagine any other way of defining the equality of two charges and still have the result that two charges each equal to a third one are equal to each other.

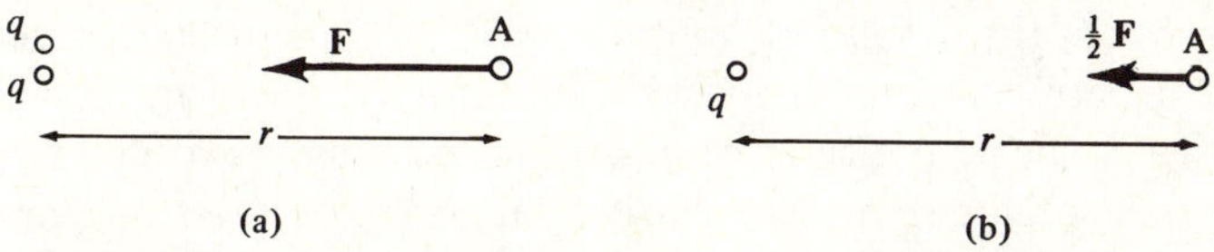

FIGURE 15.6 The force on charged particle A due to the presence of two identically charged particles side by side is twice that due to the presence of only one of them.

Consider two particles that have been prepared so that they have equal charges. They can be placed close to each other, and a third charged particle A can be brought up so that it is a distance r from both of the first particles as in Figure 15.6(a). The force on particle A is found to be twice the force that A experiences if either of the two particles is present without the other, as in (b) of the figure. If the charge on one of these particles is denoted by q, it is natural then to say that the charge on the two particles together is $2q$. Doubling the charge doubles the force, so that the electric force on A due to the presence of another charge is proportional to the size of that other charge. This can be verified for any multiple of the charge q, not just the multiple two. Two similarly charged particles side by side produce the same electric forces as a single particle carrying the sum of the two charges. This rule also applies to dissimilar charges. If a particle carrying charge q is placed beside a particle carrying the opposite charge $-q$, the electric force at great distances is almost zero, as though no charges were present. The electric force is not exactly zero unless the two particles are in exactly the same place; a pair of equal size charges of opposite sign close together constitutes a dipole, a type of object to be met with later in connection with magnetism. The cancellation of effects when two opposite charges are present justifies naming one kind of charge positive and the other kind negative. It is perfectly arbitrary that the kind of charge appearing on rubber or amber when rubbed with fur is negative. (The Greek word for amber is *elektron,* which is the origin of the words "electricity" and "electron.")

If two charged particles A and B are separated by a distance r, the force on A due to the presence of B is proportional to the charge q_B on B. Similarly, the force on B due to the presence of A is proportional to the charge on A, q_A.

$$F_{B\to A} \propto q_B, \qquad F_{A\to B} \propto q_A$$

According to Newton's third law of motion, which we can verify as applying to these electric forces, the two above forces must be equal in magnitude, so that the magnitude of the force between two charged particles must be proportional to the charges on each, that is, proportional to the product of the charges:

$$F_{A\to B} = F_{B\to A} \propto q_A q_B \tag{15.1}$$

The dependence of the force on the distance r between the charges is still to be determined. This also is an experimental matter, and one of the first relevant experiments was done in 1785 by Charles de Coulomb. He used a torsion balance shown schematically in Figure 15.7. Such a balance consists of an arm suspended by a fine elastic fiber made of bronze or fused quartz. A force applied to one end of the arm produces a torque on the arm about the point of suspension, and since the fiber is easily twisted, the arm turns in response to a very small force. The force is measured by bringing the arm back to its original equilibrium position by rotating the top support of the fiber, the angle of rotation needed being proportional to the force applied to the end of the rod. Using this device, Coulomb showed that within his experimental error the force varies as the inverse square of the distance between the charges. This has been established to much higher precision by more modern, indirect experiments. The torsion balance was used some years later by Henry Cavendish to measure the force of gravity between two spheres, thus determining the value of Newton's constant of gravitation G. The magnitude of the force between two charged particles with charges q_A and q_B may now be written as proportional to

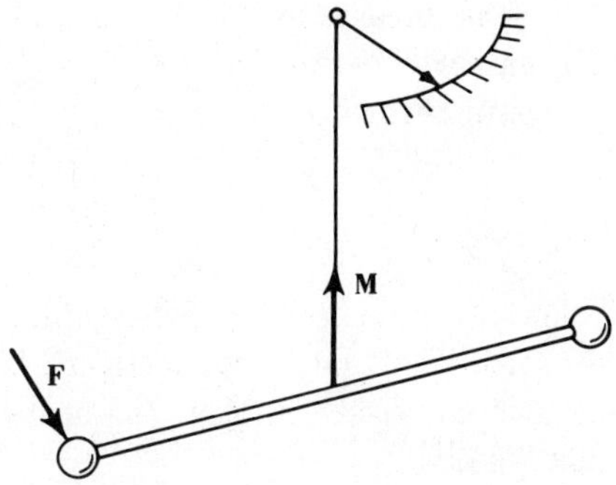

FIGURE 15.7 A schematic sketch of a torsion balance as used by Coulomb. The torque about the suspending fiber produced by the force **F** is countered by the torque caused by turning the support of the fiber until the original position of the balance arm is restored.

the product of the charges and inversely proportional to the square of the distance separating them r_{AB}.

$$F_{A \to B} = F_{B \to A} \propto \frac{q_A q_B}{r_{AB}^2} \tag{15.2}$$

This is known as Coulomb's law.

The above proportionality is converted into an equality by choosing the units in which to measure q and then inserting the appropriate constant of proportionality. A variety of choices is open to us here. There is no standard charge, as there is a standard meter or standard kilogram, with which any other charge can be compared. The difficulty of preserving a standard macroscopic charge is obviously formidable because of the tendency of charge to spread out as far as possible. One could use the electron charge as a standard, but the electron is an inaccessible standard for many purposes, its charge is microscopic, and therefore this possible standard is not used. There is no choice that is convenient for all purposes, and since our purpose is to discuss physics, let us use the units many physicists use when they are not writing elementary textbooks but are communicating with each other, the cgs Gaussian system of electrical and magnetic units. In the Gaussian system, the unit of electric charge is that which repels an equal charge at a distance of 1 cm with a force of 1 dyne. When F is measured in dynes and r in centimeters, this makes the proportionality constant needed in Eq. 15.2 unity, so that Coulomb's law takes the simple form

$$F_{AB} = \frac{q_A q_B}{r_{AB}^2} \tag{15.3}$$

This unit of charge is called the esu (electrostatic unit), or the *statcoulomb*. Other electric and magnetic quantities will be introduced later, and their units also will be given in this system. Other systems of units introduce an additional constant in Coulomb's law.

In all experiments described in this section, charge has been seen to move from place to place but never to appear out of nowhere or to disappear into nowhere. The charge that appears on a rubbed rubber rod is equal in magnitude to the charge appearing on the fur with which it is rubbed, so that the charging of both the rod and the fur can be accounted for by saying that negative charge is transferred from the fur to the rod or positive charge is transferred from the rod to the fur or both, but it is not necessary to say that any charge has been created or destroyed. Every observation made to date is consistent with the conservation of charge, and this conservation law is one of the best founded laws of physics. If the total charge that appears in any finite volume of space increases, charge must have come in through the boundary surface of that volume because it cannot be created within that volume. If the total charge in any finite volume of space decreases, charge must go out through the surface of that volume because it cannot be destroyed within that volume.

Equation 15.3 gives the magnitude of the force between two charged particles. The force is either one of attraction or of repulsion, and acts, therefore, along the line joining the particles. The vector form of Coulomb's law can be written as

$$\mathbf{F}_{A \to B} = \frac{q_A q_B}{r_{AB}^2} \times \frac{\mathbf{r}_{AB}}{r_{AB}} \tag{15.4}$$

where $\mathbf{r}_{AB} = \mathbf{r}_B - \mathbf{r}_A$ is the vector directed from A to B, as shown in Figure 15.8. The vector $\mathbf{r}_{AB}/r_{AB}$ is a unit vector which gives the direction of the force on B due to A if q_A and q_B have the same sign so that their product is positive. If the charges are unlike, the product $q_A q_B$ is negative and the force on B due to A is antiparallel to $\mathbf{r}_{AB}$. Very little use will be made of the vector form of Coulomb's law. The direction in which the force acts is usually clear without including it in the formulas.

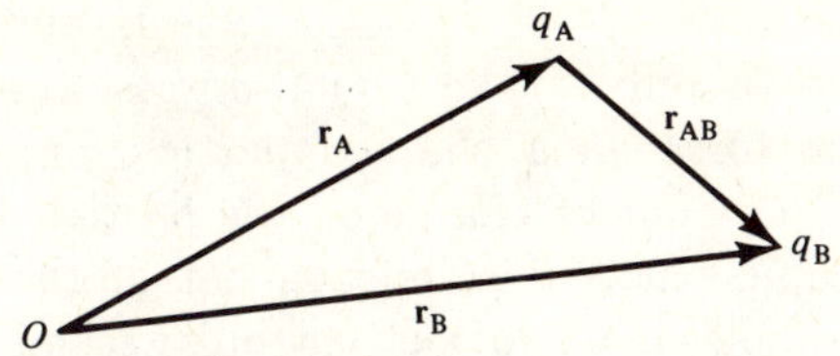

FIGURE 15.8 The vector $\mathbf{r}_{AB}$ connecting two charged particles A and B is the difference of the position vectors of these two particles.

Coulomb's law enables us to understand the first electrical effect described in this section, the attraction of a charged rubber rod for an uncharged table tennis ball. The negatively charged rod causes the side of the ball nearer the rod to become positively charged and the far side of the ball to become negatively charged. The positive charge being nearer the rod, its attraction to the rod is greater than the repulsion between the rod and the equally large negative charge, so there is a net attraction. It comes to an end if the charge on the near side of the ball can move off onto the rod when they touch. If the ball is made of an insulating material not painted with aluminum paint, it will remain in contact with the rod. This is the principle on which many dust filters operate: Dust or ash particles are attracted to charged rods and, being insulators, stick there until cleaned off. At this point one should ask why an insulating particle is attracted to the rod in the first place since charges are not free to move on insulators. The answer is that in insulators the positive and negative charges are in stable equilibrium. They can be displaced from each other to some extent by the force exerted by an external charge, but the restoring force increases with the relative displacement so that a new equilibrium is reached in which the insulating particle is a small electric dipole, with two equal size charges of opposite sign. In conductors the equilibrium is a neutral one, the motion of a negative charge from a positive one not causing these restoring forces to appear.

The form of Coulomb's law, Eq. 15.3, is very similar to Newton's law of gravitation, Eq. 10.8. There are two differences worth noting. First, Coulomb's law describes both attractive and repulsive forces, there being charges of both positive and negative sign, while Newton's law describes attractive forces only. Second, Newton's law of gravitation states that the force on a particle due to gravity is proportional to the mass of the particle, so that the mass of the particle cancels out of Newton's second law of motion when only gravitational forces are acting. Coulomb's law states that the force on a charged particle due to other charges is proportional to the amount of charge on the particle. The amount of charge on a macroscopic particle is independent of the mass of the particle, so that mass does not cancel out of Newton's second law when electric forces are involved.

Because Coulomb's law contains the distance between the particles in the same way that Newton's law of gravitation does, conclusions that depend on this distance dependence are the same for both cases. For example, the gravitational attraction between two spherical bodies is unaffected by concentrating the mass of each body at its center. Similarly, the Coulomb force between two spherical charge distributions is unaffected by concentrating the two charges at the centers of the spheres, provided the two spheres do not overlap. The Coulomb force between two charged particles is a central force, always being directed toward or away from the center of mass of the two particles. The angular momentum of two particles interacting through the Coulomb force about their center of mass therefore is conserved.

In Section 13 it was shown that a potential energy could be associated with the force of gravity. In a similar way a potential energy depending in the same manner on r can be associated with a Coulomb force. If two charged particles are separated by a distance r, there is a potential energy, measured in ergs,

$$V(r) = \frac{q_A q_B}{r} \tag{15.5}$$

associated with their nearness to each other. The argument establishing this is exactly like that given in Section 13 for the gravitational potential energy. On increasing the separation of the particles from r to $r + \Delta r$, the charge in potential energy is, for small Δr,

$$\begin{aligned}
V(r + \Delta r) - V(r) &= q_A q_B \left(\frac{1}{r + \Delta r} - \frac{1}{r}\right) \\
&= q_A q_B \frac{-\Delta r}{r(r + \Delta r)} \\
&= -\frac{q_A q_B}{r^2} \Delta r \\
&= -F_{\text{coul}} \Delta r
\end{aligned}$$

But according to Eq. 13.13, this is $-F_r\,\Delta r$, so that the potential energy $V(r)$ of Eq. 15.5 does correspond to the Coulomb force.

At this point we could discuss the motion of one charged particle in the Coulomb field of another, just as the motion of a particle moving in the gravitational field of the earth was analyzed. The force laws in these two cases are very similar, so that the orbits also would be of the same kind. This motion, however, does not take place on the macroscopic scale; it is too difficult to keep charges on a pair of macroscopic bodies large enough to make the Coulomb force the dominant force on them. The motion of electrons in atoms does follow Coulomb's law to very high accuracy, but here features connected with quantum mechanics enter, and we have yet to discuss the existence of atoms and electrons, so that we postpone treating this example until it can be given on a sounder basis and more completely, as is done in Section 39. What we can and must do next is to describe the motion of a charged particle under the influence of charges distributed in a controlled way over large bodies, because many fundamental experiments involve such motions of charged particles.

Problems

1. Describe the process of charging by induction illustrated in Figure 15.5 in detail on the assumption that only negative charges are free to move.
2. Rephrase the description of the process illustrated in Figure 15.5 so as to describe positive movable charge instead of negative movable charge.
3. Two equal positive charges are a distance $2a$ apart. The force on a positive test charge midway between them is zero. If the test charge is displaced a short distance either (a) toward one of the charges or (b) at right angles to the line joining the charges, find the direction of the force on it. Is it in stable or in unstable equilibrium at the midpoint?
4. Two opposite charges of equal magnitude are a distance d apart, as shown in the figure. (Such a pair of charges constitutes a dipole of dipole moment qd.) Find the places where the forces on a test charge due to these two charges are parallel. Is there any place where the resultant of these forces vanishes, as it did at the midpoint of the line joining the charges in Problem 2?
5. Explain why a balloon, heavy enough to fall through the air, can be made to stick to the ceiling or a wall merely by rubbing it with a dry cloth.

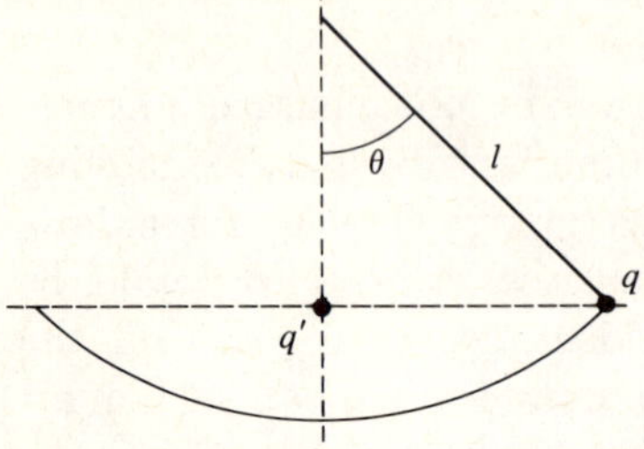

6. A ball of mass m and charge q is suspended from a string of length l. If a charge q' of the same kind as q is brought up to the position shown in the figure, find an equation which determines the angle θ shown.

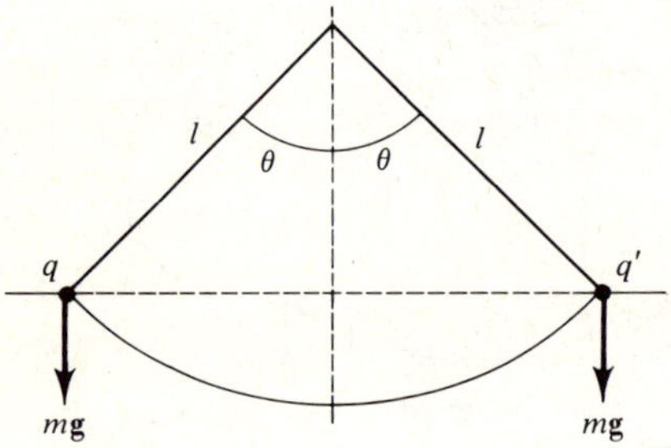

7. Two balls of mass m each are suspended from a common point by strings of length l. If one ball has a charge q and the other has a charge q' of the same kind, find an expression for θ, the angle shown in the figure. If q and q' are not of equal size, are the two angles marked θ necessarily equal? If the two balls are of different mass, are these two angles equal?
8. Give an argument that shows that the charge induced on the electroscope illustrated in Figure 15.5 cannot be larger than the inducing charge on the rod.

16. THE ELECTRIC FIELD AND ELECTRIC POTENTIAL

Coulomb's law gives an accurate description of the force on a charged particle due to the presence of another one. For the law to be directly useful, the charges must be so concentrated that there is no ambiguity in the value of r, the distance between them, or they must be distributed with spherical symmetry about two centers whose separation is then the quantity r. One often encounters distributions of charge that are neither concentrated nor spherical, so that Coulomb's law as given in Eq. 15.3 cannot be applied in any obvious way. In such cases the idea of an electric field can be very helpful. It is completely equivalent to using Coulomb's law but often is much easier to think about than the Coulomb forces themselves. The electric field produced by a set of electric charges is a generalization of the idea of a force field introduced earlier.

Consider the force on a particle P with charge q due to the proximity of another particle of charge Q. Its strength is given by Coulomb's law and is

$$F_{\mathrm{P}} = \frac{qQ}{|\mathbf{r}_{\mathrm{P}} - \mathbf{r}_{\mathrm{Q}}|^2} \tag{16.1}$$

which is, as was pointed out before, proportional to the charge q. It is useful to regard this force as the result of two things, the presence of the charge q at the point $\mathbf{r}_{\mathrm{P}}$ and the presence at that point of an electric field $\mathbf{E}$ produced by the charge Q located at $\mathbf{r}_{\mathrm{Q}}$. The electric field is considered to exist everywhere in the neighborhood of the charge Q whether or not a charged particle is present. Only its value at the position of P affects the force on P. One writes, then,

$$\mathbf{F}_{\mathrm{P}} = q\mathbf{E}(\mathbf{r}_{\mathrm{P}}) \tag{16.2}$$

When the force on P is due only to the single charge Q, the electric field produced by Q at the point $\mathbf{r}$ can be read off from a comparison of the two expressions for $\mathbf{F}_{\mathrm{P}}$. Its magnitude is

$$E(r) = \frac{Q}{|\mathbf{r} - \mathbf{r}_{\mathrm{Q}}|^2} \tag{16.3}$$

and its direction is away from $\mathbf{r}_{\mathrm{Q}}$ if Q is positive and toward $\mathbf{r}_{\mathrm{Q}}$ if Q is negative.

The concept of the electric field due to a single charge adds little to the understanding of Coulomb's law. It does help when the force on P is due to two or more other charges, say Q_1 at $\mathbf{r}_1$ and Q_2 at $\mathbf{r}_2$. Figure 16.1 illustrates the case where both Q_1 and Q_2 are positive. $\mathbf{E}_1$ is the electric field at P produced by Q_1 and $\mathbf{E}_2$ is that produced at P by Q_2. The total force on a charge q at P would be the sum of the forces due to Q_1 and Q_2 separately; it is, therefore,

$$\begin{aligned} \mathbf{F}_{\mathrm{P}} &= q\mathbf{E}_1 + q\mathbf{E}_2 \\ &= q\mathbf{E} \end{aligned} \tag{16.4}$$

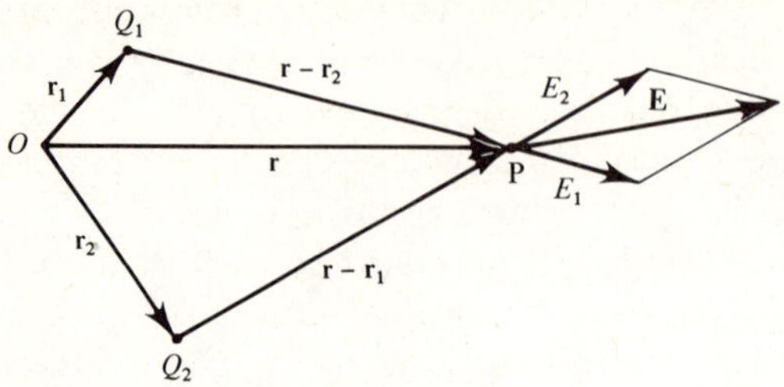

FIGURE 16.1 The electric field at P is the vector sum of the fields produced at P by charges Q_1 and Q_2 separately.

where **E** is the field at P due to both of the charges Q_1 and Q_2. The effect of both charges can be described by the single electric field **E**.

An electric field can be represented graphically. First consider the Coulomb field due to a single point charge Q. At any point the field is directed away from Q, assuming Q to be positive, so one draws lines radiating out from Q in all directions, as shown in Figure 16.2. These lines spread out in three dimensions, not just in the plane of the page. So far these lines represent only the direction of the field at any point. They also say something about the magnitude or strength of the field. To see this imagine a sphere of radius r drawn about Q as center. The surface area of this sphere is $4\pi r^2$. If the lines starting from the point Q are continued indefinitely far and no lines are started anywhere except

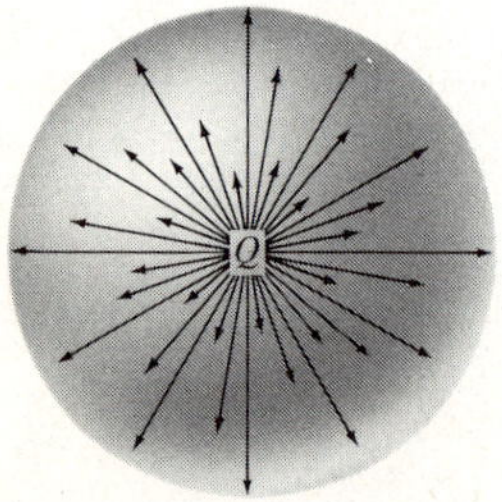

FIGURE 16.2 The electric field of a point charge can be represented by lines radiating uniformly in all directions in three dimensions.

at Q, the number of lines going through such a spherical surface will be the same whatever the radius of the sphere is. The number of lines crossing a unit area of the sphere is then the total number N of lines drawn divided by the area of the sphere, and so this density of lines can be written

$$\frac{N}{4\pi r^2}$$

The thing to notice here is the dependence on r; it is an inverse square

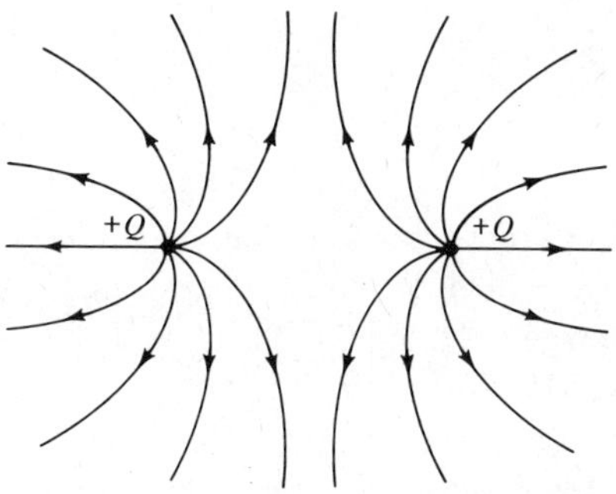

FIGURE 16.3 The electric field lines of two equal positive charges all go off to infinity.

dependence just as in Coulomb's law. The density of lines is therefore proportional to the electric field strength. A set of "field lines" in this way can represent both the direction and relative magnitude of the electric field of a point charge Q through the direction and the density of the field lines.

For a single point charge, the construction of electric field lines is hardly necessary. Figure 16.3 shows the electric field lines of the field due to two equal positive charges, and Figure 16.4 shows the electric field lines due to two opposite charges of equal magnitude. At each point the magnitude and direction of the field are found as in Figure 16.1. On the negative charge, the lines of force end rather than begin. In general lines of force begin on positive charges and end on negative charges, but all lines do not need to begin or to end. The lines in Figure

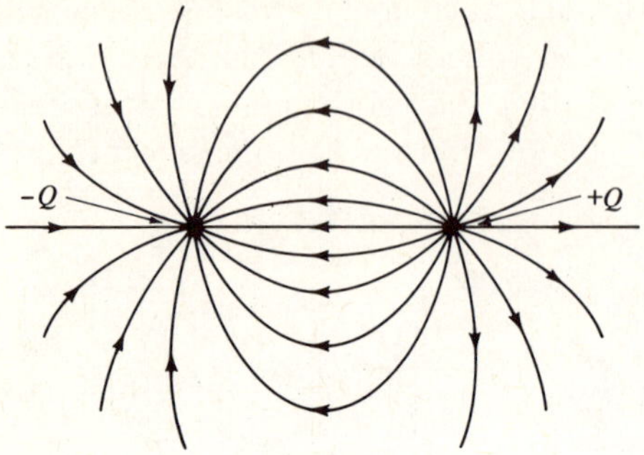

FIGURE 16.4 The electric field lines of two opposite charges of equal size start at the positive charge, and all end on the negative charge.

16.2 never end, and in the corresponding picture of a negative charge alone in the world the lines would come in from infinity, having no beginning but ending on the negative charge. The fields shown in Figures 16.3 and 16.4 are helpful in visualizing the force on a particle P with a charge q anyplace in the field. These fields are accurate representations of the physical situation as long as the charges Q_1 and Q_2 that are the source of the field are at rest. The field produced by charges at rest is called an *electrostatic field.* When the charges that are the source of the field move, magnetic fields also enter, as will be discussed later. Here only electrostatic fields are being considered.

A convenient way of speaking of field lines is to say that one line starts on a positive charge of magnitude $1/4\pi$ and one line ends on a negative charge of this same size. If this convention is adopted, then the density of field lines is not only proportional to the electric field but is numerically equal to it.

Often one is interested in electric effects in a rather limited region of space, a region where some charged particles are to be observed, but not elsewhere. Under these circumstances, the idea of the electric field is extremely helpful because one can think of the field existing without bothering about what charges are necessary to produce it. Eventually, if what is talked about is to be realized in the laboratory, the field has to be produced by properly placed charges. This may be a nontrivial matter, but it is one that can be ignored here.

The presence of an electric field and its magnitude and direction can be determined experimentally by introducing a charged particle and measuring the force on it. A particle used in this way is called a *test particle*. If the test particle has unit charge, the force on it gives the electric field directly. It is important not to use a test particle with too large a charge on it, as it then might exert big enough forces on the charges producing the field to disturb those charges and so change the field being measured. A measurement should not disturb the system being observed.

A conductor is defined as a body in which there are charges that are able to move if forces are exerted on them, so if an electric field existed in a conductor there would be motion of these charges, as in a wire carrying an electric current. In normal conductors there is some friction associated with the motion of charge, so when the conductor is of finite size and when no extraneous devices such as batteries or electric generators are part of the system, the current cannot flow forever and eventually the electric field inside the conductor will vanish. As an example of how this comes about, imagine a conducting rod on which there is zero total charge to materialize suddenly with its axis parallel to an electric field $\mathbf{E}_0$. Immediately charge would start to move along the rod, as shown in Figure 16.5, with positive charge accumulating at the right and negative charge accumulating at the left. This charge would not oscillate back and forth like a mass on a spring because there is friction in the form of electrical resistance to the motion of charge through the material. In (a) of the figure, before any charges have moved, a positively charged test particle near the middle of the rod would be urged to the right by the applied field $\mathbf{E}_0$. In (b), after charges have moved some, this test particle would be urged to the right by

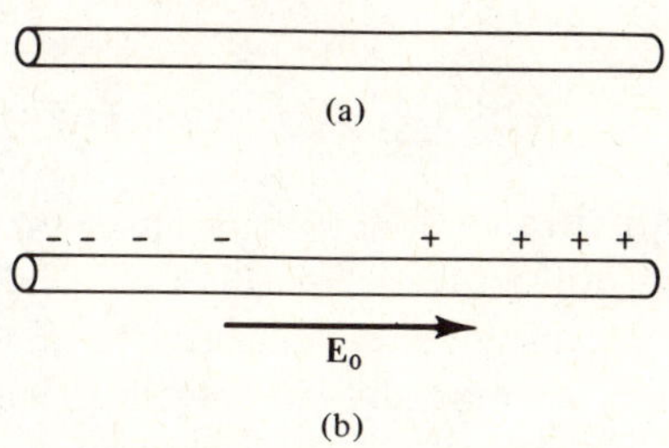

FIGURE 16.5 An uncharged conductor placed in an external electric field becomes charged positively at one end and negatively at the other, the total charge remaining zero.

the applied field $\mathbf{E}_0$ as before but would be repelled by the positive charges already at the right and attracted by the negative charges at the left. If the influence of $\mathbf{E}_0$ were still dominant, positive charges would still be urged to the right, increasing the positive charge at the end so that a test particle in the middle would be more strongly repelled. This would go on until so much positive charge has accumulated that its repulsion cancels the force caused by the applied field. In the end, then, a test particle anywhere along the rod feels no force in either direction, and the electrostatic field inside the conductor is zero. While the charges are moving to their equilibrium positions the field is not static, but in the end it becomes so. There exist conductors in which there is no resistance to the motion of charge. These are called superconductors, and they are becoming of great practical and theoretical interest. The argument just given does not apply to them, but in them, too, the electric field is eventually zero.

Associated with the Coulomb force between two charged particles is a potential energy. If the particles have charges q and Q and are located at positions $\mathbf{r}$ and $\mathbf{r}_Q$, this potential energy is, according to Eq. 15.5,

$$V = \frac{qQ}{|\mathbf{r} - \mathbf{r}_Q|} \tag{16.5}$$

If the charge Q is held fixed at a certain position and the position of the particle with charge q is allowed to vary, this may be regarded as the potential energy of q at the position $\mathbf{r}$. If q moves in the neighborhood of several charges, its potential energy is the sum of the potential energies due to those charges considered separately. Thus, in the presence of two charges Q_1 and Q_2 located at $\mathbf{r}_1$ and $\mathbf{r}_2$ as in Figure 16.1,

$$V = \frac{qQ_1}{|\mathbf{r} - \mathbf{r}_1|} + \frac{qQ_2}{|\mathbf{r} - \mathbf{r}_2|} \tag{16.6}$$

The potential energy of a particle at $\mathbf{r}$ is proportional to its charge q. As in the case of the electric field, it is useful to think of this potential energy V as the result of two factors, the charge q of the particle and something characteristic of the state of affairs at $\mathbf{r}$ due to the presence of the charges Q_1 and Q_2. This something is called the *electrostatic potential* at the point $\mathbf{r}$ and is usually denoted by φ (the Greek letter phi).

$$V(\mathbf{r}) = q\varphi(\mathbf{r}) \tag{16.7}$$

$\varphi(\mathbf{r})$ is then the potential energy that a unit charge placed at $\mathbf{r}$ would have due to charges Q_1 and Q_2. In the above situation, this electrostatic potential is

$$\varphi(\mathbf{r}) = \frac{Q_1}{|\mathbf{r} - \mathbf{r}_1|} + \frac{Q_2}{|\mathbf{r} - \mathbf{r}_2|} \tag{16.8}$$

The electrostatic potential (often shortened to potential) is a scalar, and the addition in Eq. 16.8 is ordinary algebraic addition rather than the vector addition appearing in Eq. 16.4.

The electrostatic potential φ is related to the electric field **E** just as the potential energy of a particle is related to the force on the particle. Using Eq. 13.13, which gives this relation for energy and force, we see that

$$\varphi(\mathbf{r} + \Delta\mathbf{r}) - \varphi(\mathbf{r}) = -E\,\Delta r \cos\theta \tag{16.9}$$

From this follows the important result that all points of a conductor in equilibrium are at the same potential, as inside such a conductor **E** vanishes. If there is an electric field at the surface of a conductor, it must be perpendicular to that surface as it can have no component parallel to it. Such a component would do work on a charge moved along the surface of the conductor, leading to a difference of potential between two points of the conductor.

If a set of charged particles P′, P″, . . . with charges q', q'', is within a region of space, and if a closed surface is drawn enclosing this region, one can ask how many field lines enter or leave the region. If the total charge within the region is zero, for every field line that starts in the region there is another one that ends, so the total number of lines crossing the boundary, the number leaving the region minus the number entering, is zero. This does not mean that no lines of force cross the surface; it means that for every one that crosses in one direction there is another that crosses in the other direction. The converse of this is also true: If the total number of field lines crossing a closed surface is zero, the total charge inside the surface is zero. Figure 16.6 shows schematically the field due to some charges whose sum is zero. Surface A encloses all the charges, and so the numbers of lines entering and leaving the enclosed region are equal. Surface B contains a net positive charge, and more lines leave it than enter it.

It is now possible to show that there can be no net charge within a conductor, or more accurately, that no net charge can stay within a conductor indefinitely. It is possible to put a charge in a conductor, but if the conductor is a good one such as copper, it will remain inside

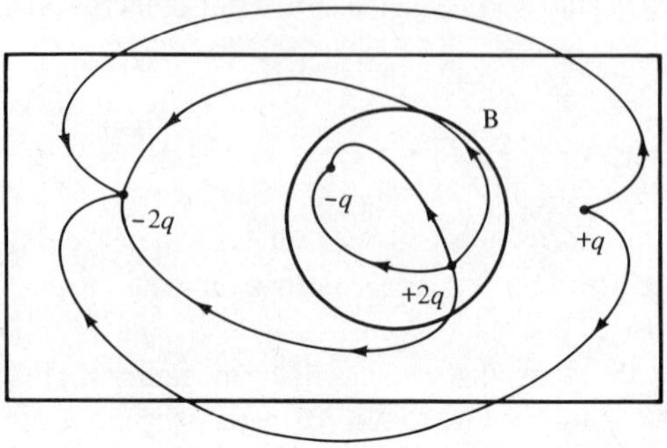

FIGURE 16.6 The electric field due to four charges that add up to zero is shown. Surface A contains zero total charge, and all lines that leave it reenter it. Surface B contains total charge q, and not all lines leaving it reenter it.

only for as long as it takes light to travel the length of the conductor, 10^{-10} sec for a 3-cm sphere. In poorer conductors it will stay longer, but in the end it will have left. We are interested in the equilibrium situation when the charge has stopped moving. In equilibrium the electric field within a conductor is zero, as has been seen, so that there are then no lines of force within a conductor. Any region whose boundary lies within a conductor therefore has zero total charge within it, as no lines of force enter or leave the region. This is a special case of a general theorem known as Gauss' law, which states that the number of lines of force leaving a region in all directions is 4π times the total charge within the region.

If a hollow conducting sphere is mounted on an insulating stand, the following experiment can be done: The sphere is made neutral by touching it with a grounded wire for a moment. A door in the sphere is opened by means of an insulating handle, and a charged particle is thrown in through the door, which is then closed, as shown in Figure 16.7. Is there a net charge inside the conductor? It has just been shown that there cannot be one, so there is none, but how this happens may not be clear. If the particle thrown in is itself a conductor, the charge on it may simply be transferred to the inside of the sphere and then move quickly to the outside surface of the sphere, leaving no charge behind. If the thrown particle is an insulator, the charge cannot leave it and the mechanism given above is not sufficient to describe what happens. In this case a charge equal in magnitude but opposite in sign to the

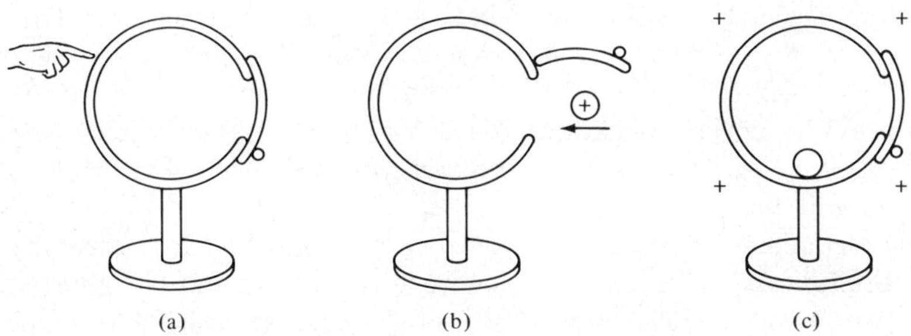

FIGURE 16.7 A charged object is thrown into a hollow conducting sphere that originally is uncharged. The sphere then has a charge on its outer surface equal to the charge thrown in.

charge on the particle must be induced on the inner surface of the sphere. The sphere itself is uncharged, so a charge equal to the particle's charge then must be left on the outside surface of the sphere. These two cases are illustrated in Figure 16.8. The size of the charge that appears on the outside surface of the sphere is the same whether the injected particle is a conductor or an insulator; it is equal to the charge on that particle. Only the situations inside the sphere are different.

If some of the charge introduced into the conducting sphere by a conducting particle remained within the sphere and did not appear on the outside, it would follow that Gauss' law is not correct and that the representation of the electrostatic field by lines of force starting

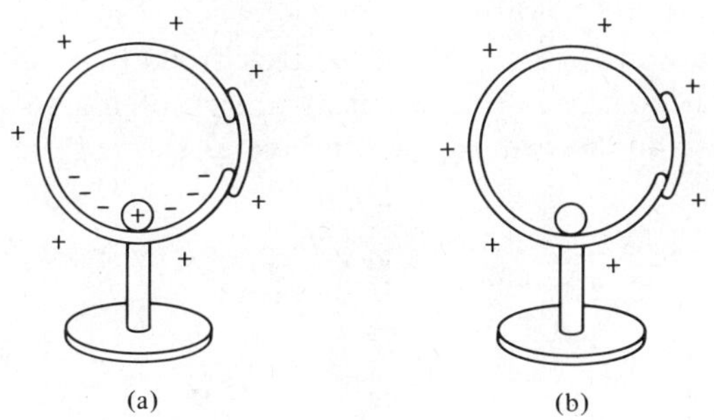

FIGURE 16.8 The situation inside the conducting sphere of Figure 16.7 depends on whether the thrown particle is a conductor or an insulator. (a) Insulating particle and (b) conducting particle.

on positive charge and ending on negative charge is not correct. This would mean that Coulomb's law is wrong. It has been very accurately verified that when a conductor is charged, no charge appears inside it, which provides a verification of Coulomb's law. This way of checking Coulomb's law is capable of much greater precision than experiments with a torsion balance.

Uniform or almost uniform electric fields are useful in controlling the motion of particles. A uniform electric field is most easily achieved by the use of two parallel conducting plates whose separation is small compared with their length and breadth. Such a pair of plates is called a *capacitor* and is illustrated in Figure 16.9. The field lines start on positive charges on the upper plate and end on negative charges on the lower plate. They must meet the plates at right angles because the plates are conductors, and except near the edges the lines remain essentially straight all the way across because there is no more reason for them to bend one way than another. If the plate separation is d and the potential difference between the plates is $\Delta\varphi$, the electric field strength between the plates is

$$E = \frac{\Delta\varphi}{d} \tag{16.10}$$

and the potential at any height y above the bottom plate is

$$\varphi(y) = Ey \tag{16.11}$$

if the potential of the bottom plate is chosen to be zero. This is very like the situation in a uniform gravitational field, but here both the direction and magnitude of the field are at the experimenter's disposal, which is not true for the gravitational field.

If a particle of charge q (positive) is released at rest from the top plate of the capacitor where its energy is all potential energy of magnitude $q\ \Delta\varphi$, it will be accelerated downward and will arrive at the bottom

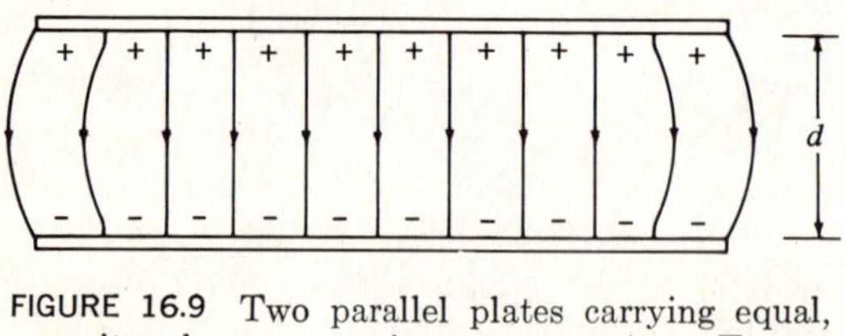

FIGURE 16.9 Two parallel plates carrying equal, opposite charges constitute a capacitor. Except near the edges, the electric field in a capacitor of this kind is very nearly uniform.

plate with kinetic energy K given by

$$K = \tfrac{1}{2}mv^2 = q\,\Delta\varphi \tag{16.12}$$

in view of energy conservation. We have treated the contribution mgd due to gravitation as being negligible, which is not always permissible. In the Millikan oil drop experiment, described in Section 19, the effect of gravity is essential, but in most experiments its effect is far too small to be detected. If a hole is made in the lower plate, the accelerated particle can pass out of the capacitor into a region free of electric field, and we have a device for providing charged particles of a given kinetic energy. To obtain negatively charged particles instead of positively charged ones, one needs only to reverse the signs of the charges on the plates of the capacitor. This method of generating charged particles with a specified kinetic energy is usually applied to atomic particles whose charge is the same as the electron's in magnitude, or some small multiple of it.

In the Gaussian system of units used here, the unit of electric field is the dyne per statcoulomb and the unit of electrostatic potential is the erg per statcoulomb, or statvolt. These units are seldom used in practice. The unit of potential most used for numerical work is the volt (abbreviated to V), which is related to the Gaussian unit by

$$\begin{aligned} 1\ \text{statvolt} &= 2.9979 \times 10^2\ \text{V} \\ &\approx 300\ \text{V} \end{aligned} \tag{16.13}$$

The most-used unit of electric field is the volt per centimeter instead of the dyne per statcoulomb:

$$1\ \text{dyne/statcoul} \approx 300\ \text{V/cm} \tag{16.14}$$

The reader should verify that this equation is dimensionally correct. When E and φ appear in formulas, however, they represent the electric field and potential as measured in Gaussian units.

The kinetic energy of atomic particles is most often measured in electron volts, 1 electron volt (eV) being the energy acquired or lost by a particle whose charge is equal to that of an electron on passing through a potential difference of 1 V. The connection between the electron volt and the erg clearly depends on the measured value of the electron charge. A determination of this is described in Section 19. The conversion factor is

$$1\ \text{eV} = 1.60 \times 10^{-12}\ \text{erg} \tag{16.15}$$

A uniform field may be used to deflect charged particles. If a particle

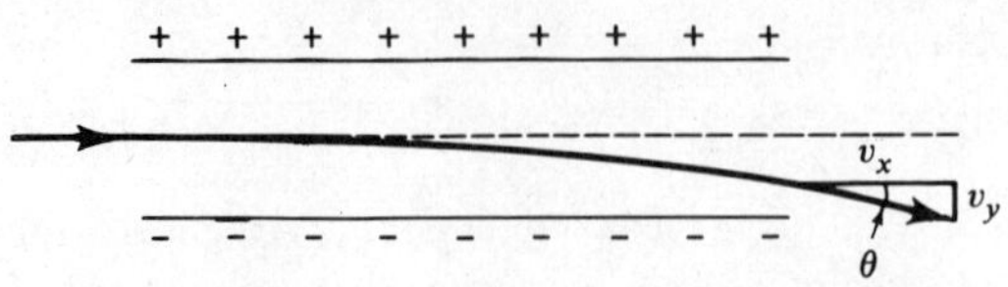

FIGURE 16.10 A positively charged particle moving through the uniform electric field of a charged capacitor is deflected from its original straight-line motion, the angle of deflection θ depending on the original speed, the electric field strength, and the length of the capacitor as well as on the charge and mass of the particle.

of mass m and charge q enters the space between the plates of a capacitor with velocity $\mathbf{v}$ perpendicular to the field $\mathbf{E}$, the motion inside the capacitor will be like the motion of a projectile in a uniform gravitational field. If we take $\mathbf{E}$ to be downward along the negative y-axis, the x-component of the velocity will remain unchanged, while the y-component changes according to

$$ma_y = -qE$$

It has been assumed that initially v_y is zero, so at time t

$$v_y(t) = a_y t$$

$$= -\frac{qE}{m}t$$

$$y(t) = -\frac{1}{2}\frac{qE}{m}t^2$$

If the length of the capacitor is L, the total time T needed to traverse it is L/v_x. The particles will emerge from between the plates going at an angle θ with its initial direction of motion, as can be seen in Figure 16.10, the tangent of the angle being

$$\tan\theta = \frac{-v_y}{v_x} = \frac{qEL}{mv_x^2} \tag{16.16}$$

and displaced downward by the amount

$$y = -\frac{1}{2}\frac{qE}{m}\left(\frac{L}{v_x}\right)^2 \tag{16.17}$$

Both uses of uniform electric fields described here, the imparting of kinetic energy and the deflection of charged particles, are combined in the cathode ray tube, similar to a television tube. At one end of the evacuated tube is an "electron gun" consisting of a source of relatively low energy electrons and two plates made of mesh or something with holes in it. The electrons are accelerated to a rather high energy by the potential difference between these plates, say 2000 V, and then pass successively between two pairs of deflecting plates, one to produce vertical and one to produce horizontal deflections. The electron beam then strikes the wide end of the tube, which is coated with a phosphor that emits light when struck by energetic electrons. The visible spot where the electrons strike the phosphor moves when the potential difference between either pair of deflecting plates is changed. The tube is shown in Figure 16.11. In television tubes the deflection is produced by magnetic rather than electric fields.

It is interesting to note that if electric fields are used both to accelerate and to deflect charged particles, the resulting deflection is independent of both the charge and the mass of the particles involved. Equation 16.16 may be written, with the help of Eq. 16.12, as

$$\tan\theta = \frac{EL}{2\,\Delta\varphi_{\text{acc}}} \tag{16.18}$$

where $\Delta\varphi_{\text{acc}}$ is the potential difference between the accelerating plates. Similarly Eq. 16.17 may be written as

$$y = \frac{EL^2}{4\,\Delta\varphi_{\text{acc}}} \tag{16.19}$$

In neither of these does the charge q or the mass m appear. The speed

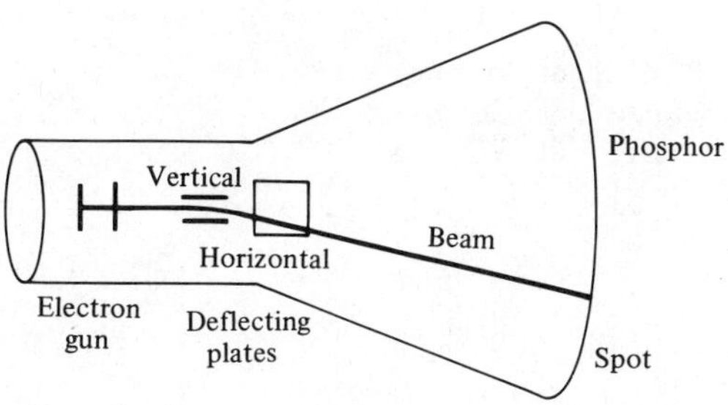

FIGURE 16.11 In an electrostatic cathode ray tube, the electron beam produced by the electron gun is deflected vertically and horizontally by two sets of capacitor plates and produces a spot on the phosphor on the wide end of the tube.

of the particle does depend on q and m, but this cannot be determined just by looking at the angle of deflection or the amount of vertical displacement. As we shall see in the next section, the deflection of a charged particle in combined electric and magnetic fields does depend on the ratio of the charge to the mass of the particle and can be used to determine its speed.

Problems

1. A table tennis ball of mass 2.5 g and radius 1.9 cm is charged to a potential of 3000 V = 10 statvolts. What is the charge on the ball? How large an external electric field is needed to hold this ball up in the gravitational field at the surface of the earth?
2. To what potential would the ball of Problem 1 need to be charged if the mass of the ball to be supported is unchanged but the radius is cut in half?
3. Four charges of equal size, two positive and two negative, are arranged at the corners of a square in the two ways shown. For each arrangement draw the lines along which the electric potential vanishes. [The configuration shown in part (a) of the figure consists of two dipoles side by

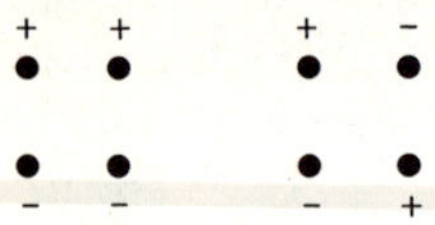

(a) (b)

side and in the same direction, so the effect is almost that of a single dipole of double dipole moment. In the configuration shown in (b) the two dipoles are in opposite directions and tend to cancel each other out. What is left is called a *quadrupole*.]
4. Two charges of equal magnitude q but of opposite signs are separated by a distance d form an electric dipole of dipole moment qd. Show that the electric potential of such a dipole at a point P as shown in the following figure is given approximately by

$$\varphi = \frac{qd}{r^2}$$

when the distance r is much larger than d. Note that this decreases faster with increasing r than a Coulomb potential does.

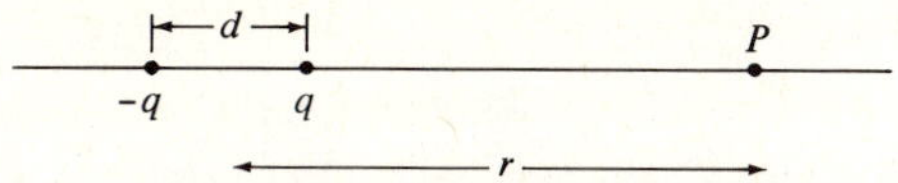

5. Apply the vector addition of electric fields to show that the electric field of an electric dipole is at right angles to the plane that bisects the line joining the charges perpendicularly.
6. Imagine a world in which like charges attract each other and unlike charges repel one another. Describe how a neutral conducting rod in which there are equal amounts of movable positive and negative charge might have its charge distributed. Would the electric field in such a conductor be zero?
7. A beam of charged particles is accelerated by moving through a potential difference of 24,000 V, as the electrons in a television tube are. Find the transverse electric field, expressed in volts per centimeter, that will deflect these particles by 0.2 cm in a 5-cm path length. If the deflecting plates are 0.5 cm apart, what potential difference between them is needed to produce this transverse field?
8. Charge is distributed along a line with a constant charge density μ, that is, there are μ statcoul/cm along the line (see the figure). By applying Gauss' law to a cylindrical surface of unit length drawn about the wire, show that the electric field has the magnitude given by $E = 2\mu/r$. (If the line is very long, the distribution of charge on the two sides of one of the end surfaces are almost identical so that the component of the field parallel to the wire coming from the charges on the two sides will cancel and so there are no field lines crossing the ends of the cylinder.)

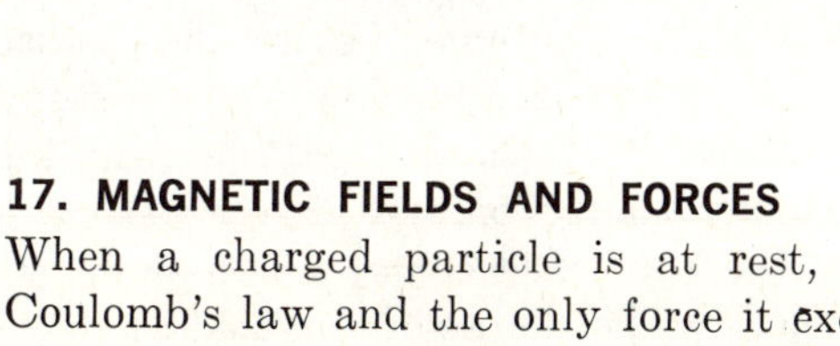

17. MAGNETIC FIELDS AND FORCES

When a charged particle is at rest, its electric field **E** is given by Coulomb's law and the only force it exerts on another particle of charge

q, moving or at rest, is the force

$$\mathbf{F} = q\mathbf{E} \tag{17.1}$$

(We neglect the force of gravity between these particles as being much too small to affect any experiment contemplated.) When the charged particle that is the source of this electric field is in motion the situation is observed to become more complicated and the force exerted on another charged particle is no longer completely given by Eq. 17.1. A magnetic force is produced by this moving charge that acts in addition to this electric force. Here we want to see what this new force is like and how it can be described with the help of a magnetic field. Later, in Section 29, the relation between electric and magnetic fields will be explored to some extent.

Magnetic effects are most simply studied when electric effects are absent and the only forces acting are those due to magnetism, as when a charged particle moves near a current-carrying wire made of a very good electrical conductor. In such a wire a current can be made to flow by means of a battery, and if the wire is made of a good conductor the electric field needed to keep the current flowing is small. The wire itself is electrically neutral because it contains equal amounts of positive and negative charge. We shall need to be able to specify the current that flows, so we must define a unit of current. In the Gaussian units we are using, 1 statampere of current consists of a flow of 1 statcoulomb of charge per second past a fixed point on the conductor. The practical unit of current is the ampere, which is related to the statampere by the equation

$$\begin{aligned} 1 \text{ amp} &= 2.9979 \times 10^9 \text{ statamp} \\ &\approx 3 \times 10^9 \text{ statamp} \end{aligned} \tag{17.2}$$

We shall write the formulas describing magnetic effects using Gaussian units because they are simpler in these units than in any other. In working problems it must be remembered to express all currents in statamperes, no matter what units are used in stating the problem.

Let the current I flow in a circuit, part of which consists of a long, straight wire of good conductor so that, as mentioned before, the electric field needed to cause the current to flow is negligible. We want to study the effect of this current on the motion of a charged particle near the current-carrying wire. There is no net charge on the wire, so there is no electrostatic field produced by the wire. The charge on the current-carrying particles is exactly cancelled by the charge of the fixed conduct-

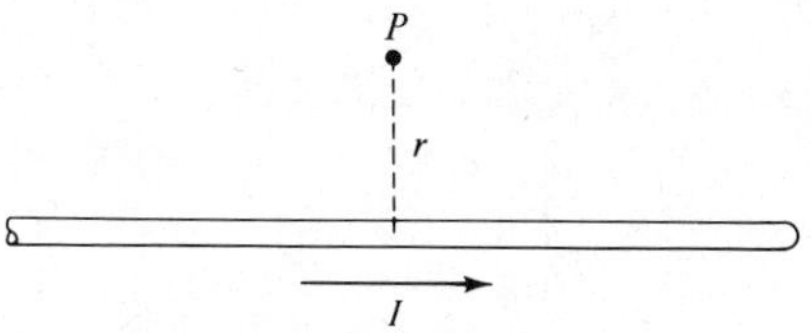

FIGURE 17.1 A current I flowing through a wire influences the motion of a charged particle at P a distance r from the center of the wire by producing a magnetic field around the wire.

ing material; the relative motion of these two kinds of charge constitutes the current I.

Figure 17.1 shows a length of the current-carrying wire with the current flowing from left to right. Such a current would be caused by positive charges moving to the right or by negative charges moving to the left. We are concerned with the effect of this current on the motion of a particle P with positive electric charge q located at a distance r from the wire. Unlike electric forces that do not depend on the velocity of the particle acted upon, magnetic forces are found to depend on the velocity of the particle affected. Several different situations may be described:

1. If no current is flowing through the wire, so that $I = 0$, then the charged particle P moves like a free particle. The wire exerts no force on it.
2. If there is a current flowing but the particle P is at rest, it remains at rest and no force acts on it to accelerate it.
3. If there is a current I in the wire and the particle P is moving parallel to it with speed v, there is a force on the particle proportional to I and to v and directed toward the wire. If the particle is moving antiparallel to I, then the force on it is directed away from the wire. This is illustrated in Figure 17.2(a) and (b).
4. If there is a current I and the particle is moving perpendicular to the wire, there is no force on the particle, as in (c) of the figure.
5. Whether or not there is a force on the particle, its speed or kinetic energy is not changed but remains constant.

It is clear from this description that the force on a charged particle caused by a current is harder to specify than the force due to charges at rest. This new force depends on the direction of the current and

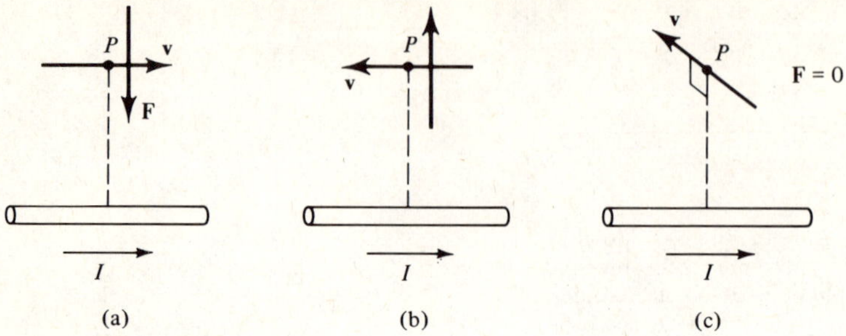

FIGURE 17.2 The force on a positively charged particle at P due to the current I in the wire depends on the direction of the velocity of the particle relative to the direction of the current as shown.

on the direction of motion of the particle. Despite its complication, it can be specified with the help of the magnetic field **B** produced by the current.

The magnetic field produced by a current I flowing in a long, straight wire can be described by field lines, just as the electric field is described by lines of force. The direction of the lines is the direction of the field, and the density of the lines is a measure of the strength of the field. The magnetic field lines are circles about the current-carrying wire as center, as illustrated in Figure 17.3. The direction of the field is defined

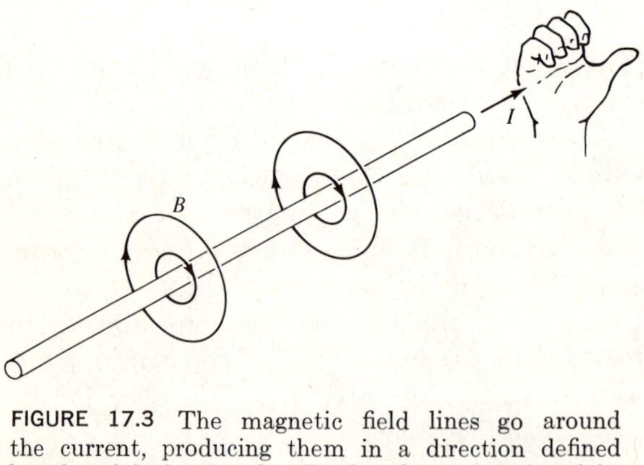

FIGURE 17.3 The magnetic field lines go around the current, producing them in a direction defined by the right-hand rule: If the thumb of the right hand points in the direction of the current, the fingers of that hand can be wrapped around the current and point in the direction of the field.

by a right-hand rule just as the direction of the angular momentum vector in Section 12 was defined. If the wire is grasped (figuratively) with the right hand so that the thumb points along the current, the fingers go round in the direction of the field **B**. This is an arbitrary convention, but it is universally used.

Saying that the magnetic field lines go around the current in circles is not an arbitrary convention. The direction of a magnetic field can be determined by use of a magnetic compass needle, and a graphic way of illustrating the magnetic field produced by a current is to place iron filings on a plastic sheet perpendicular to the wire and to tap the sheet while the current is flowing. Each filing acts like a compass needle and lines up with the magnetic field at its location. A photograph of such a set of filings is shown in Figure 17.4. In the picture the two directions around the circles are, of course, equivalent. The positive direction is arbitrarily defined by the right-hand rule.

The magnitude of the magnetic field is proportional to the current I and is inversely proportional to the distance away from the wire r. This proportion can be made into an equality by introducing a propor-

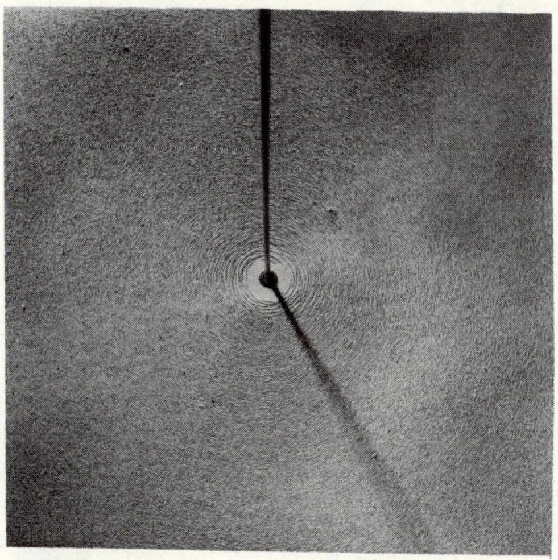

FIGURE 17.4 The magnetic field produced by a current flowing through the wire is illustrated by the iron filings which align themselves with the magnetic field. The decrease in the strength of the field is indicated by the decreasing exactness of the filings' alignment. (From *PSSC Physics,* D. C. Heath, Lexington, 1965.)

tionality constant, so we write

$$B = \frac{2I}{cr} \tag{17.3}$$

with $2/c$ as the proportionality constant. The magnitude of c is discussed later.

Having specified the magnitude and the direction of the magnetic field **B** produced by the current I flowing in a long, straight wire, we must now relate the force on the particle to **B** and to the particle velocity **v**. First we note that according to point 5 above, the speed of the particle is not changed by these magnetic forces. This means, as we know from Section 7, that the force is perpendicular to the particle velocity **v**. If we also assume **F** to be perpendicular to **B**, the observations under point 3 are partly accounted for, since when **v** is parallel to the current, a vector perpendicular to **v** and to **B** is either toward or away from the wire.

The magnitude of the force **F** is given by the expression

$$F = q\frac{v}{c}B\sin\theta \tag{17.4}$$

This says that F is proportional to the particle's charge q, its speed v, the magnetic field through which it is moving, and the sine of the angle θ between the velocity vector **v** and the magnetic field vector **B**. It is zero when **v** is parallel or antiparallel to **B**, which happens when **v** is perpendicular to the wire, as can be seen by combining Figures 17.2(c) and 17.3. The constant c here is the same as the c in Eq. 17.3.

In which direction **F** points along the line perpendicular to both **v** and **B** is determined by another application of the right-hand rule. If the fingers of the right hand point in the direction of **v** and can be bent into the direction of **B**, then the right thumb points in the direction of **F**. Application of this to the situations shown in Figure 17.2(a) and (b) provides a check. If the charge on the particle is negative, the direction of the magnetic force is the reverse of that just given.

It is important to notice that in determining the direction of the force on the charged particle caused by the current I, the right-hand rule is applied twice; once it defines the direction of **B** relative to I, and once it defines the direction of **F** in terms of the directions of **B** and **v**. No physical result would be changed if both of these right-hand rules were replaced by left-hand rules. As far as electromagnetism is concerned, nature is completely symmetrical between right and left. This symmetry is not universal; there are effects in the radioactive decay of elements in which right and left are distinguished from one another. A little is said about this in Section 28.

A few words must be said about the constant c that is introduced in Eqs. 17.3 and 17.4. Suppose the particle P is moving parallel to the current I. Substituting from Eq. 17.3 for B into Eq. 17.4 for F we obtain

$$F = \left(q\frac{v}{c}\right) \times \frac{2I}{cr}$$

$$= \frac{2qvI}{c^2 r} \tag{17.5}$$

Here $\sin\theta = \sin 90° = 1$. The force F can be measured and so can the charge q, the current I, the speed v, and the distance r. c^2 therefore can be determined experimentally. When force is measured in dynes, charge in statcoulombs, and r in centimeters, it turns out that

$$c = 2.9979 \times 10^{10} \text{ cm/sec} \tag{17.6}$$

It is easy to check that c has the units of speed. On the right side of Eq. 17.5 the units of the various quantities appearing are

q: statcoulombs

I: statcoulombs per second

v: centimeters per second

r: centimeters

so that the units of $2qvI/r$ are $(\text{statcoulombs})^2(\text{second})^{-2}$. From Coulomb's law,

$$F = \frac{qq'}{r^2} \quad \text{or} \quad qq' = Fr^2$$

we know that the units $(\text{statcoulomb})^2$ are those of $(\text{force}) \times (\text{centimeters})^2$ so that

$$\frac{2qvI}{r}\text{: (force)} \times \left(\frac{\text{centimeters}}{\text{second}}\right)^{-2}$$

According to Eq. 17.5, however, we see that

$$\frac{2qvI}{r} = c^2 F$$

so the units of c^2 are $(\text{centimeters per second})^2$.

The magnitude of the speed c, determined here by measuring the force on a moving charge near a current-carrying wire, turns out to be the same as the measured speed of light. This remarkable fact was one of the early supports for the idea that light is an electromagnetic wave. More is said about this in Section 31. The unit of magnetic field resulting from inserting c in the relation Eq. 17.3 is the gauss. This unit for magnetic field is the most commonly used one in physics. The constant appearing in the definition of the ampere in Eq. 17.2 is just one-tenth of c.

Magnetic fields are not always produced by currents flowing along straight wires. Mathematical expressions for the magnetic field due to other shapes of wires can be written down, but we do not need them so they are omitted. It is possible to wind coils of wire that produce almost uniform magnetic fields over a considerable volume of space. There are magnetic materials such as iron with very complicated magnetic properties that are useful in making magnetic fields of desired kinds. We ignore these matters because they are too complicated to go into here; indeed they are far from completely understood by anyone at the present time. We shall often state that a particle is moving through a uniform magnetic field without saying how this field is produced.

There is one source of magnetic fields other than currents about which something should be said: the permanent magnet. This is a piece of steel or of some alloy that has been treated in such a way as to magnetize it. The magnetic field of a permanent bar can be mapped out using the iron filing technique mentioned earlier, and shown in Figure 17.5. The magnetic field of Figure 17.5 looks remarkably like the electric field produced by two equal but opposite electric charges, an electric dipole. A magnet of this form is called a magnetic dipole, and the ends of the magnet are its poles, labeled N (north) and S (south). These names originate from the fact that the earth is a magnet with poles near the geographic poles. Note that the magnetic pole near the north geographic pole is a south pole, as it attracts the north poles of compass needles.

It is possible to isolate an electric charge of one sign, but it has so far proved impossible to isolate a magnetic pole. If one cuts an electric dipole in two, one gets one positive charge and one negative charge, as shown in Figure 17.6. If one cuts a magnetic dipole in two, one gets two smaller magnetic dipoles but no analog of the electric charge. No magnetic *monopole* or isolated pole has ever been observed, although such things have been searched for in many ways.

The influence of a magnetic field on the motion of charged particles can be seen directly with the help of a bar magnet and the cathode ray tube described in Section 16. The electron gun at one end of the

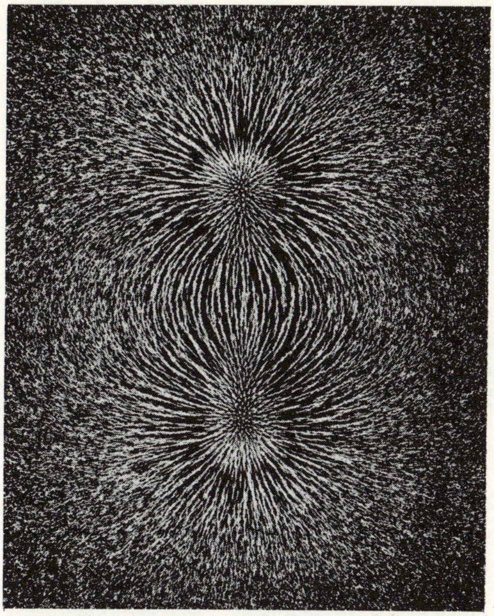

FIGURE 17.5 The magnetic field of a permanent magnet is shown by the iron filing technique. Near the poles of the magnet the filings stand on end because here the magnetic field is not parallel to the sheet on which the filings are distributed. (Courtesy Education Development Center.)

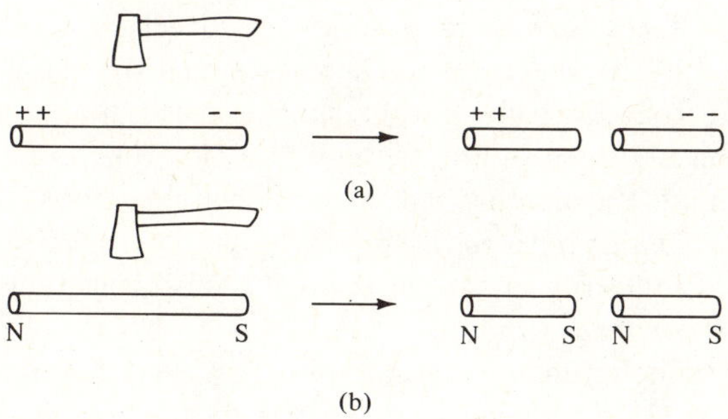

FIGURE 17.6 (a) Chopping an electric dipole in two results in a positive and a negative charge. (b) Chopping a magnetic dipole in two results in two magnetic dipoles and no isolated magnetic pole.

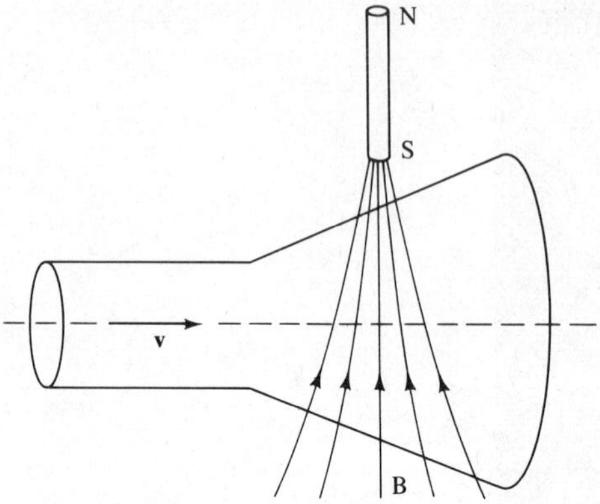

FIGURE 17.7 The magnetic field of a permanent magnet will deflect the beam in a cathode ray tube, the acceleration of the particles being perpendicular both to the magnetic field and to the particle velocity and in the direction expected for negatively charged particles.

tube provides a narrow beam of electrons, and the phosphor at the other end allows us to see where this beam goes; the deflecting plates are not needed in this experiment and are not shown in Figure 17.7. When no magnet is present the electron beam travels down the axis of the tube and produces a spot in the center of the screen at the wide end. Now let a magnet be placed above the tube as shown in the figure. The magnetic field produced by this magnet in the region traversed by the electrons has its largest component in the upward direction. The components of the field parallel to **v** can be made very small by using a strong magnet placed far from the tube rather than a weak magnet placed close as shown. In the presence of the magnet the electron beam no longer strikes the screen at the center, but it is deflected *into* the page. The deflection is perpendicular both to the direction of the magnetic field **B** and to the direction of motion **v** of the electrons. This is consistent with what was said above, as the reader should verify.

The charged particles on which the magnetic force acts need not be particles free of other forces; they may be the particles that carry the current through a second wire. If there are two parallel wires carrying currents in the same direction, one may regard the first of these currents as producing a magnetic field through which the charged particles carrying the second current are moving. The discussion just given, then, shows that these two currents should attract each other, as in fact they do. The charge carriers of the second current experience a magnetic force toward the first, but since they are not free to leave their wire this force is transmitted to the wire as a whole, the mechanism by which this happens being the subject of Section 43. It does not matter which current is considered to generate the field in which the other one moves because Newton's third law requires the forces on either of the wires caused by the other to be equal in magnitude. Large currents can be produced in wires. If one tries to produce a large current of free charged particles as with an oscilloscope beam, a difficulty is encountered because the charged particles in the beam repel each other with Coulomb forces and the beam spreads out. In a wire there is no net charge, current being carried by the motion of particles whose charge is neutralized by that of fixed particles. Later it will be shown that the moving charge is usually negative and consists of electrons and the positive charge is that of the stationary ions that make up the structure of the metal. The force between two current-carrying wires is purely magnetic because the total charge on each is zero; the force between two parallel beams of charged particles is a combination of a Coulomb repulsion and a magnetic attraction, of which the repulsion is always the greater.

Magnetic fields are used in many practical devices, usually to exert forces on currents. Where large forces are needed, as in electric motors, the magnetic fields are most often produced by currents flowing through coils of wire wound around iron cores. The magnetic field induced in the iron reinforces that produced directly by the current in the coil. Where small forces are sufficient as in loudspeakers, permanent magnets are commonly used.

Problems

1. A current of 3 statamperes is carried by charges moving 5 cm/sec along a tube. How much charge per unit length must be moving to produce this current? What is the strength of the magnetic field 15 cm from the current? What is the strength of the electric field produced by these charges at this distance? (See Problem 8, Section 16.)
2. Two billiard balls of mass 500 g each and of radius 3 cm are in contact. What is the gravitational attraction between them according to Newton's law of gravitation, Eq. 10.8, using the value of G in Eq. 10.14? How much charge would each ball have to have distributed uniformly over its surface (or concentrated at its center) for the electric repulsion just to cancel this attraction? Does your answer depend on the separation of the balls?

18. MOTION IN A UNIFORM MAGNETIC FIELD

We now want to consider the motion of a charged particle in a uniform static magnetic field. In such a field both the magnitude and direction of **B** are everywhere the same, but this does not mean that the force exerted on a charged particle is fixed in either direction or magnitude because this force depends on the velocity of the particle as well as on **B**. A magnetic field is not, therefore, a "force field" in the same sense that gravitational and electric fields are. The magnetic force is always perpendicular to the velocity **v** of the particle, so it is impossible to change the kinetic energy of a charged particle by the use of a steady magnetic field alone; a steady magnetic field can do no work on a current. (A time-varying magnetic field produces an electric field that can change the kinetic energy of a charged particle, an effect discussed in Section 31.) If the particle starts out with **v** perpendicular to the uniform field **B**, then θ in Eq. 17.4 is a right angle and $\sin\theta = 1$, so the magnitude of the magnetic force is

$$F = q\frac{v}{c}B \tag{18.1}$$

To obtain the motion of the particle under the influence of such a force, we note that **F** is of constant magnitude and perpendicular to **v**, which is itself of constant magnitude, so that the circumstances are just those encountered in uniform circular motion. One therefore can equate the magnetic force to the centripetal force needed to produce circular motion, obtaining

$$q\frac{v}{c}B = \frac{mv^2}{r} \tag{18.2}$$

or

$$mv = \frac{qBr}{c} \tag{18.3}$$

The radius r of the path of a charged particle in a uniform magnetic field **B** is a direct measure of the momentum of the particle. This provides a very important way of measuring the momentum of particles whose charges are known.

The length of time it takes a charged particle to complete one revolution in its circular motion in a uniform magnetic field, the period τ of its motion, is equal to the circumference of the circle divided by the speed. From Eq. 18.3 one obtains

$$\tau = \frac{2\pi r}{v}$$

$$= 2\pi \frac{mc}{qB} \qquad (18.4)$$

All the quantities on the right are constants independent of the motion of the particle, so that the period τ depends only on the mass and charge of the particle and the magnitude of the magnetic field, not on the speed of the particle. A faster-moving particle moves in a larger circle; the increased circumference of this circle just compensates for the increased speed, leaving the period unchanged. Another way of saying this is that the angular velocity ω of the particle in its circular motion does not depend on the radius of the circle or, therefore, on the momentum of the particle. We have by the definition Eq. 7.7 of angular velocity

$$\omega = \frac{2\pi}{\tau} = \frac{v}{r}$$

$$= \frac{qB}{mc} \qquad (18.5)$$

This angular velocity or angular frequency of the particle is called the *cyclotron frequency* for reasons that appear in Section 24.

For very high speeds when v is not very small compared with c, the speed of light, m cannot be considered to be independent of the speed v but increases with increasing speed. This means that for particles moving with such high speeds the period τ increases with increasing speed or the cyclotron frequency decreases. This is a relativistic effect described in Section 22.

If the initial velocity of the particle is not perpendicular to **B**, the motion is not circular motion in a plane perpendicular to **B** but is along a helix whose axis is parallel to **B**. The component of **v** along the direction of the field is unaffected by the magnetic force and is therefore constant in time. This motion was described at the end of Section 7 and is illustrated in Figure 7.4. It consists of uniform circular motion in a plane perpendicular to **B** together with a uniform linear motion of the plane of the circle parallel or antiparallel to the field **B**.

Problems

1. A table tennis ball has a radius of 1.9 cm and a mass of 2.5 g. What electric charge must be placed on this ball in order to have it go around in a uniform magnetic field of 12,000 gauss with an angular velocity of 20 rad/sec? (This is about 7200 rpm, top speed for an automobile engine.) What is the electric field at the surface of this table tennis ball? To what potential must the ball be charged?
2. An atomic particle is observed to move with angular velocity 2×10^6 rad/sec in a uniform magnetic field of 12,000 gauss. What is the charge-to-mass ratio of this particle? How does it compare with that for an electron, an alpha-particle?
3. If a table tennis ball like that of Problem 1 is charged to a potential of 3000 V = 10 statvolts, how fast must it move across a horizontal magnetic field with $B = 5000$ gauss in order that the magnetic force just cancels the force of gravity on the ball?
4. If a table tennis ball like that of Problem 1 is charged to a potential of 3000 V = 10 statvolts, how fast must it move across a horizontal magnetic field of 500 gauss in order that the magnetic force just cancels the electric force on the ball produced by a uniform electric field of 300 dynes/statcoul? Does your answer depend on the charge or on the mass of the ball?

CHAPTER 4
The Atom and Its Constituent Particles

19. THE ELECTRON

The demonstration that the electron exists as a charged material particle was a turning point in physics, the electron being the first "elementary particle" to be identified. The history of this part of physics is long and complicated and involves the work of many people. There were ideas of the nature of charge and of electromagnetic phenomena that now, in the light of many years of work, seem far-fetched but at the time were widely accepted. These ideas had to be overcome and to be replaced with better ones. There is no point in describing them here; there are enough difficult concepts that have survived until now without adding to the list ideas that are no longer relevant. The brief account we give here of the history of the electron follows a path that became clear only after the events described were all fitted together and that was almost totally obscure at the time.

The first phenomena we describe are those which led to the idea of the *cathode rays*. This name has survived mainly in connection with the *cathode ray oscilloscope*. The cathode rays were observed in the study of the conduction of electricity through gases, where they appear when the gas pressure gets very low. Therefore, they could not be discovered before good vacuum pumps were developed, and this happened in the last third of the 1800s.

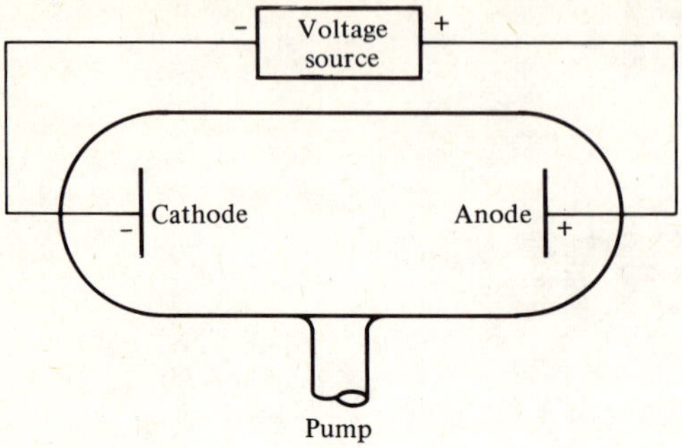

FIGURE 19.1 A tube for showing types of electrical conduction in gases has electrodes at the two ends connected to a source of potential difference and is connected to a pump that can exhaust most of the air or other gas from the tube.

The conduction of electricity through gases can be investigated by the use of a glass tube through whose ends wires are introduced that are connected to small metal disks. The tube can be connected to a pump to reduce the pressure to any desired value, and the wires can be connected to a source of potential difference. Such sources were available in the form of "electrostatic machines" such as the Wimshurst machine (which could charge bodies to very high potentials provided the bodies were well insulated), batteries, and various other devices. With the sources available only small currents could be used. The source of the potential difference need not concern us beyond noting its existence. The apparatus to be used is shown in Figure 19.1. The metal disks inside the tube are called *electrodes*. The electrode connected to the negative terminal of the power supply (source of potential difference) is called the *cathode*, the other one connected to the positive terminal being the *anode*.

When the pressure of the gas in the tube is about the same as that in the atmosphere, the conduction of electricity between the electrodes takes place in the form of sparks like lightning on a small scale. As the pressure is reduced by removing gas from the tube, a stage is reached in which there are narrow, continuous arcs or streamers stretching from one electrode to the other. It is impossible by inspection to see in what direction electricity is moving, that is, whether positive charge moves to the left or negative charge moves to the right. On further reduction of the pressure the streamers grow broader and eventually the whole tube is occupied by a continuous glow. This happens when the pressure is about one three-hundredth of atmospheric pressure. The color of the glow depends only on the gas in the tube: Air gives a purple glow, neon a red one, and argon a blue one. This glow is that seen in "neon" signs.

The pressure in the tube can be lowered still more. The glow ceases to be uniform, and dark spaces appear. In particular a dark space appears near the cathode, and this dark space becomes larger and larger as the pressure is further reduced. Eventually it reaches the walls of the tube, which then fluoresce with a color that depends on the kind of glass the tube is made of, usually a greenish or bluish color. This color does not depend on the kind of gas that is in the tube. A typical pressure at which this fluorescence appears is between 10^{-5} and 10^{-6} atmospheres. The fluorescence of the glass is produced by something coming from the cathode and bombarding the glass; this something is the cathode rays. If the pressure is reduced still further, the conduction of electricity through the gas ceases, there not being enough gas left to carry charge in observable amounts from one electrode to the other.

Let us consider the pressure to be kept in the range where the cathode rays have a visible effect on the tube walls. The first thing to be demonstrated is that the cathode rays do come from the cathode. This can be done by interposing an obstacle, say a star-shaped piece of aluminum, between the cathode and the glass envelope of the tube. For this experiment a tube like that shown in Figure 19.2 is convenient. The anode is off to the side and out of the way of the cathode rays, which seem to come off the cathode at right angles to its surface. The metal star is on a hinged support and can be raised as shown or can lie flat on the bottom of the tube. When the star is on the bottom, the right end of the tube fluoresces when the potential difference is applied, but when it is raised its shadow is clearly seen on the end of the tube. This proves

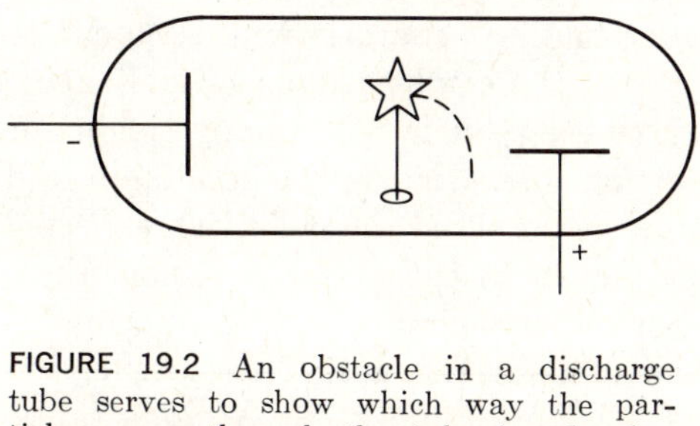

FIGURE 19.2 An obstacle in a discharge tube serves to show which way the particles move through the tube by showing where its shadow appears.

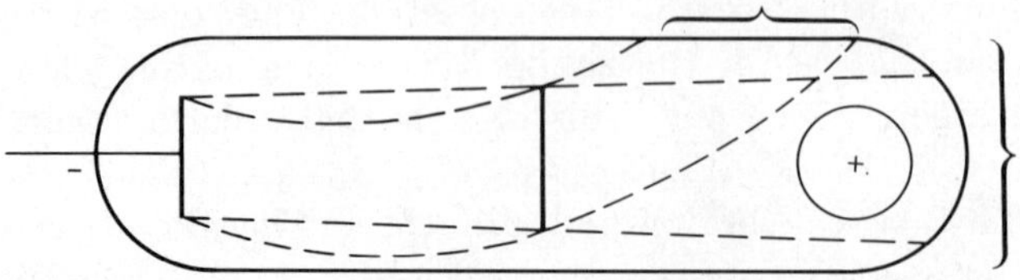

FIGURE 19.3 Top view of the tube of Figure 19.2 shows how the position of the shadow is affected by the presence of a magnetic field out of the page. The direction of the deflection shows that the particles moving are negatively charged.

conclusively that the cathode rays do come from the cathode and that they travel in straight lines.

The next observation is that the cathode rays are deflected by a magnetic field. If a magnet is placed with its north pole below the star in Figure 19.2 and its south pole rather far away, the shadow of the star is displaced *into* the page. A view of the tube from above in Figure 19.3 shows what happens. The magnetic field lines are coming up out of the page. The shadow is observed to be moved to the left side of the tube, looking along the beam, when the magnet is present. The paths of the particles are bent by the magnetic field in the same way that the paths of negatively charged particles moving from the cathode would be bent by this field, as an application of the rules for the magnetic force given in Section 17 will show.

Are the cathode rays streams of negatively charged particles? If so, they should be deflected by an electric field directed across the tube. If electrodes are placed above and below the tube and charged, as in the capacitor of Figure 16.8, no displacement of the shadow of the star is observed. This was regarded as strong evidence against the charged particle model of cathode rays, but it was eventually realized by J. J. Thomson that since the gas is conducting, charges will accumulate on the inside of the glass walls opposite the condenser plates and cancel out the electric field inside the tube, as the electric field in any conductor is canceled. When he put the deflection plates inside the tube and used a low enough gas pressure in the tube, he observed the deflection expected on the basis of the charged particle picture of cathode rays. Under these circumstances the gas is part of an electric circuit and equilibrium is not reached, so there is an electric field in the gas. This verification of the particle nature of cathode rays is only qualitative. A quantitative study of their motion confirms this picture in a way we describe directly.

In Section 16 it was pointed out that a charged particle can be accelerated by an electric field and that when a particle of charge q starting from rest is accelerated by an electric field, by the time it has moved through a potential difference $\Delta\varphi$, it has acquired a kinetic energy given by

$$\tfrac{1}{2}mv^2 = -q\,\Delta\varphi \tag{19.1}$$

If q is positive, $\Delta\varphi$ must be negative in order to accelerate the particle, and if q is negative, $\Delta\varphi$ must be positive. We write the charge on a cathode ray particle as $-e$ so that e itself is positive. From this we see that

$$\frac{e}{m} = \frac{v^2}{2\,\Delta\varphi} \tag{19.2}$$

The charge-to-mass ratio can be determined by measuring the speed v of a particle that has been accelerated from rest by a potential difference $\Delta\varphi$. Cathode ray particles cannot be seen individually, so that no direct measurement of their speed is possible. This speed can be measured indirectly, however, and one very elegant method of doing this is to compare the deflections of a beam of cathode rays produced by a transverse electric field and by a transverse magnetic field. The force an electric field exerts on a charged particle does not depend on the speed of the particle, but the magnetic force does depend on this speed, so a comparison of these forces can yield the value of speed. We analyze this in detail.

In Section 16 we saw that an electric field in the y-direction will deflect a beam of charged particles traveling in the x-direction. According to Eq. 16.13 the deflection produced by an electric field of strength E acting over a length L is given by

$$\tan\theta = \frac{qEL}{mv_x^2} = \frac{(-e)EL}{mv_x^2} \tag{19.3}$$

If the electric field in the y-direction is now replaced by a magnetic field of strength B in the z-direction, as shown in Figure 19.4, the particles will again be deflected in the y-direction. If the deflection is small we may neglect the change in the magnitude of v_x as the particle moves through the field. The magnitude of the y-component of the magnetic force is given by

$$F_y = q\frac{v_x}{c}B \tag{19.4}$$

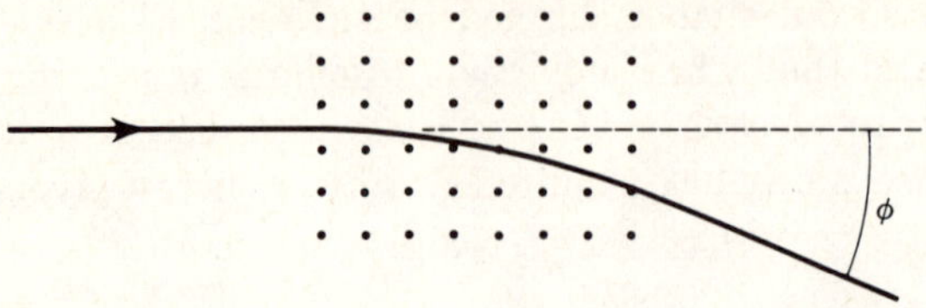

FIGURE 19.4 A charged particle moving through a region with a uniform magnetic field into the page is deflected through an angle ϕ whose size depends on the field strength, the length of the path through the field, and the charge, mass, and speed of the particle.

and the vertical component of velocity produced has the magnitude

$$\begin{aligned} v_y &= a_y t \\ &= \frac{F_y}{m} \times \frac{L}{v_x} \\ &= \frac{qBL}{mc} \end{aligned} \tag{19.5}$$

The final velocity vector has components v_x, v_y, and therefore makes an angle φ with the direction of the x-axis such that

$$\begin{aligned} \tan\varphi &= \frac{v_y}{v_x} \\ &= \frac{qBL}{mcv_x} = \frac{(-e)BL}{mcv_x} \end{aligned} \tag{19.6}$$

In his classic experiment of 1897, J. J. Thomson carried out the steps just described. He deflected a beam of cathode rays by means of a transverse electric field acting over a path length L. He then introduced a magnetic field also acting over the path length L. The strength of this magnetic field was adjusted to produce the same deflection as that produced electrically, but in the opposite direction so that the net effect of the electric and magnetic fields was no deflection of the beam at all. With $\varphi = \theta$, Eqs. 19.6 and 19.3 lead to

$$\frac{eBL}{mcv_x} = \frac{eEL}{mv_x^2}$$

or

$$\frac{v_x}{c} = \frac{E}{B} \tag{19.7}$$

This determines the speed of the particles.

The value of the speed just determined could be used in Eq. 19.2 to determine e/m, but this requires a separate measurement of the accelerating potential $\Delta\varphi$. It is easier to insert this back into Eq. 19.6 and to measure the deflection angle φ, which is by adjustment the same as θ. The result is

$$\frac{e}{m} = \frac{c^2 E\theta}{B^2 L} \tag{19.8}$$

All the quantities on the right are measureable, so that this gives the value of e/m for the cathode ray particles. Thomson's value for this was about 2.3×10^{17} statcoul/g. The presently accepted value for this quantity is more than double Thomson's original value and is 5.2728×10^{17} statcoul/g. There are several important things to say about this experiment and its result. The value of e/m is independent of the kind of metal the emitting cathode is made of and of the gas with which the tube is filled. This suggests that the cathode ray particles are constituents of all kinds of matter. The value obtained is very much higher, almost 2000 times larger, than that corresponding to charged atoms (ions). The ratio e/M for ions can be measured when electrolysis takes place. When an electric current is passed through water containing a small amount of acid, the water is decomposed with gaseous hydrogen being released at one electrode and gaseous oxygen at the other. The mass of hydrogen released is proportional to the total electric charge passed through the water, and both the mass and the charge can be measured. If the hydrogen is released in the form of identical atoms, as is known from chemical evidence to be the case, one can measure the charge-to-mass ratio for the hydrogen ion or charged atom. It is 2.879×10^{-20} statcoul/g. This ratio can be measured for other elements too, and it depends strongly on the element involved; it has its greatest value for hydrogen. The cathode ray particle, the electron, is therefore a very light constituent of atoms.

The Thomson experiment does not, in fact, establish the existence of a unique kind of cathode ray particle, although it makes it very probable that such a particle exists. The results of the experiment would be unchanged if there were many different kinds of particles as long as they all have the same ratio of charge-to-mass. The matter can be settled by determining that there is a smallest unit of charge and that all charges are multiples of this unit. An experiment to show this was done by R. A. Millikan: the famous oil drop experiment.

The problem is to measure a very small charge. The charge on a particle can be found if the force on that particle in an electric field can be measured. It will not do to observe the acceleration of the charged particle in the electric field unless the mass of the particle is known separately, and for the cathode rays this is not known, only the ratio e/m being determined from Thomson's experiment. The electric force on the particle must be compared to some nonelectric force that can be measured. Millikan used the force of gravity as the comparison force.

To measure the force on a particle directly, a visible particle is needed, and to measure a small force, a small particle is needed. A small drop of oil from an atomizer spray can be seen through a microscope, especially when it is illuminated from the side, just as very small dust particles are visible in a ray of sunlight in a darkened room. An oil drop is round, and if the right kind of oil is used it is almost constant in mass because evaporation is very slow. A small oil drop is light enough to be influenced by the force due to a very small charge when the charged drop is in an attainable electric field. The idea of the oil drop experiment is to make the electric force on the drop just cancel the gravitational force on the drop so that the drop does not fall.

The experimental arrangement is sketched in Figure 19.5. Oil drops are introduced between a pair of capacitor plates. They are viewed through a microscope, which has a scale built into the eyepiece. The oil drops are charged when they enter the space between the plates because of the effect of friction during their formation. Their charge can be altered by exposing them to X-rays. (This process had long

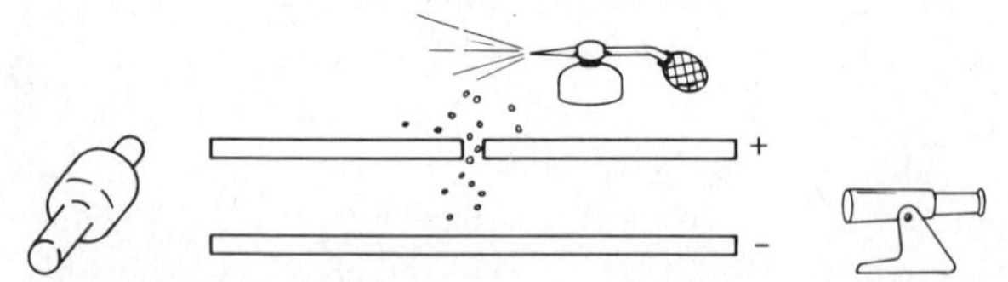

FIGURE 19.5 The essential ingredients of the Millikan oil drop apparatus are a capacitor to provide the electric field, an atomizer to provide oil drops, an X-ray tube to charge the drops, the earth to cause the drops to fall, and a microscope to see them with.

been known to make gases conductors of electricity, it now being known that X-rays eject electrons from the atoms of the gas leaving them charged. In the same way they eject electrons from an oil drop, or the drop may pick up an electron ejected from some other atom or drop. This is a random process, and one just has to look for a drop that is charged.) The electric field between the capacitor plates is then adjusted to hold the drop stationary, so that if the buoyant force of the air is neglected,

$$qE = mg \tag{19.9}$$

If we knew m we could then find the charge q on the drop. Obviously the drop cannot be placed on the pan of a balance and weighed. A less direct method must be used.

When the electric field is turned off, the drop does not fall with acceleration **g** like a free particle, but it falls with a constant speed called the *terminal speed.* This is because it is falling through air, and motion through air involves frictional forces. The frictional force on a sphere moving with speed v through a fluid such as air is known to be proportional to the speed v and to the radius of the sphere a. Its magnitude depends on the fluid and is written as

$$F_{\text{vis}} = 6\pi\eta\, av \tag{19.10}$$

A frictional force proportional to v is called a *viscous force,* and the constant η (the Greek letter eta) is called the coefficient of viscosity. The terminal speed for a sphere falling through a viscous fluid is that speed for which F_{vis} is equal to mg, so that the total force on the sphere is zero and the sphere is not accelerated. η can be measured by observing the fall of a large sphere through the fluid, a sphere that can be weighed, or by equivalent but more accurate methods. This gives the needed method of determining m. If ρ (the Greek letter rho) is the density of the oil, then

$$m = \tfrac{4}{3}\pi a^3 \rho \tag{19.11}$$

since the volume of a sphere is $(4\pi/3)a^3$. The terminal speed of fall is determined by

$$mg = F_{\text{vis}}$$

again neglecting the buoyant force of the air for simplicity, or

$$\tfrac{4}{3} a^3 \rho g = 6\pi\eta a v \tag{19.12}$$

in which a is the only unknown. Solving this for a, the result is used in Eq. 19.11 to determine the mass of the drop m. When m is known, Eq. 19.9 determines the charge on the drop. This analysis is a beautiful example of the impossibility of separating physics into little specialties that have nothing to do with each other.

The result of the experiment, done by Millikan in 1910 and repeated and refined many times since, is that q is always an integer multiple of a charge e,

$$q = ne \tag{19.13}$$

where n may be either positive or negative and e has the value

$$\begin{aligned} e &= 4.80325 \times 10^{-10} \text{ statcoul} \\ &= 1.6022 \times 10^{-19} \text{ coul} \end{aligned} \tag{19.14}$$

The interpretation of the result is that there is a smallest unit of charge, the charge on the electron, and that all other charges are integer multiples of this charge. The charge on the electron is negative, as we know from the cathode ray experiments, so we have written it as $-e$ making e a positive charge. Taking $-e$ as the electron charge, the mass of the electron is

$$m_e = 9.1096 \times 10^{-28} \text{ g} \tag{19.15}$$

Let us summarize briefly the ideas involved in this discussion of the electron: The observation of the cathode rays led to the concept of a stream of negatively charged particles emerging from the cathode when electricity is conducted through a gas at low pressure. The Thomson experiment, using electric and magnetic deflection of these rays, established that they act like charged particles moving with a definite speed that is determinable by the experiment and having a definite charge-to-mass ratio much larger than that met with in electrolysis. The value of e/m is independent of any materials used in the measuring apparatus. The oil drop experiment established the existence of a smallest unit of charge e, of which other observed charges are integral multiples. On identifying this charge with the magnitude of the charge on the cathode ray particles, one is led to the existence of the electron, a constituent of all forms of matter, with charge $-e$ and mass m_e, both of which are known experimentally. Notice that the existence of the electron has been inferred by assuming that the laws of motion of large, directly visible particles apply equally well to small, invisible particles. As far as we know, this assumption is very accurate as long as the paths followed by the particles are long enough and straight enough to be directly visible. We shall see later that when the paths followed get small, as when an electron moves around in an atom, this assumption no longer is valid. The new rules that apply to these small orbits are the subject of quantum mechanics, about which something will be said quite a bit later in this book.

Problems

1. The numerical results of this section are given in Eqs. 19.14 and 19.15, which give the charge and mass of the electron. Compare the values of e^2 and of Gm^2, G being Newton's gravitational constant, and thereby compare the magnitude of the electric repulsion of two electrons with their mutual gravitational attraction. Does this make gravitation seem unimportant for atomic particles?
2. There are $N_0 = 6.02 \times 10^{23}$ atoms in a gram-atom of an element. (A gram-atom of an element is a mass of that element whose measure in grams is the atomic weight of the element.) What fraction of the atoms in a copper sphere containing 1 g-atom of copper can lose one electron each before this sphere will repel a similar one at a distance of 10 cm with a force of 10^7 dynes?
3. How long does it take an electron that has been accelerated from rest by a potential difference of 2000 V to travel 40 cm down a cathode ray tube and strike the screen? By what factor is this time shortened if the potential difference is quadrupled to 8000 V?
4. If the length of Thomson's apparatus from electron gun to phosphorescent screen were 20 cm and if the speed of the particles were $0.1c$ how far would they fall under the influence of gravity while traversing the apparatus? Would this have a visible effect on the location of the light spot caused by the electron beam?
5. How fast would a particle have to be moving if it is to fall 1 mm under the conditions of Problem 4?
6. In principle, what would be the effect of gravity if the Thomson experiment were done using an electron beam moving vertically upward?

20. THE NUCLEAR ATOM

By the end of the nineteenth century an enormous amount of evidence had accumulated that matter is made up of atoms and that each of the chemical elements is composed of atoms of a distinct kind. The evidence was chemical in nature. Elements combine to form compounds in which the masses of the various elements concerned always bear simple ratios to one another, and this can be easily understood if the elements occur in identical units that can combine to form molecules. This kind of evidence does not, however, tell one much about the individual atoms because the experiments always involve very large numbers of them. The ratio of the mass of an atom of hydrogen to the mass of an atom of any other element could be accurately determined, but the mass of a single atom of hydrogen could not be measured. Thus as late as 1910 there were eminent chemists and physicists who considered the atomic hypothesis a convenient one for describing chemical behavior but who were highly skeptical of the actual existence of atoms as material particles. Only when many experiments dealing with individual particles had been done were these skeptics convinced.

The electron was the first particle really identified as a material constituent of atoms. The experiments of Thomson, and others of a very different nature too complicated to describe here, established the value of e/m for electrons; the oil drop experiment established the value of e. At about the same time another particle was being identified, one much heavier than the electron and called the *alpha particle.* Because the alpha particle played such a central role in the establishment of the currently accepted model of the atom, something must be said about it here.

In 1896 A. H. Becquerel discovered that certain kinds of elements spontaneously emit radiation that can darken a photographic plate even when the plate is wrapped in lightproof paper. Such elements are said to be *radioactive.* It was found shortly afterward that this emission involved a transformation of the emitting element into another chemical element. Radium was isolated by the Curies in 1898; radium is a very radioactive element that decays at such a rate that half of the initial amount has emitted radiation and transmuted into another element after a period of 1690 years, a period called the half-life of the element. The element resulting from the decay is also radioactive, and in fact these elements are members of a long chain of elements that decay one into another with various half-lives.

The radiation emitted by radioactive elements was shown to be of three kinds: alpha, beta, and gamma radiation. The alpha radiation is not very penetrating, being stopped entirely by a thin metal foil. The beta radiation is more penetrating, and the gamma radiation is the most penetrating of all. These different forms of radiation are differently affected by a magnetic field: The alpha rays are deflected as though they were positively charged particles, the beta rays as though they were negatively charged, and the gamma rays are not deflected at all. In 1906 Ernest Rutherford showed by using electric and magnetic deflection of alpha rays from radioactive sources that the charge-to-mass ratio of these particles is very nearly one-half that of hydrogen as measured in electrolysis and one four-thousandth that for electrons, so that here particles of atomic mass and not electrons are involved. Three years later he showed that these particles are helium atoms that have lost two electrons each—doubly charged helium ions. This he showed by providing a glass bulb with a very thin portion of wall through which alpha particles from his source could pass. He found that traces of helium, not detectible before the bombardment, were present in the bulb after the thin window was bombarded by alpha particles.

The beta and gamma rays do not concern us at the moment. By electric and magnetic deflection experiments it was shown early that the beta rays are streams of particles whose value of e/m is that of electrons, and hence that these most likely were electrons. The gamma rays were shown to be radiation like ordinary light, but with much shorter wavelength. We shall come back to these radiations in later sections.

By applying the known properties of electric and magnetic fields to the study of the motion of alpha particles, the nature of these particles was firmly established; they are doubly charged ions of helium. The electric and magnetic fields used in the experiments proving this are those due to charged capacitor plates and large coils of wire carrying currents. Knowing this about alpha particles it was possible to use them to find out what forces act within atoms. This one can do by getting an alpha particle inside an atom and finding the force on it. Of course one cannot place the alpha particle on the end of a rod and stick it into the atom, but one can do scattering experiments that accomplish almost as much. Because so much of our knowledge of atomic and subatomic systems comes from scattering experiments, what they involve must be described.

The simplest kind of scattering experiment involves two particles: the target particle and the bombarding particle or projectile. Usually one arranges things so that the target particle is initially at rest, and one propels the bombarding particle at it with a certain momentum $\mathbf{p}$. When this particle comes near the target particle, it is acted on by a force due to the target that causes its momentum to change. Finally the two particles are again far apart and move like free particles. The bombarding particle has final momentum $\mathbf{p}'$, and the target particle has final momentum $\mathbf{p}''$. Conservation of momentum in the collision requires that the total initial momentum equal the total final momentum,

$$\mathbf{p} = \mathbf{p}' + \mathbf{p}'' \tag{20.1}$$

The change in momentum of the bombarding particle

$$\Delta\mathbf{p} = \mathbf{p}' - \mathbf{p} \tag{20.2}$$

is called the *momentum transfer* during the collision. A measurement of this momentum transfer gives information about the force acting on the projectile during the collision.

An atom is a small target, so one cannot aim an atomic projectile accurately enough to control how close it gets to the target. A source

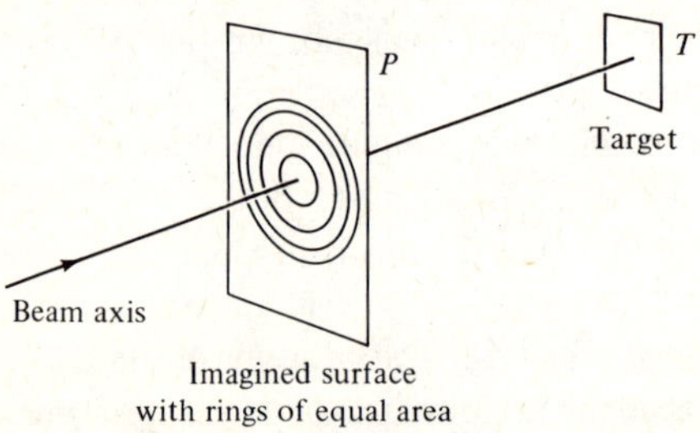

FIGURE 20.1 Beam particles move parallel to the beam axis in such a way that equal numbers of particles go through ring-shaped areas of equal area on the imagined surface P and go on to strike the target.

of atomic projectiles cannot pitch "high and inside" to an atom. The best it can do is to provide a beam of projectiles that uniformly covers a target area very much larger than an atom, so that equal numbers of projectiles go through equal size parts of that target area. This tells something about the probability of a projectile being aimed a certain distance from a target atom. In Figure 20.1 S is the source of projectiles and T is the target atom. The target area is P. We divide P up into circular rings of equal area, so that equal numbers of projectiles go through each ring. If the inner radius of a particular ring is b and its width is Δb, a projectile going through that ring would pass the target T at a distance between b and $b + \Delta b$ if it were not deflected by the force field of the target. This projectile is then said to have the *impact parameter* b for the collision.

If the force field of the target were known, it would be possible to predict what the path of a particle moving with a given speed and a given impact parameter b would be. The angle between the initial projectile momentum vector and the final projectile momentum vector, shown in Figure 20.2, is called the scattering angle θ. This angle can be anything between 0 (no change in direction) and 180° (reversal of direction, back scattering). In the scattering experiment it is known how many particles have an impact parameter between b and $b + \Delta b$; it is the number going through a ring of radius b and width Δb. Each of these particles is scattered through an angle between θ and $\theta + \Delta\theta$. It is possible, then, to calculate for a given force field how many particles

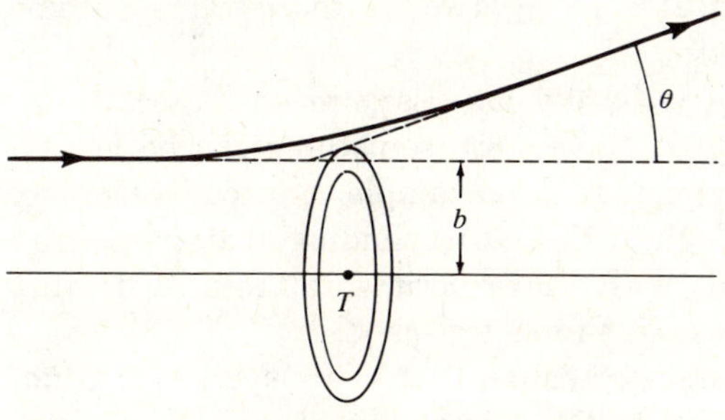

FIGURE 20.2 The path of a projectile past the target T with impact parameter b is deflected by the amount θ.

are scattered through angles between θ and $\theta + \Delta\theta$. One speaks of the *cross section* for this scattering, this being the area of the corresponding ring.

Scattering cross sections can be measured experimentally. On bombarding the target with a beam of particles, one counts the number of particles that are scattered through angles between θ and $\theta + \Delta\theta$ and also counts the total number of bombarding particles crossing the unit area of the plane P. Then the ratio

$$\frac{\text{number of particles scattered through angles between } \theta \text{ and } \theta + \Delta\theta}{\text{number of particles per unit area that could have been scattered}} \tag{20.3}$$

is simply related to the desired cross section. This is called a *differential cross section* because it differentiates between different scattering angles. The total scattering cross section σ (the Greek letter sigma) is defined by

$$\sigma = \frac{\text{number of particles scattered through any angle between 0 and 180°}}{\text{number of particles per unit area that could have been scattered}} \tag{20.4}$$

An actual target contains more than one target atom. Usually the target can be made thin enough so that a given bombarding particle passes close to only one target atom. The effect of having N target atoms is then just to multiply the probability for scattering through any angle by N. If the target is thick so that a given bombarding particle has a good probability of being scattered by two or more target atoms, then things get very complicated.

If the force field is known, it is hardly necessary to do the scattering experiment, but if the force field is not known, something can be learned about it from scattering experiments. For example, one can assume a certain kind of force acts, calculate the corresponding scattering cross section, and compare the result with the experimental results. If they agree one is fairly sure that the correct force was chosen.

Rutherford was one of the first to realize that the behavior of alpha particles bombarding a very thin gold foil might be the effect of single collisions between an alpha particle and a single target particle. (Gold foils were used because gold can be beaten or rolled into very thin sheets, and gold does not corrode.) It was known that alpha particles are scattered by such a gold foil, but most of them are deviated only through a very small angle from their original direction of motion by the foil. When only small angle scattering occurs, it is easily interpreted as the cumulative effect of many very small, independent scatterings by many target atoms, but in the case of alpha particles there are always a few scatterings through large angles that cannot be accounted for as the cumulative effect of many small angle scatterings. Independent scatterings are subject to the laws of probability, and when applied to this problem these laws give a clear and unambiguous result: Large angle scatterings occur practically not at all. This gave Rutherford the hint that single scattering events could be important.

To have large angle scattering take place, two requirements must be met, both of which are fairly obvious once the problem is presented: First, there must be very large forces acting, in this case electrical forces, so there must be strong electric fields somewhere in a gold foil. Second, the source of electric field must be heavier than the alpha particle, or else the alpha particle would knock it out of the way and go on in almost its original direction. The extreme values of both the possible electric field strength and the mass of the target particle are obtained if one assumes that the entire mass of the gold atom (except for the mass of some electrons) is concentrated in a small sphere carrying a positive electric charge. Near this sphere, the *atomic nucleus,* the electric field is given by Coulomb's law and is very great if the value of r to be inserted in this law is very small. If these nuclei are very small compared with the radius of a gold atom, they are far apart and most alpha particles would not get near even one in penetrating a thin foil. Those that did happen to get near one, however, would be scattered through a large angle.

It is possible to calculate the cross section for the scattering of an alpha particle of given energy through an angle between θ and $\theta + \Delta\theta$ on passing through the Coulomb field of a point charge, and this was done by Rutherford. The theoretical results agreed with the experimental results both in their dependence on scattering angle and alpha particle energy provided the Coulomb field on the point charge was taken to be

$$E = \frac{Ze}{r^2} \tag{20.5}$$

where the number Z is about half the atomic weight of gold. Z is now interpreted as the number of protons in the nucleus, each proton having a positive charge of the same size as the electron charge. For gold, $Z = 79$. Z is also the number of the element in the periodic table of the chemical elements.

This experiment alone establishes the nuclear model of the atom. There is a massive, positively charged nucleus that is highly concentrated, the main volume of the atom being occupied by electrons. This nuclear model is now completely established because the electronic structure of atoms is known both experimentally and theoretically to conform to it completely. We return to this subject later.

The scattering experiments carried out in Rutherford's laboratory by H. Geiger and E. Marsden could not establish that the gold nucleus is a point, but they did put an upper limit on its size. If the radius of the nucleus is R, then inside the nucleus the field is no longer given by Coulomb's law with r measured from the nuclear center, just as Newton's law of gravitation with r measured from the center of the earth is not correct beneath the surface of the earth. If the scattered alpha particles got closer than a distance R to the center of the gold nucleus, they would no longer be scattered as though moving through a Coulomb field. One also would expect something like a contact force between the alpha particle and the nucleus that would modify the scattering. No such modification was seen in these early experiments. The closest an alpha particle with positive charge $2e$ can get to a gold nucleus with positive charge $79e$ depends on the energy of the alpha particle. It can certainly not get any closer than the distance r at which all its kinetic energy is converted into potential energy, a distance determined by

$$\frac{1}{2} m_\alpha v_\alpha^2 = \frac{2e \times 79e}{r} \tag{20.6}$$

Actually only the relative kinetic energy can be converted to potential energy, as was shown in Section 14, but here there is not much difference between the total and the relative kinetic energies of the alpha particle because the gold nucleus is so heavy (cf Eq. 14.7). The most energetic alpha particles used by Geiger and Marsden could penetrate, according to this idea, to within about 10^{-12} cm of the center of the nucleus, so that the nuclear radius is no larger than this.

The alpha particle scattering experiments do not give any information about the electrons in the atom except their number. There must be enough electrons to neutralize the positive nuclear charge, but electrons are too light to scatter an alpha particle, so no information on their location in the atom is given.

One can ask what happens to the target nucleus in Rutherford scattering. Can it be detected as it recoils under the impact of the scattered alpha particles? Unfortunately it cannot, but the reasons are not hard to understand. The gold nucleus is much heavier than the alpha particle (atomic weights 197 and 4, respectively), so that even in a head-on collision it will not acquire much kinetic energy although it acquires nearly twice the alpha particle's momentum in such a collision (cf Eq. 13.24). This small amount of kinetic energy is soon dissipated among the other gold atoms in the foil. We shall study many scattering events later in which the recoil of the target particle can be and is observed.

The existence of the nucleus of an atom can also be established by studying the scattering of electrons by atoms. There are several reasons why the alpha particle scattering experiments were the first to show this. The most important is the large mass of the alpha particle. When a charged particle moves through a foil of gold, it experiences Coulomb forces due to the electrons in the atoms as well as those due to the nuclei. If the moving particle is an alpha particle with a mass some 7500 times that of an electron, Newton's third law tells us that the acceleration of the alpha particle by an electron will be only 1/7500 of the acceleration of the electron by the alpha particle. The passage of the alpha particle thus will affect the motion of the electrons drastically, but its own motion will be affected only slightly. This fact is used to detect the presence of the charged particles, as will be described in Section 24. The only large deflections of the alpha particles are those produced by the Coulomb force due to the nucleus, which is much heavier than the alpha particle.

If the moving particle is an electron, it can be deflected through large angles in collisions with other electrons and the nuclear scattering is

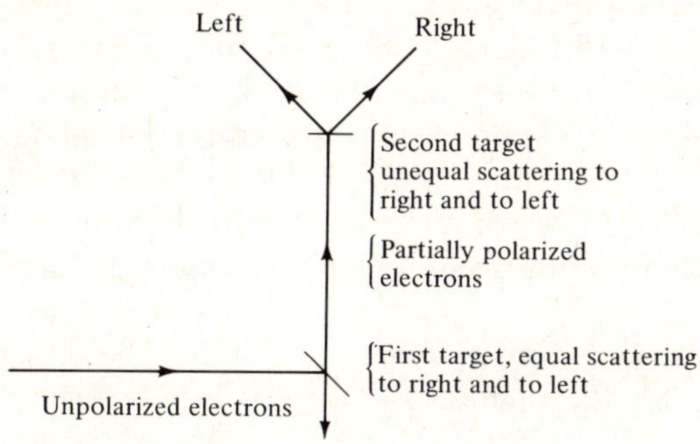

FIGURE 20.3 In a double-scattering experiment, the first scattering produces partially polarized electrons that have been scattered to the left and the second scattering proves the existence of the polarization by giving unequal numbers of electrons scattered to the right and to the left.

not so clearly distinct from this other scattering. If the electron is moving very fast, according to the relativistic mechanics developed in Section 22 its mass is greater than that of an electron at rest. In this case the deflection of the incident electron by the atomic electrons is small and the nuclear scattering again can be differentiated from scattering by electrons. By the use of extremely energetic electrons the distribution of electric charge within the nucleus can now be measured.

The scattering of electrons by nuclei also can reveal something about the electron, showing that it is not a simple point particle with a charge e. Suppose one has a beam of electrons that have been scattered to the left through 90° by the gold nuclei in a thin foil, as shown in Figure 20.3. These electrons then can be allowed to strike a second gold foil and can be scattered a second time. A question to ask is whether in this second scattering the electrons are scattered to the right and to the left with equal probability. The experimental answer is no, the two directions of scattering are not equally probable. We must see what this implies for the properties of the electron.

If one throws a table tennis ball without giving it a spin, it will follow a trajectory that lies in a vertical plane and will not curve to the right or to the left. If the ball is given a spin about a vertical axis when thrown, its trajectory will no longer lie in a vertical plane but will bend to the right if the spin is in one direction and to the left if the spin is in the other direction. One could say that a spinning ball is "scattered" by the air. If a stream of balls emerges from a throwing machine and if the trajectories of these balls bend to the right as often as to the left, one would conclude that the balls are spinning but that they are equally likely to be spinning in either direction about a vertical axis. If they were not spinning there would be no left or right deflection, and if they tended to be spinning in one direction oftener than in the other, the deflection would occur oftener to one side than to the other.

A spinning ball has angular momentum due to its spin. According to the definition given in Section 12, angular momentum is represented by a vector pointing along the axis of rotation, the direction along the axis being determined by the right-hand rule. We can say, therefore, that the behavior of a table tennis ball depends upon its spin angular momentum. If this angular momentum points up, as in Figure 20.4, the ball will be deflected to the left, while if it points down the deflection will be to the right. The reason it happens this way is not obvious at all, and it is not worth going into here. What matters to us here is that there is a deflection produced by the spin whose direction depends on the spin direction.

The behavior of the electrons in the double-scattering experiment described above is somewhat similar to the behavior of spinning table tennis balls. The initial electron beam contains electrons with positive and negative vertical components of spin angular momentum in equal numbers; the electron gun does not discriminate between spin directions, so the initial electron beam is called "unpolarized." In the first scattering those electrons with "spin up," that is, with a positive upward vertical component of spin angular momentum, are more likely to be scattered to the left than to the right, but because the beam is unpolarized equal numbers of electrons are scattered in each direction. (That "spin up" gives a leftward deflection for both table tennis balls and electrons scattered from nuclei is a pure coincidence.) Thus even though the initial

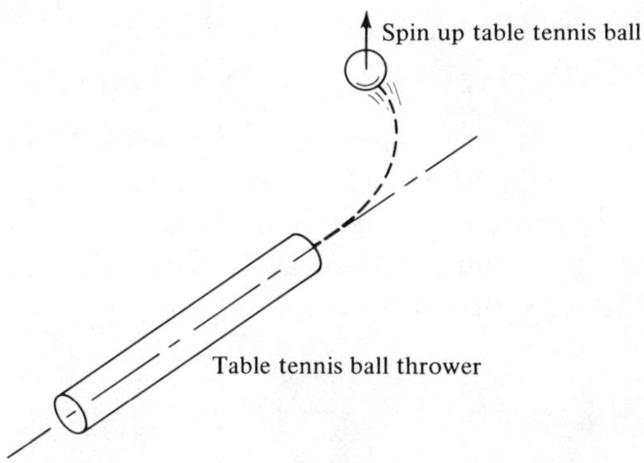

FIGURE 20.4 A table tennis ball spinning so its angular momentum has an upward component will follow a path through air that veers off to the left.

beam is unpolarized, a beam composed of particles that have been scattered to the left contains more electrons with spin up than with spin down; it is partially polarized. When this once-scattered beam is scattered again, the numbers scattered to the left and to the right are unequal because what is now the incident beam is partially polarized. This double-scattering experiment demonstrates that electrons have spin angular momentum and provides a standard means for detecting the polarization of electrons.

The spin angular momentum of an electron is analogous to but not identical with the spin angular momentum of a ball. Atomic particles can be described by Newtonian mechanics only when moving in very large orbits, as in Thomson's apparatus. When the orbits get small, as when electrons move around an atomic nucleus, the description furnished by Newtonian mechanics breaks down and quantum mechanics must be used. One of the rules of quantum mechanics is that a component of angular momentum is "quantized." This means that its value can change only in units of a definite size, just as the electric charge on a body is quantized, and can change only by an integral number of electron charges. Because the electron charge is small, this quantum nature of charge is not noticed until very small charges are investigated. The quantum of angular momentum is denoted by $\hbar$ (h with a line through it read as "aitch-bar"). It is a very small unit, having the value

$$\hbar = 1.054 \times 10^{-27} \text{g cm}^2 \text{ sec}^{-1}$$

Further analysis, which is too mathematical to go into, shows that the electron can have only two values for a component of angular momentum, $\pm\frac{1}{2}\hbar$, which differ, as they must, by an integer multiple of $\hbar$, namely, $\hbar$ itself. For this reason the electron is said to have "spin $\frac{1}{2}$." A particle of spin 1 would have a maximum value $\hbar$ for an angular momentum component, and two other values would be possible, 0 and $-\hbar$. Not all particles have spin; the alpha particle, for example, has no spin angular momentum, and it is called a particle of spin zero.

Problems

1. Gold can be rolled or beaten into very thin foils, such as the gold leaf used in stamping the titles of books on the covers, which is one reason gold targets were used by Rutherford. In a target foil that is 1000 atoms thick, about what fraction of the area of the foil is occupied by nuclei, assuming the nuclear radius is 10^{-4} of the atomic radius? What has this to do with the probability for a given alpha particle being scattered through a large angle by each of two nuclei?
2. Devise a method for determining the radius of a gold atom. The method may draw on the results of several experiments. (One method uses the Millikan experiment, the electrolysis of water, and the measurement of the density and atomic weight of gold.)
3. Evaluate the distance of closest approach of 6.2 MeV ($= 10^{-6}$ ergs) alpha particles to a uranium nucleus with $Z = 92$, neglecting the nuclear recoil. (Recall that 300 V = 1 statvolt of potential difference.) How close could electrons of this energy get?
4. Evaluate the maximum kinetic energy that an alpha particle of 6.2 MeV can transfer to a gold nucleus, using the results of Section 13. Do the same for an incident 6.2 MeV electron.
5. It is known from various scattering experiments that the radius of the gold nucleus is 5.9×10^{-13} cm. What is the value of the electric field due to the nuclear charge just outside the nuclear surface? Express the answer in volts per centimeter. (It is difficult in practice to maintain an electric field of 10^5 volts/cm outside a macroscopic charged body without causing a spark.)

21. THE NUCLEONS, PROTON AND NEUTRON

The nuclear model of the atom has been established by experiment and theory. The successes of the model in accounting for the behavior of atoms of all kinds are so numerous and so complete that this model can be counted as correct beyond doubt. According to this model two very different kinds of constituents are present in any atom, the atomic nucleus and the electrons. Experimentally electrons from one kind of atom are indistinguishable from electrons from any other kind of atom, so the constituent that differs between atoms of one element and those of another is the nucleus. The chemical identity of an element depends on the number of electrons in a neutral atom of that element, but this in turn depends on the electric charge of the nucleus.

Atomic nuclei can be studied experimentally in a number of ways, perhaps the most important one being by scattering experiments. We have already seen that Rutherford's early experiments provided an upper limit to the size of a nucleus. By the use of bombarding particles of higher energy than described earlier the projectile can be made to reach the nuclear surface, and when it does various interactions take place. In general these are very complicated, and they are not completely understood yet, so we do not try to describe such reactions or the picture of nuclear structure that has emerged. All we do here is to show how the particles of which the nucleus is made are identified and how their properties are determined.

The lightest atom has the lightest nucleus, so this nucleus should be the simplest one also. The lightest chemical element is hydrogen, so the nucleus of hydrogen is the one to study first. Chemically hydrogen behaves very simply, having chemical valence one in all simple compounds. This was accounted for early in this century by saying that hydrogen has one "valence" electron that it can share with other atoms in a molecule. In the case of hydrogen this valence electron is the only electron in the atom. Hydrogen gas is electrically neutral; if it were not, the flow of hydrogen gas would involve the flow of electric charge, and this has been looked for and found to be too small to detect if it exists at all. Within very narrow limits, then, the charge on the nucleus of the hydrogen atom must be $+e$, the exact opposite of that on the electron.

The mass of the hydrogen atom can be measured by the process of electrolysis. When a small amount of acid is added to pure water, the water becomes an electrical conductor. While a current is passing through the water, hydrogen gas appears at the electrode where the current leaves the liquid and the mass of gas evolved is proportional to the total charge (current $\times$ time) that passes through. The model of this process that accounts for many aspects of the phenomenon is that some of the hydrogen coming from the acid is present in the liquid as positive hydrogen ions, that is, as hydrogen atoms that have lost their electrons. The rest of the acid molecule is present as a negative ion carrying the hydrogen's electron in addition to their own. Thus, if the acid is hydrochloric acid, the acid molecule in water is dissociated according to

$$HCl \rightleftharpoons H^+ + Cl^- \tag{21.1}$$

Not all the molecules dissociate: The larger the fraction that does, the stronger is the acid.

When an electric field is applied to such an ionized solution, the positive ions move in the direction of the field and the negative ions move in the opposite direction. When the positive hydrogen ion reaches the negative electrode it acquires the electron it needs to become a whole, neutral atom and is freed from the solution as hydrogen gas. That two atoms combine to form a hydrogen molecule is irrelevant here. The model describes clearly why the mass of gas evolved is proportional to the charge transported, since one ion yields one atom and carries one charge e. This proportionality is observed experimentally to be exact within the accuracy of measurement.

There is now at our disposal a method of determining the mass of any atom that can be freed by electrolysis, in particular the mass of the hydrogen atom. One has only to pass a known charge through an acid solution and measure the mass of the hydrogen evolved. The number N of hydrogen atoms is known, because if Q is the charge used in the experiment, then

$$Q = Ne \tag{21.2}$$

If M is the mass of hydrogen evolved, the mass m_{H} of a hydrogen atom is

$$m_{\mathrm{H}} = \frac{M}{N} \tag{21.3}$$

Finally, the mass m_{p} of the hydrogen nucleus is the mass of the atom minus the mass of the electron.

$$m_{\mathrm{p}} = m_{\mathrm{H}} - m_{\mathrm{e}} \tag{21.4}$$

Combining the results of the Thomson experiment to measure e/m for the electron, the Millikan experiment to measure the electron charge e, and the electrolytic experiment just described, one can find the mass of the hydrogen nucleus.

The nucleus of the hydrogen atom is called the *proton.* (There are hydrogen atoms with nuclei two or three times as massive as the typical one. We shall be able to understand how this happens later in this section.) The proton is one of the building blocks of atomic nuclei. Like the electron, the proton has spin angular momentum, being a particle of spin $\frac{1}{2}$. It also has a magnetic moment, but this is much smaller than that of the electron. These properties of the proton show up particularly clearly in the properties of the hydrogen atom, but this is beyond the scope of our discussion here. Protons also can be studied by the methods used for the electron. When an electric discharge takes place in hydrogen gas, free electrons and free protons are produced. The protons will be accelerated by any applied electric field, and a proton beam can be deflected by electric and magnetic fields so that the charge-to-mass ratio can be measured directly. The results agree with those obtained from electrolysis.

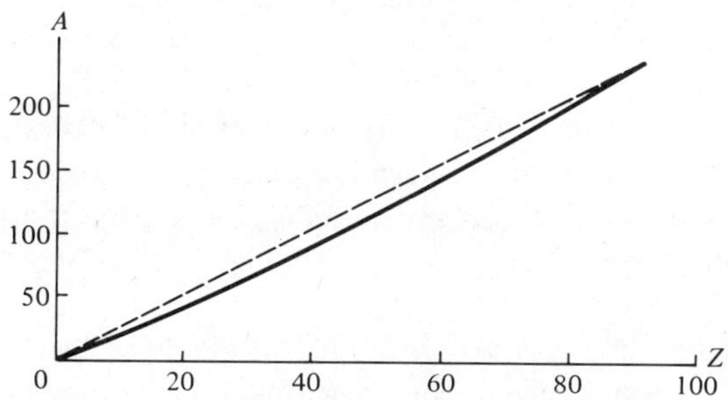

FIGURE 21.1 The chemical atomic weights of the elements are plotted against their atomic numbers. If the nuclei of all elements were made of a single kind of particle, the graph should be a straight line. It is not because of the varying ratio of neutron number to proton number in nuclei.

It was mentioned in Section 20 that Z, the atomic number of an element in the periodic table, is the charge on the nucleus of an atom of that element measured in units of e, the magnitude of the electron charge or the proton charge. This now can be interpreted as the number of protons in the nucleus of the atom. If nuclei consisted of nothing but protons, the atomic weight of all elements would be equal to the atomic number Z, but this is not the case. A plot of the chemical atomic weight of the elements as a function of Z is shown in Figure 21.1. It is not even a straight line, so there must be something else in the nucleus besides protons.

The other constituent of nuclei is the neutron. It was first discovered in 1932 by James Chadwick. The discovery involved applying the principles of mechanics to certain collisions. When the element beryllium (Be) with $Z = 4$ is bombarded by alpha particles from a natural radioactive source such as radium, a very penetrating radiation results with which no electric charge is associated. Calling the unknown neutral particle that is produced n, the reaction can be written as

$$\mathrm{Be}^9 + \alpha = \mathrm{C}^{12} + \mathrm{n} \tag{21.5}$$

The superscripts are a modern notation; they give the mass number or *baryon number* of the atom, that is, the number of heavy particles such as protons and neutrons in the atom. The term "baryon" is very useful in high energy physics, as are the terms "lepton" and "meson." The leptons are light particles; electrons belong to this class, so that one could say that Z is the number of leptons in the neutral atom. The mesons are a class of particles with masses intermediate between baryons and leptons. Not all leptons are electrons, and not all baryons are neutrons and protons. We shall meet other members of these classes soon. This notation did not exist in 1932, naturally, because no experimental basis for it had been discovered. The problem then was to find out as much as possible about the neutral particle produced in the reaction of Eq. 21.5, the neutron.

An uncharged, or neutral, particle can go through solid matter losing very little energy because it does not act on the electrons, and its chance of hitting a nucleus is small. In Section 13 the head-on collision of two particles was studied; a projectile of mass m_1 and kinetic energy K_1 collides with a target of mass m_2 at rest. It was shown in Eq. 13.24 that the kinetic energy K_2' given to the target in a head-on collision depends on the masses of the projectile and target according to

$$K_2' = \frac{4m_1m_2}{(m_1 + m_2)^2} K_1 \tag{21.6}$$

If collisions of the neutrons with charged particles are observed, something can be learned about the mass of the neutron. This was the method used by Chadwick. The projectile is the neutron, and the target is the nucleus of an atom of the material through which the neutron is passing. It was noticed that when the material is paraffin, which contains lots of hydrogen, recoil protons of kinetic energy as high as 5.3 MeV are produced. If the projectile were about as massive as a proton, it would require only 5.3 MeV to produce these protons, but if it were much lighter or much heavier, it would require considerably more energy. For example, if the neutron mass were one-tenth of the proton mass, the neutron would require an energy of

$$K_1 = \frac{(m_1 + m_2)^2}{4m_1m_2} K_2'$$

$$= \frac{121}{40} \times 5.3$$

$$= 16 \text{ MeV}$$

Because such high energies had never been encountered in nuclear transitions at that time, and for other reasons, it was inferred that the neutron mass is comparable to the proton mass. Shortly after the existence of such a neutral particle was established, it was suggested by Werner Heisenberg that it was a constituent of nuclei. It is now known that

$$m_p = 1.67221 \times 10^{-24} \text{ g}$$

$$m_n = 1.67450 \times 10^{-24} \text{ g} \qquad (21.7)$$

Neutrons being neutral, they cannot be accelerated by electric fields or deflected by magnetic fields. Their study therefore is more complicated than that of protons or other charged particles. Their energy, momentum, and angular momentum must be inferred from the amount of these quantities they can transfer to charged particles that can be observed directly, assuming that the laws of the conservation of energy, momentum, and angular momentum are valid in these collisions.

The neutron and the proton are, aside from their charge difference, very much alike. The spin of the neutron is $\frac{1}{2}$, like that of the proton, and even though it is electrically neutral, it has a small magnetic moment as does the proton. It is appropriate to think of these two kinds of particles as two different varieties of a single kind of particle called the *nucleon*. The two kinds of nucleons play similar roles in the formation of atomic nuclei. The number of protons in the nucleus determines the atomic number Z, thereby determining the number of electrons in the neutral atom and the chemical identity of the atom. The mass or baryon number A is the total number of nucleons in the nucleus, and this determines the atomic weight of the atom. A given number of protons can combine with different numbers of neutrons to form different nuclei with the same charge. In this way atoms of the same chemical element may differ in atomic weight. An element found in nature is generally a mixture of various "isotopes" with different atomic weights but necessarily with the same atomic number. The atomic weight of an element as determined by chemical means is the average over the isotopes occurring in the sample analyzed. The variation in the abundances of various isotopes from sample to sample gives information about the source of the sample—information that is often useful especially in geological problems.

From what has been said above, one would conclude that the mass of any nucleus should be given by the formula

$$m = Zm_p + (A - Z)m_n - \Delta m \tag{21.8}$$

with $\Delta m = 0$, but this is not true. The *mass defect* Δm appearing in this equation always makes the value of m less than one would expect, except in the case of hydrogen, where $m = m_p$. The nucleus is less massive than the parts of which it is made. This is a reflection of the fact to be developed in the next section that there is a relation between energy and mass, namely, that a system of mass M has a total energy of Mc^2, where c is the speed of light. A nucleon in a nucleus cannot be removed from the nucleus without work being done from the outside. Performing work increases the energy of the system and thereby increases its mass, so that the neutron and the residual nucleus from which it came are together more massive than the original nucleus. Because in conventional units c^2 has such a large numerical value, it takes a large number of ergs of energy to produce a mass change of 1 g. The energy change required to remove a neutron from a nucleus is large enough to yield a measurable mass change. The energy change involved in removing an electron from an atom is much smaller and does not result in mass change large enough to be detected experimentally as yet.

Problems

1. Suppose you were told that the mass of the neutron is twice the mass of the proton. Suppose also that you have a source of neutrons that give protons an energy of 1 MeV in a head-on collision.
   (a) What would the energy of these "neutrons" have to be to do this?
   (b) How much energy would one of these "neutrons" give to a lithium nucleus (mass about that of six protons) in a head-on collision? Does this provide means for an experimental check on the statement concerning the mass of the neutron?
2. If the charge on the proton were one part in 10^8 larger in magnitude than the charge on the electron, what would the net charge on 100 g of hydrogen in a pressure tank be? Show that if the hydrogen is allowed to escape from

the tank, the electric potential of the tank will change if it is electrically insulated. If the tank is a sphere of radius 25 cm, by how much will its potential change? (This forms the basis for an extremely accurate measurement of the neutrality of hydrogen done by J. G. King in 1960. He proved the equality of the electron and proton charges to one part in 10^{20}.)

3. Give an argument based on comparing the escape of ordinary hydrogen gas (hydrogen nucleus the proton) and of heavy hydrogen (deuterium with a nucleus containing one proton and one neutron) from pressure tanks that would lead to an experimental test of the lack of any charge on the neutron.

CHAPTER 5
Fast-Moving and Subatomic Particles

22. RELATIVISTIC MECHANICS

Newton's laws of motion have been applied to many systems so far in this book, from the motion of planets around the sun to the motion of alpha particles in Rutherford's scattering experiments and of electrons in Thomson's e/m experiment. It has been mentioned from time to time that these laws as stated apply only to motions at speeds much less than the speed of light. This limitation has no practical effect as far as macroscopic particles are concerned because the speed of light is so much greater than any speed now attainable by a large body, but it does have an effect in the study of atomic particles. Atomic particles are identified as such and have their masses and charges measured in low speed experiments like the Thomson and Millikan experiments and in the electrolysis of water. Once the charge and mass of an electron are known, for example, it is possible to use fast electrons to see if the Newtonian laws of motion remain valid at high speeds. The experimental result is that they do not remain valid, but that seemingly slight changes in their form are sufficient to restore their accuracy. We shall see that these slight changes in the basic laws are not so slight in their consequences and that the view of the world resulting from these laws is different from the view suggested by the original form of Newton's laws of motion. In this section we examine the motion of high speed particles and arrive at a relativistic mechanics of particles. In a later section this same theory is approached from a more usual point of view involving the properties of light. In both ways we come to the special theory of relativity introduced by Albert Einstein in 1905.

Newton's first law of motion can be interpreted as asserting the existence of inertial coordinate systems, an inertial coordinate system being one in which any free particle moves with constant velocity. Deviations from motion with constant velocity are ascribed to external influences acting on the particle, making it no longer a free particle. There is nothing in this one needs to change even for high speed motions. A free particle is recognized as free by its constant velocity, so even a fast-moving free particle should have constant velocity. We retain Newton's first law of motion in its original form.

The second law of motion states that the deviation from motion with constant velocity is to be ascribed to a force acting on the particle, this force being defined by

$$\mathbf{F} = m\mathbf{a} \tag{22.1}$$

Force was defined in this particular way because Newton's third law of motion is especially simple with this definition. It states that when two particles interact with each other, the force exerted by the first on the second is the opposite of the force exerted by the second on the first. Any definition of force differing from Eq. 22.1 in an essential way would require a change in this statement of the third law. One of the most immediate consequences of the second and third laws is the conservation of momentum. Defining momentum by

$$\mathbf{p} = m\mathbf{v}$$

we have

$$\begin{aligned} \Delta\mathbf{p} &= m\,\Delta\mathbf{v} = m\mathbf{a}\,\Delta t \\ &= \mathbf{F}\,\Delta t \end{aligned} \tag{22.2}$$

so the total change in momentum caused by two opposite forces is zero.

These laws are more specific in form than the first law and can be modified without changing the main conceptions behind them in any obvious way. For example, the mass can be imagined to depend on the speed of the particle so that the momentum becomes

$$\mathbf{p} = m(v)\mathbf{v} \tag{22.3}$$

If force is then still defined by

$$\mathbf{F} = \frac{\Delta \mathbf{p}}{\Delta t} \tag{22.4}$$

then force is no longer given by $m\mathbf{a}$ because a change in $\mathbf{p}$ can arise from a change of $m(v)$ as well as from a change in $\mathbf{v}$ itself. Newton's third law might still be valid with this modified force; this is a matter for experimental determination rather than speculation. In fact, modifications of just this kind are required by experimental observation. Let us look at a little of the evidence.

One of the early experiments to show the need of modifying Newton's laws of motion was an experiment much like Thomson's experiment, which measured the charge-to-mass ratio of the electron. In this experiment electrons are accelerated to a speed v in an electron gun and are then deflected by a transverse electric field or a transverse magnetic field, the relative strengths of these two being adjusted to give the same angle of deflection θ. The speed of the electrons is determined by Eq. 19.7, which states that

$$\frac{v}{c} = \frac{E}{B}$$

The charge-to-mass ratio is then that in Eq. 19.8,

$$\frac{e}{m} = \frac{c^2 E\theta}{B^2 L}$$

if L is the path length through either of the deflecting fields. When v is small, which is the case when the potential difference across the electron gun is small, the value of e/m obtained is independent of this speed. As the electron speed is increased, however, the value of e/m begins to decrease, and its value is accurately represented by the expression

$$\left(\frac{e}{m}\right)_v = \left(\frac{e}{m}\right)_0 \sqrt{1 - \frac{v^2}{c^2}} \tag{22.5}$$

where c has the known value 3×10^{10} cm/sec. Thus either the charge e or the mass m of the electron depends on the speed v.

There are good reasons for believing that the charge of an electron does not depend on its speed. One argument against such a speed dependence is this: Atoms of all the chemical elements are electrically neutral to extremely high accuracy. All the electrons in all kinds of atoms cannot possibly be moving with the same speeds, so if the electron charge depended on the electron speed, not all electrons in an atom would have the same charge. It would be an incredible coincidence if the motion of the protons in all kinds of nuclei were just such as to cause the protons' charges to vary so as to exactly cancel out the change in charge of the electrons caused by their motion. The only simple explanation of this electrical neutrality is the constancy of the charge, so we assume e to be independent of the speed of the particle of which it is the charge. This implies that m does depend on the particle speed according to the formula

$$m(v) = \frac{m_0}{\sqrt{1 - \frac{v^2}{c^2}}} \tag{22.6}$$

This variation of electron mass with speed does not lead to any difficulties such as the one encountered above if the charge is assumed to be variable. m_0 is the value of the mass when the particle is at rest: the *rest mass* of the particle. For what is considered a high speed for a macroscopic particle $m(v)$ is effectively m_0. Take, for example, the escape speed of a particle from the gravitational field of the earth, which was found in Section 13 to be 25,000 mph or 1.12×10^6 cm/sec. For this value of v one finds

$$\frac{v}{c} = 3.7 \times 10^{-5}$$

and then

$$\begin{aligned} m(v) &= \frac{m_0}{\sqrt{1 - 1.4 \times 10^{-9}}} \\ &= 1.0000000007 m_0 \end{aligned}$$

Clearly this does no violence to everyday experience. The variable mass $m(v)$ is called the *relativistic mass* of the particle.

The mass of a particle is a measure of the effort needed to accelerate the particle; the larger the mass, the less acceleration produced by a given force. The relativistic mass increases with increasing speed, so the acceleration produced by a given force decreases with increasing particle speed. This makes it impossible to accelerate a particle all the way up to the speed c because as v approaches c the denominator in the expression for $m(v)$ approaches zero and $m(v)$ gets large without limit. Any speed less than c is attainable in principle. Experimental evidence supporting this conclusion exists, and some of it will be described a little further on.

The relativistic momentum of a particle is defined by

$$\mathbf{p} = m(v)\mathbf{v} = \frac{m_0\mathbf{v}}{\sqrt{1 - (v^2/c^2)}} \tag{22.7}$$

This definition reduces to the nonrelativistic Newtonian definition when v^2/c^2 is small enough to be neglected compared to unity. The relativistic force is defined by

$$\mathbf{F} = \frac{\Delta\mathbf{p}}{\Delta t} \tag{22.8}$$

where $\Delta\mathbf{p}$ includes the change in $\mathbf{p}$ caused by a change in $m(v)$ as well as by a change in $\mathbf{v}$. When two particles interact directly with each other, Newton's third law of motion applies to this force and the total relativistic momentum of the two particles is conserved. When two charged particles interact with each other through electric and magnetic fields, some momentum can be associated with those fields and the momentum of the particles alone may not be conserved. This extra complication need not concern us here. We will be able to understand how it arises a little later.

The relativistic momentum of a charged particle can be measured by measuring the radius of curvature r of the path of the particle in a uniform magnetic field, just as nonrelativistic momentum is measured. The magnetic force on the charged particle is perpendicular to the velocity of the particle and therefore causes no change in the speed of the particle. This being so, the relativistic mass does not change so the motion is described by the same form of equation as in the nonrelativistic motion, which is

$$\frac{m(v)v^2}{r} = \frac{evB}{c}$$

This leads to

$$p = m(v)v = \frac{eBr}{c} \tag{22.9}$$

The old formula automatically gives the relativistic momentum.

The next thing to look at is the kinetic energy. The nonrelativistic expression for this is $\frac{1}{2}mv^2$. The most obvious thing to do here is to replace m by $m(v)$, but this does not lead to conservation of energy when the particle is not free but has a potential energy $V(\mathbf{r})$. A potential energy is energy of position by definition and should not depend on the particle's speed but only on the particle's environment. V is therefore assumed to be the same in relativistic motion as in slow, nonrelativistic motion. An expression for the kinetic energy K that in combination with a potential energy does lead to energy conservation is this:

$$K = \frac{m_0c^2}{\sqrt{1 - (v^2/c^2)}} - m_0c^2 \tag{22.10}$$

This expression looks very formidable, but we can see that it reduces to the old, nonrelativistic result when v is much less than c if we recall the approximate formula

$$(1 + x)^n \approx 1 + nx$$

which is the more accurate the smaller the value of x. For example,

$$(1 + x)^2 = 1 + 2x + x^2 \approx 1 + 2x$$

$$(1 + x)^3 = 1 + 3x + 3x^2 + x^3 \approx 1 + 3x$$

when x is so small that x^2 and x^3 can be neglected. This "binomial expansion" is now used in connection with

$$\frac{1}{\sqrt{1 - (v^2/c^2)}} = \left(1 - \frac{v^2}{c^2}\right)^{-1/2} \approx 1 + \frac{1}{2}\frac{v^2}{c^2}$$

On inserting this into Eq. 22.10, we get

$$K \approx m_0c^2\left(1 + \frac{1}{2}\frac{v^2}{c^2}\right) - m_0c^2$$

$$= \tfrac{1}{2}m_0v^2$$

The proof that $K + V$ is a constant when the force acting is related to the potential energy by

$$F\,\Delta s \cos\theta = -\Delta V$$

as in Eq. 13.13 involves the use of calculus and will not be given here. Inspection of the expression for K shows that it is zero when $v = 0$ and that the first term is always greater than the second when v is not zero, so that K is never negative. It gets large without limit as v approaches the limiting speed c.

Expression 22.10 gives K as the difference of two terms, and it suggests that it might be useful to associate an energy $E = m(v)c^2$ with a particle. When the particle is at rest this energy has the value m_0c^2, called the *rest energy* of the particle. When the particle is moving, its mass increases and so does its energy; the *increase* in its energy is the kinetic energy of the particle. At first this seems to be doing nothing more than adding a constant to the kinetic energy, something we said earlier that should not be done, but more than that emerges. There are reactions in which the rest masses of particles change, which means that particles change their identity. In such reactions conservation of energy is valid only if rest energy is included among the possible forms of energy. The production of power in nuclear power plants involves processes of this kind, the energy released corresponding to a reduction in the rest mass of the fuel. It takes a very small change in rest mass to produce a large amount of energy because the factor c^2 appearing is so large.

There is a very useful relation among the energy, momentum, and rest mass of a free particle that is simple in form if by energy we mean the total energy E including the rest energy and not just the kinetic energy K. It is

$$E^2 = p^2c^2 + m_0^2c^4 \tag{22.11}$$

or

$$E = +\sqrt{p^2c^2 + m_0^2c^4} \tag{22.12}$$

The positive square root is always chosen to make the energy of a free particle positive. A particle in a bound orbit such as a planet moving around the sun may have a negative value of what we called "mechanical energy," $K + V$, but here E is the total energy of a free particle including the rest energy—and with $V = 0$ because the particle is free. The correctness of Eq. 22.11 is verified by inserting the expressions for E and p as functions of m_0 and v into the equation. This gives

$$E^2 = \left[\frac{m_0c^2}{\sqrt{1 - (v^2/c^2)}}\right]^2 = \frac{m_0^2c^4}{1 - (v^2/c^2)}$$

and

$$\begin{aligned} p^2c^2 + m_0^2c^4 &= \frac{m_0^2v^2c^2}{1 - (v^2/c^2)} + m_0^2c^4 \\ &= \frac{m_0^2v^2c^2 + m_0^2c^4[1 - (v^2/c^2)]}{1 - (v^2/c^2)} \\ &= \frac{m_0^2c^4}{1 - (v^2/c^2)} \end{aligned}$$

which are equal.

Another useful equation connects the energy, momentum, and speed of a particle. It is

$$v = \frac{pc^2}{E} \tag{22.13}$$

This is easy to check.

We have collected a lot of formulas. Now we want to see what they mean and what they are good for. What they should be good for is to find the motion of even a fast-moving particle that is subject to known forces, and they are good for this. Unfortunately the equations are rather more complicated than the original form of Newton's laws because the speed v enters in an awkward way, so the only motions we discuss are the motion of a free particle and the motion of a charged particle in a uniform magnetic field. These are all that we really need for the purposes of this book. The meaning of these modified equations is very interesting and important, and some of their consequences are far from obvious.

The speed c is a limiting speed that a particle can approach but cannot attain. Presumably the experiments that determine c are done in some laboratory, and that laboratory specifies the reference frame in which c is given. What happens if the same experiment is done in another laboratory that is moving uniformly relative to the first? Is the same value of c obtained, or does c depend on the motion of the laboratory in which the experiment is done? The experimental result is that the value of the limiting speed does not depend on the motion of the laboratory in which it is measured, but the same value is obtained in all laboratories regardless of their motion. We cite one kind of experimental evidence for the correctness of this statement.

Electron accelerators have been designed and built on the assumption that electrons follow the modified form of Newton's laws of motion given above, including the assumption that the limiting speed c has the value 2.9979×10^{10} cm/sec. In these accelerators the electrons are kept in circular orbits by a magnetic field and are given increased energy by electric fields. The energy attained by an electron in some accelerators is as high as 10 GeV (1 GeV = 10^9 eV = 10^3 MeV). The rest energy of an electron, m_0c^2, is 0.51 MeV, so the total energy of a 5 GeV electron is 10^4 times its rest energy. The value of $1/\sqrt{1 - (v^2/c^2)}$ is therefore 10^4, and the corresponding value of v/c is

$$\frac{v}{c} = 0.999999995$$

These accelerators operate just as they are supposed to. In particular the time it takes an electron to go around its orbit once is never less than the time it would take light to travel that same distance, so the existence of the limiting speed c is directly verified, at least in the reference frame provided by the earth.

The earth is well known not to provide an inertial reference frame. The most direct evidence for this is perhaps the behavior of a spinning object such as a gyroscope. If a spinning gyroscope is mounted so that it is free to turn in any direction and so that the force of gravity does not tend to cause it to turn, the axis of rotation or the direction of the angular momentum of the gyroscope should remain fixed in an inertial reference frame. It does not remain fixed relative to the earth, but as closely as can be determined it remains pointed at the same place in the sky. Thus the distant stars seem to provide a very good approximation to an inertial reference frame relative to which the earth rotates. The earth itself is a pretty good gyroscope, and its polar axis always points very close to the north star, Polaris. The motion of the earth around the sun also makes the earth a noninertial frame, as at any time its velocity of motion around the sun is opposite to what it was 6 months earlier. The speed of this motion is 2.97×10^6 cm/sec $\approx 10^{-4}\ c$.

The fact that the earth does not provide an inertial reference frame is not important for the operation of something like an electron accelerator. The total time required to accelerate an electron from rest to its final energy is less than 1 sec, so that the change in velocity of the machine due to the acceleration of the earth is very small during this time. What is worth noticing is that while during any 1 sec the accelerator itself can be regarded as an inertial reference frame, it is a very different inertial frame at one time than it was 6 months earlier because of the orbital motion of the earth, as seen in Figure 22.1. If all inertial reference frames were not equivalent, the accelerator ought to have very different characteristics in the spring and fall and less different but still distinguishable characteristics in the morning and evening. The complete absence of such effects is a strong indication that all inertial reference frames are equivalent and that in them all the limiting speed is $c = 2.9979 \times 10^{10}$ cm/sec.

That the measured speed of light is independent of the motion of the laboratory in which the speed is measured was firmly established by optical experiments long before electron accelerators were built. The most important of these optical experiments was that of A. Michelson and E. Morley. The Michelson-Morley experiment is described in Section 31, where wave motion, including light waves, is the subject.

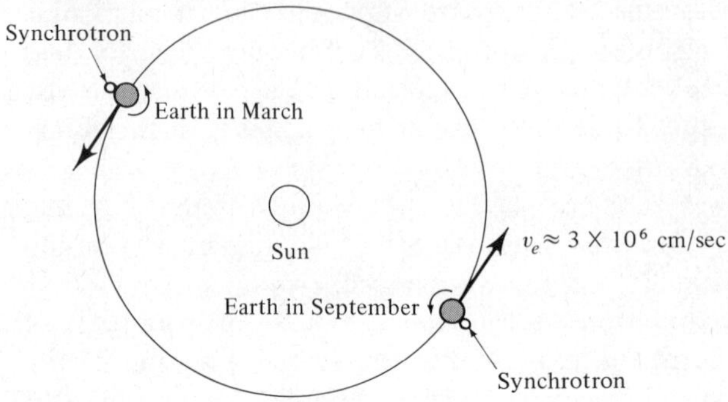

FIGURE 22.1 A synchrotron on the earth is moving relative to the fixed stars very differently in March and in September, but the limiting speed of electrons in the synchrotron is always the same.

The existence of a limiting speed in a given reference frame may seem strange, but to have that limiting speed the same in all reference frames at first seems stranger still. To see what is involved, suppose a particle is moving with a speed $\frac{2}{3}c$ along the $+x'$-axis in a coordinate system S' and that S' is moving with a speed $u = \frac{2}{3}c$ along the x-axis of another coordinate system S, as shown in Figure 22.2. What is the speed of the particle relative to S? The answer based on what has been developed so far in this book is that $v = v' + u$, but this cannot be

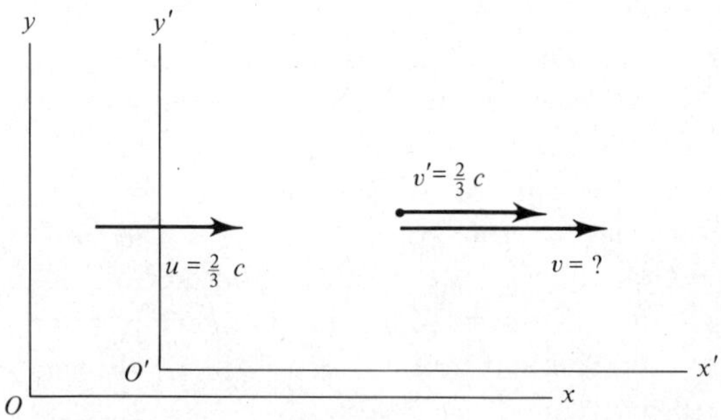

FIGURE 22.2 Reference frame S' is moving with speed u as seen from reference frame S, with $u = \frac{2}{3}c$. A particle moving with speed $v' = \frac{2}{3}c$ as seen from S' might be thought to have speed $u + v' = v = \frac{4}{3}c$ as seen from S, but it cannot have speed greater than c. What is v?

right because $\frac{2}{3}c + \frac{2}{3}c = \frac{4}{3}c > c$ and this would make the particle be moving faster than the limiting speed in S. The low speed rule used up to here for adding velocities must be modified. We must try to write down a connection between the energy and momentum of a particle as seen in one system S' in terms of the energy and momentum of the same particle in another coordinate system S in a way that avoids this difficulty.

There is a connection among the energy and the momentum and the mass of a particle given in Eq. 22.11 that can be written

$$E^2 - p^2c^2 = m_0^2c^4 \tag{22.14}$$

The right-hand side of this equation contains nothing that depends on any special reference frame, so it must have the same value in all reference frames. Such a quantity is called a *relativistic invariant*. There is a transformation that relates the energy and momentum of a particle as given in the coordinate system S to the energy and momentum of the same particle as viewed from coordinate system S' and that leaves the value of $E^2 - p^2c^2$ unchanged. This is called a *Lorentz transformation*. We write it down for the two coordinate systems shown in Figure 22.1:

$$\begin{aligned} p_x' &= \frac{p_x - (u/c^2)E}{\sqrt{1 - (u^2/c^2)}} \\ p_y' &= p_y \\ p_z' &= p_z \\ E' &= \frac{E - up_x}{\sqrt{1 - (u^2/c^2)}} \end{aligned} \tag{22.15}$$

The primed momentum components and energy are those of the particles measured in S', the others are those measured in S, and u is the speed of S' as seen from S.

To see that this is a possible rule for going from one coordinate system to another, we look at some special cases:

1. A particle is at rest in S, so that

$$p_x = p_y = p_z = 0, \qquad E = m_0c^2$$

What are its energy and its momentum as seen from S'? In S' it is moving with speed u in the negative x-direction. We insert the above values in Eq. 22.15 and get

$$\begin{aligned} p'_x &= \frac{-um_0}{\sqrt{1-(u^2/c^2)}} \\ p'_y &= p'_z = 0 \\ E' &= \frac{m_0c^2}{\sqrt{1-(u^2/c^2)}} \end{aligned} \qquad (22.16)$$

The value of $\mathbf{p}'$ is just what it should be for a particle moving with speed u in the negative x'-direction in S', and the value of E' is also correct. The Lorentz transformation gives the desired results here.

2. A particle is moving with speed v' along the x'-axis of S'. What is its speed v as seen from S? Is it $(u + v')$ or not? To answer this question we use the expression Eq. 22.13 for v, namely,

$$v = \frac{p_xc^2}{E}$$

and insert the expression for p_x and E in terms of p'_x and E', which are obtained by solving Eq. 22.15. This solution is

$$\begin{aligned} p_x &= \frac{(p'_x + (u/c^2)E')}{\sqrt{1-(u^2/c^2)}} \\ p_y &= p'_y \\ p_z &= p'_z \\ E &= \frac{(E' + up'_x)}{\sqrt{1-(u^2/c^2)}} \end{aligned} \qquad (22.17)$$

which is just the same as the original Lorentz transformation except for the reversal of the sign of u. This is an agreeable result, since if S' is moving with speed u in the positive x-direction in S, then S is moving with speed u in the negative x'-direction in S'. Using this, we obtain

$$\begin{aligned} v &= \frac{p'_xc^2 + uE'}{E' + up'_x} \\ &= \frac{(p'_xc^2/E') + u}{1 + u(p'_x/E')} \\ &= \frac{v' + u}{1 + (v'u/c^2)} \end{aligned} \qquad (22.18)$$

In the second line both the numerator and denominator have been divided by E', and in the third line $v' = p'_x c^2/E'$ has been used.

The result obtained is called the *Einstein rule for adding velocities.* It was developed by Einstein in a different way, which will be described in Section 31, where a different approach to relativity is given. The Einstein law differs from the nonrelativistic law only by the presence of the denominator, but when either u or v' is large, this makes a big difference. Consider the case mentioned above, when $u = \frac{2}{3}c$ and $v' = \frac{2}{3}c$. Inserting these into the Einstein rule gives

$$v = \frac{\frac{2}{3}c + \frac{2}{3}c}{1 + \frac{4}{9}} = \frac{12}{13}c$$

which is less than the limiting speed. If $v' = c$, then

$$v = \frac{c + u}{1 + u/c} = \frac{c(1 + (u/c))}{1 + (u/c)} = c$$

so the limiting speed in S' is also the limiting speed in S.

The subject matter of this section can be summarized quite briefly: Experimentally it is found that particles cannot be given speeds exceeding a limiting speed c, which is the speed of light, and that the numerical value of this limiting speed is the same in all inertial reference frames. The existence of a limiting speed requires a modification of Newton's laws of motion to allow for a variable mass. Thus m depends on the speed v according to

$$m(v) = \frac{m_0}{\sqrt{1 - (v^2/c^2)}}$$

The relativistic momentum is defined to be

$$\mathbf{p} = m(v)\mathbf{v} = \frac{m_0\mathbf{v}}{\sqrt{1 - (v^2/c^2)}}$$

and the relativistic force is the time rate of change of this momentum:

$$\mathbf{F} = \frac{\Delta\mathbf{p}}{\Delta t}$$

The total energy of a free particle is related to its momentum and mass by

$$E^2 = p^2c^2 + m_0^2c^4$$

and the speed of a particle is given by

$$v = \frac{pc^2}{E}$$

The experimental fact that the limiting velocity has the same value in all inertial coordinate systems is accounted for by assuming that the components of the momentum and energy of a particle as measured in one coordinate system are related to these quantities as measured in a second system by a Lorentz transformation as given by Eq. 22.15 or 22.17 depending on which way one wants to go, from S to S' or from S' to S.

These results are absolutely basic to the subject of the next section and several subsequent sections. The topics of these sections will serve as illustrations of the use of relativistic mechanics.

Problems

1. A particle moves in a circle with speed $v = 0.9c$, as seen from the reference frame S. If reference frame S' is moving with speed u in S, find the maximum and the minimum speed of the particle as seen from S' when u is (a) $0.01c$, (b) $0.4c$, and (c) $0.9c$.
2. How long would a rocket have to experience an acceleration of 10^3 cm/sec^2 (just a little larger than the gravitational acceleration g) in order to reach a speed equal to half the speed of light? How far would it travel in the process?
3. Compare the Newtonian value $\frac{1}{2}mv^2$ for the kinetic energy of a body moving at half the speed of light with the relativistic expression for this kinetic energy. By what percentage does the second exceed the first? What does this percentage become if the speed is $0.99c$?
4. A particle at rest decays into two particles identical to each other whose combined mass is 10 percent less than the mass of the original particle. In what relative directions do these product particles move off and with what speed?
5. The mass of an electron at rest is 9.1×10^{-28} g. What is its relativistic mass after it has been accelerated by going through a potential difference of 25,000 V, as in the electron gun of a television tube?
6. A particle is known to have a total energy E (including the rest energy) and a momentum p. Can you determine the rest mass of the particle from this information?
7. Two particles are approaching each other, one moving to the right with speed $\frac{4}{5}c$ and the other moving to the left with the same speed. How fast would the other particle be observed to be approaching an observer moving with the first one?

23. PARTICLES AND ANTIPARTICLES

The particles with which Newtonian mechanics has to do are usually considered to be permanent parts of the system in which they exist. When a meteorite strikes the earth's atmosphere and burns up, it is true that its identity as a particle is lost, but in even such a catastrophic event one feels that the atoms of which the meteorite was made still exist. This permanent existence of particles does not always persist down to the atomic and subatomic scale of events. Particles can appear or disappear; they can be created and annihilated. These events cannot occur in arbitrary ways: The conservation laws for energy, momentum, angular momentum, and charge all still are valid in these processes provided rest energy is included as a kind of energy, and some other conservation laws also apply to these events. There is, however, no law of conservation of identity. In the kind of process being described a particle present after a reaction cannot always be identified with one particular particle present before the reaction.

In the last section the idea of the rest energy associated with a particle was introduced on rather artificial grounds; it was suggested by the form of the expression for the kinetic energy of a particle moving at relativistic speed, and its inclusion in the energy made the Lorentz transformation of momentum and energy look simple, but these are frivolous reasons. The solid reason for introducing it is that its introduction saves the law of conservation of energy in many processes, including those to be described here.

There are processes in which all the rest energy of a particle can be converted into other forms of energy. A little bit of the rest energy of a uranium nucleus is turned into heat in a nuclear reactor, but only about one part in a thousand. What form of energy is there that involves no rest energy? If a particle's rest energy is completely converted into other forms of energy, no particles of the usual kind can be products of the transformation because such particles themselves have mass and hence have rest energy. The only kind of particle without rest energy is one without rest mass, and such particles do exist. They are referred to as *massless particles.*

If a particle has zero rest mass (is massless), the relation among energy, momentum, and mass given in Eq. 22.12 becomes simply

$$E = pc \tag{23.1}$$

The energy of such a particle is proportional to its momentum. The speed of the particle can now be found using Eq. 22.13,

$$v = \frac{pc^2}{E} = \frac{pc^2}{pc}$$
$$= c \tag{23.2}$$

A massless particle always moves with the speed of light, no matter what its energy may be. There is no reference frame in which such a particle is at rest because, as we saw in the last section, the limiting speed c is the same in all inertial reference frames.

There are two kinds of massless particles known to exist in nature. The first is the photon, a particle associated with light, as its name implies. The second is the neutrino, a particle involved in what are called the weak interactions, to be studied later. A beam of light has a dual nature, for some purposes being best described as an electromagnetic wave and for other purposes as a stream of photons. Both kinds of description are only aspects of light, the full description of which requires the quantum theory. When the photon aspect is used, the photons must be treated as particles of zero rest mass so that they move with the speed of light. Photons are denoted by γ (the Greek letter gamma) since the gamma rays mentioned in Section 20 consist of photons.

If a particle is to convert all its energy into other forms of energy not involving rest mass, it must transform itself into photons or neutrinos. Here we consider only photons.

There are a few particles that can transform into photons when they are by themselves. One such particle is a meson, the so called π^0 (read as "pi zero"), which we shall discuss in some detail later. A π^0 cannot decay into a single photon because this would violate either energy conservation or momentum conservation or both. If such a decay were possible, it would be written $\pi^0 \rightarrow \gamma$, but since it does not occur one writes

$$\pi^0 \not\rightarrow \gamma$$

Suppose the π^0 to be at rest before the assumed decay. Its momentum $\mathbf{p}$ would then be zero and its energy would be $m_\pi c^2$. To conserve energy in the decay process, the photon must have energy $E = m_\pi c^2$, and according to Eq. 23.1 it must then have momentum of magnitude $p = E/c = m_\pi c$. But the initial momentum was zero, so momentum conservation would be violated if energy is conserved in this process. The process is never seen to happen. What does happen is the decay

$$\pi^0 \rightarrow \gamma + \gamma \tag{23.3}$$

in which two photons are created. If the π^0 is initially at rest, the two photons must be emitted in opposite directions with momenta of equal magnitude, as illustrated in Figure 23.1. The laws of energy and momentum conservation are satisfied if each photon has an energy

$$E_\gamma = \frac{m_\pi c^2}{2}$$

Charge conservation is also satisfied, neither the π^0 nor the photons having any charge.

The π^0 particle is one we have not encountered before, so it is unfortunate to use it for the introduction to this subject, but it is the commonest particle that can decay into photons when alone. Electrons, which we have studied extensively, cannot decay into photons because charge would not be conserved in the process even though energy and momentum could be. This does not prevent electrons from being converted into photons under any circumstances; it merely prevents an electron from doing this by itself. It appears that in nature types of particles occur in pairs, called *particle-antiparticle* pairs, and that the two members of such a pair can decay into massless particles without violating any conservation laws. There are a few special cases such as the π^0, which can decay in this way without an antiparticle being present. Such particles are their own antiparticles.

The antiparticle corresponding to the electron is the *positron,* denoted by e^+. It is a particle of spin one-half like the electron; it has the same rest mass m_e and the same size charge although of positive sign. If a positron-electron pair meet while moving so slowly that the kinetic energy and momentum of each is negligibly small, the total relativistic energy of the pair is $2m_ec^2$, the total momentum is zero, and the total charge is zero. The process

$$e^+ + e^- \rightarrow \gamma + \gamma \tag{23.4}$$

can occur, and does. As in the decay of the π^0, the two photons are

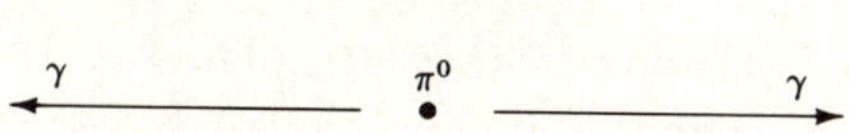

FIGURE 23.1 A π^0 meson at rest can decay into two photons of equal but oppositely directed momenta, each photon having half the rest energy of the π^0 meson; both momentum and energy are conserved in this decay process.

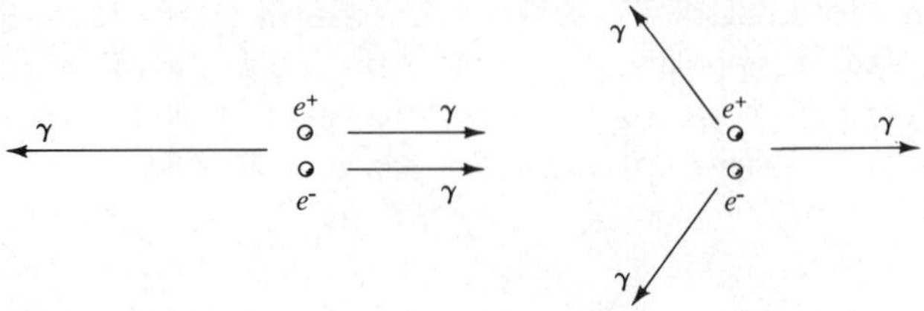

FIGURE 23.2 An electron-positron pair can decay into three photons in infinitely many ways and still conserve both momentum and energy. Two possibilities are illustrated. The three photons must always lie in a plane so that their momenta can add up to zero, if the pair is at rest when it decays.

emitted in opposite directions with equal energies of m_ec^2 each; in this way energy and momentum are conserved. Other decays of a positron-electron pair at rest are also possible, for example, a decay into three photons. When three photons are emitted their momenta must lie in a plane so they can add vectorially to zero, but they are not uniquely determined relative to one another as are those of a two-photon decay. Two possible three-photon decays are shown in Figure 23.2, and there are infinitely many other ways for it to happen.

The positron's existence was hinted at by a quantum theory of the electron proposed by P. A. M. Dirac in 1928, but at that time the prediction of new particles on theoretical grounds was not taken seriously. It was discovered experimentally in 1932 by C. D. Anderson in the cosmic rays that continually bombard the earth from above. Anderson observed the first positron by the use of a cloud chamber, a device whose principle of operation is described in Section 25. The cloud chamber track of a particle moving in a magnetic field and passing through a metal plate in the chamber is seen in Figure 23.3. The direction of curvature of the track requires the particle either to have positive charge and be moving upward or to have negative charge and be moving downward. The larger curvature of the path above the plate shows that the momentum here is less than the momentum below the plate, and this determines the direction of motion of the particle along the track. A charged particle going through matter loses energy by transferring energy to particles with which it collides; it cannot gain energy by collision with nearly stationary particles. The particle in Anderson's picture, therefore, must be traveling upward through the cloud chamber and have a positive charge. It cannot be a proton because a proton of this momentum (as determined by the curvature of the track) leaves a much heavier track than an electron or a light particle like an electron.

There are almost no positrons in the world because there is a surplus of electrons with which they can annihilate. The study of positrons requires their manufacture either by radioactive decay of unstable nuclei, by collisions of cosmic ray particles with particles of the atmosphere, or by artificial means.

Electron-positron pairs can in principle be made by having two sufficiently energetic photons collide because, just as with chemical reactions, particle reactions can go in either direction. These photons would need to have energies at least as great as 0.51 MeV, and it is at present completely impossible to produce enough such photons in a laboratory to make collisions between them at all probable. Photons necessarily move with the speed of light, so a target foil of photons does not exist. Intense sources of photons of visible light, such as lasers, are no use because the energy of the photons associated with visible light ranges from 1.5

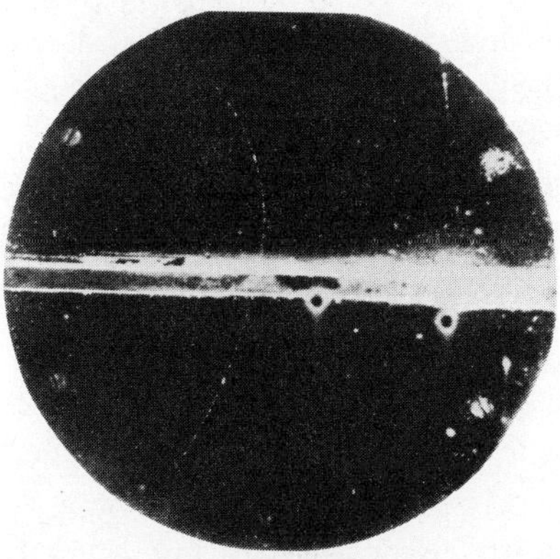

FIGURE 23.3 A charged particle that goes through a metal plate leaves a track in a cloud chamber. The track could be made by a negative particle moving downward through the magnetic field in the chamber or by a positive particle moving upward. The latter must be the actual case because the particle is going slower above the plate than below it. This is the first observation of the antielectron, the positron. (C. D. Anderson, California Institute of Technology.)

to 3 eV and is much too low. A single photon cannot by itself create a pair, again because this necessarily violates energy or momentum conservation. The most practical way to create electron-positron pairs is to bombard a target containing heavy nuclei with energetic photons. The reaction can be written as

$$\gamma + \mathrm{A} \rightarrow \mathrm{A} + \mathrm{e}^{+} + \mathrm{e}^{-} \qquad (23.5)$$

The presence of the nucleus A permits the energy-momentum balance to be maintained because even though its identity is maintained in the reaction, its momentum can change. In chemical terms A is a catalyst. This reaction is so probable when the nucleus has a large charge Ze that it provides a practical method for detecting the presence of energetic photons.

A thin foil of lead is placed in a photon beam. The photons create pairs whose momentum can be measured by the use of a magnetic field. The photon itself is unaffected by the magnetic field since it carries no charge. This method of detecting photons has been superseded in practice by other more accurate and efficient methods. (Figure 27.5 will show a photograph of the tracks of a pair created in a bubble chamber.)

The processes in which the antiparticles corresponding to other, heavier particles are produced and annihilated are discussed in Section 27.

Particle-antiparticle pairs are produced in the laboratory in collisions between particles one of which is at rest and the other of which is moving with high kinetic energy. This is because a large number of target particles at rest can be provided in the form of a thin foil, and the bombarding particle can produce the desired reaction in association with any one of these target particles. The reference frame in which the target particle is at rest is called the *laboratory frame.* The energy required to have a reaction take place is, in most cases, the energy as measured in the laboratory frame. We saw in Section 14 how this is related to the energy required in the center-of-mass reference frame when the energies involved are nonrelativistic. Here we must redo this calculation in such a way that particle creation is allowed for and the relativistic form of the energy is used.

The calculation is very simple if the idea of relativistic invariance, that is, the equivalence of all inertial reference frames, is used. The energy and momentum of a single particle, E and $\mathbf{p}$, combined as in Eq. 22.18 form a relativistic invariant, a quantity whose numerical value is the same in all inertial coordinate systems:

$$E^2 - p^2c^2 = m_0^2c^4$$

The value of this invariant is the square of the rest energy of the particle.

The same is true for the energy E and momentum $\mathbf{p}$ of a system of particles. Let

$$\begin{aligned} E &= E_1 + E_2 + \cdots \\ \mathbf{p} &= \mathbf{p}_1 + \mathbf{p}_2 + \cdots \end{aligned} \tag{23.6}$$

where the sum includes all the particles in the system. Then

$$s = E^2 - p^2c^2 = M^2c^4 \tag{23.7}$$

is a relativistic invariant. The M appearing can be defined to be the rest mass of the entire system of particles. It is greater than the sum of the rest masses of the particles constituting the system unless these are all at rest relative to one another or unless there are attractive forces between the particles as between the nucleons in a nucleus. Here we are considering particles far from one another, as before or after a reaction has taken place, so they are free particles. Any reaction that takes place between the particles constituting the system conserves the total energy E and the total momentum $\mathbf{p}$ of the system, so that M or s has a value that is unaffected by any reaction within the system itself, although of course this value can be changed by external forces applied to the system.

Suppose that we want to produce some particles A, B, C, ... by means of a collision between particles a and b. The smallest amount of energy with which we can do this is the sum of the rest energies of A, B, C, ... This amount of energy will produce the particles at rest, nothing being left over for kinetic energy, so the momentum of each particle produced by this minimum energy is zero, and so is the total momentum. A reference frame in which the total momentum of the system of particles is zero is the *center-of-mass* frame, so in the center-of-mass reference frame the minimum value of s called s_{th}, which allows particles A, B, C, . . . to be produced, may be expressed as

$$\begin{aligned} s_{\text{th}} &= (m_A + m_B + m_C + \cdots)^2c^4 \\ &= M^2c^4 \end{aligned} \tag{23.8}$$

This is called the *threshold* value of s. Our problem is to see what is needed in the laboratory frame before the reaction so that the reaction is just possible.

In the laboratory frame s has the same value as in the center-of-mass frame because it is a relativistic invariant. It has the same value before the reaction as afterward because it is constructed out of quantities, energy and momentum, which are conserved in any reaction. Therefore, we may evaluate s_{th} in the laboratory frame before the reaction when only particles a and b are present. Let us take b to be the stationary target particle, with energy $m_b c^2$ and zero momentum, and a to be the bombarding particle, with nonzero momentum. Then in the laboratory frame

$$\begin{aligned} s &= (E_a + E_b)^2 - (\mathbf{p}_a + \mathbf{p}_b)^2 c^2 \\ &= (E_a + m_b c^2)^2 - p_a^2 c^2 \end{aligned} \tag{23.9}$$

We use the relation

$$E_a^2 - p_a^2 c^2 = m_a^2 c^4$$

to eliminate $p_a^2 c^2$ from the above equation and obtain

$$\begin{aligned} s &= (E_a + m_b c^2)^2 - E_a^2 + m_a^2 c^4 \\ &= 2E_a m_b c^2 + (m_a^2 + m_b^2) c^4 \end{aligned}$$

This may be solved for E_a, the total energy of particle a in the laboratory frame, yielding

$$E_a = \frac{s - (m_a^2 + m_b^2)c^4}{2m_b c^2} \tag{23.10}$$

Inserting the threshold value $M^2 c^4$ for s then gives the threshold value of E_a:

$$E_{a_{th}} = \frac{M^2 - (m_a^2 + m_b^2)}{2m_b} c^2 \tag{23.11}$$

Another quantity of interest is the threshold kinetic energy particle a must have for the reaction to proceed. This is

$$\begin{aligned} K_{a_{th}} &= E_{a_{th}} - m_a c^2 \\ &= \frac{M^2 - (m_a + m_b)^2}{2m_b} c^2 \end{aligned} \tag{23.12}$$

and is the least kinetic energy that must be provided by an accelerator to make the reaction go at all.

Let us apply the result of Eq. 23.12 to two cases of importance:

1. Production of an electron-positron pair in a photon-nucleus collision
The production reaction is

$$\gamma + \mathrm{A} \rightarrow \mathrm{A} + e^- + e^+ \tag{23.13}$$

Because $m_\gamma = 0$, here

$$m_a = m_\gamma = 0, \qquad m_b = m_A$$
$$M = m_A + m_{e^-} + m_{e^+}$$
$$= m_A + 2m_e$$

The nucleus is not altered in structure during the process, but its presence is necessary so that it can take off the excess in momentum that the pair cannot. Then

$$K_{a_{th}} = E_\gamma = \frac{(m_A + 2m_e)^2 - m_A^2}{2m_A} c^2$$
$$= \frac{4m_A m_e + 4m_e^2}{2m_A} c^2$$
$$= 2m_e c^2 \left(1 + \frac{m_e}{m_A}\right) \tag{23.14}$$

If the nuclear mass is very large compared with the electron mass, as it is even for hydrogen where the ratio

$$\frac{m_{nuc}}{m_e} = 1836$$

the second term in the parenthesis is unimportant and we see that an electron-positron pair can be created by a photon of energy just a shade greater than the rest energy of the pair. This is not surprising.

2. Production of a proton-antiproton pair in a collision between a moving and a stationary proton An antiparticle is denoted by a bar over the symbol for the particle so the antiproton is written $\overline{\text{p}}$. The reaction is

$$\text{p} + \text{p} \rightarrow \text{p} + \text{p} + \text{p} + \bar{\text{p}} \tag{23.15}$$

Here, then,

$$m_a = m_b = m_p$$
$$M = 4m_p$$

we obtain

$$K_{a_{th}} = K_p = \frac{(4m_p)^2 - (2m_p)^2}{2m_p} c^2$$
$$= 6m_p c^2 \tag{23.16}$$
$$= 5628 \text{ Mev}$$

This result is surprising. In order to create a single proton-antiproton pair in a proton-proton collision, three times the rest energy of the pair is required in the form of kinetic energy of the bombarding proton. The extra energy is associated with the center of mass motion and is not available to help the reaction take place. Because of the relativistic increase of mass with speed, in the collision of two similar particles the moving one has the greater relativistic mass, which means that more than half the energy is energy associated with the motion of the center of mass. In nonrelativistic mechanics, in the collision of two similar particles one of which is initially at rest, half of the kinetic energy is associated with the center of mass motion and half with the relative motion, as shown in Eq. 14.7. It is this relativistic effect that makes it necessary to have particle accelerators producing protons with energies as high as 33 GeV at Brookhaven National Laboratory and to make the construction of a 500-GeV proton accelerator a matter of great importance for the study of the particles that are produced in collisions between nucleons.

We have seen that all the kinetic energy of two particles is available to produce a reaction if the energy is measured in the center of mass reference frame. If the particle accelerator could be made to operate in this reference frame instead of in a frame where one of the particles, the target particle, is at rest, much more effective use could be made of the energy it produces. The difficulty with operating in the center-of-mass frame is that of providing enough fast-moving particles to constitute a reasonable target in which an observable number of reactions take place. A stationary target can be very dense and contain many target particles, perhaps 10^{18} per square centimeter, for the accelerated particles to react with. Such a dense target cannot be accelerated as a whole to a speed large enough to make any difference at all. What can now be done is to accumulate a large number of energetic particles moving in circular orbits in "storage rings" and then to let two such stored beams intersect each other. If the number of particles in each beam is large enough, an observable number of reactions will take place in the intersection volume, and those particles that do not react go around again and have another chance.

Storage rings for protons are under construction at the CERN laboratory in Geneva. There the Proton Synchrotron, a machine whose operation is described in Section 24, can fill the storage rings with protons of 28 GeV of kinetic energy. For purposes of discussion we will call the total relativistic energy of these protons "30 GeV," so the available energy in the intersecting beams is 60 GeV. We can find what energy is needed when one proton is at rest and only the other is moving to achieve the same available energy. From Eq. 23.10 with $s = (60 \text{ GeV})^2$ we find that (taking $m_pc^2 = 1$ GeV)

$$E_a = \frac{3600 - 2m_pc^2}{2m_pc^2}$$
$$\approx 1800 \text{ GeV}$$

This is so large that the rest energy of the proton does not matter.

There are storage rings for electrons and positrons in operation in Novosibirsk, USSR, and at Frascati, Italy, so that very high energy reactions of electrons can be studied. Other storage ring facilities are being constructed both here and abroad.

Problems

1. A photon is a particle with zero rest mass, and it moves with speed c in any reference frame. If in reference frame S a photon is moving with momentum p in the positive x-direction and therefore has energy $E = pc$, what is its energy as measured in a reference frame S' moving with speed v in the negative x-direction of S? Is there any upper limit to this energy as v gets closer and closer to c?
2. Repeat the calculation of Problem 1 for a photon moving in the negative x-direction in S' with momentum p. Is there any lower limit to the energy E' as measured in S as v gets closer and closer to c?
3. Show that the nonreaction

   $$e^- \not\rightarrow e^- + \gamma$$

   is forbidden by the laws of energy and momentum conservation when the electron is initially at rest.
4. Show that a moving electron cannot emit a single photon and slow down because this process violates the laws of energy and momentum conservation. How is your answer to this question related to your answer to Problem 3?
5. A photon can be scattered by an electron. By applying the laws of energy and momentum conservation find the energy of a photon that has been back-scattered by an electron as in the figure. This is a special case of what is called *Compton scattering.*

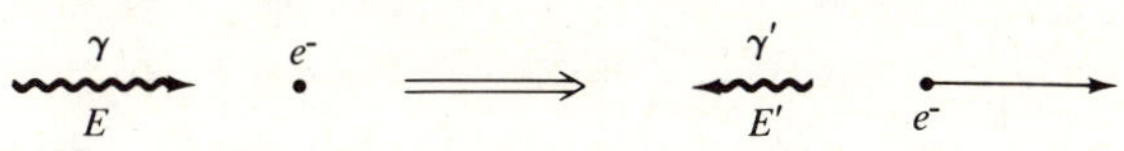

6. If the electron in Problem 5 is moving to the left with total energy E_e before the collision, find the energy of

the scattered photon. This provides a method for producing energetic photons; photons from a laser with energy of about 2 eV are scattered from electrons of energy $E_e = 20$ GeV. The back-scattered photons then have much more energy than the original ones.

7. On the basis of the reaction of Eq. 23.13, how can one conclude that the photon is a particle of integer spin? Its spin value is actually unity.

24. PARTICLE ACCELERATORS

The identification of the electron as a particle was made by Thomson using a beam of electrons accelerated in the electric field produced by a battery. The discovery of the atomic nucleus by Rutherford was made using a beam of alpha particles from radioactive elements. The discovery of the positron by Anderson was made using fast particles bombarding the earth from space, the cosmic rays. These examples show the usefulness of beams of fast particles in studying the structure of matter in its most fundamental forms. An important aspect of experimental physics is the development of accelerators that can furnish controlled beams of particles of specific kinds, because with such beams experiments can be done that would be impossible without them. Much of our knowledge of the constitution of atoms and nuclei and of the properties of the subatomic "fundamental" particles has come from experiments using beams of particles.

There are many types of particle accelerators suitable for different purposes. In atomic physics one may want electron beams ranging from less than 1 eV energy up to 1 million eV or more. In nuclear physics beams of protons and other nuclei with energies from a few million electron volts up to a few hundred million electron volts are needed. In high energy physics beams of protons with the highest possible energy can be used to advantage. The highest energy accelerator now available is one of 70 GeV in Serpukhov in the USSR. Here we shall describe three types of particle accelerators, the electrostatic or Van de Graaff accelerator, the synchrotron, and the cyclotron. Of these the synchrotron can provide the highest energy particles.

One way of obtaining energetic positively charged particles is to produce them at rest in a place that is at an electric potential higher than that of the ground and to let them move from there to ground. If the electric potential where the particles of charge q are produced is φ, the potential energy of one of the particles is $q\varphi$. The particles can then be accelerated by the electric field existing between that place and ground. When they reach ground (zero) potential, energy conservation requires them to have a kinetic energy K given by

$$K = q\varphi \tag{24.1}$$

The maximum energy obtainable in this way depends on the charge on the particle and on the maximum potential relative to ground that can be obtained. When one wants to accelerate particles of a definite kind, not much can be done about the charge on the particle, so the problem of obtaining fast particles of a given type is the problem of obtaining high potentials.

In order to keep a body at high potential, charge must be prevented from leaving it too rapidly; in other words, the body must be insulated from the ground. Insulators are made of matter, and matter is composed of atoms that contain electrons. If an insulator is placed in an electric field strong enough to detach electrons from atoms of the material, the insulator breaks down and becomes a conductor. This happens in dry air at atmospheric pressure when the electric field is about 10^4 volts per centimeter. It is necessary, therefore, to avoid very strong electric fields in the construction of a high voltage system.

To see what is implied by the requirement that strong fields be avoided, let us consider the problem of raising a conducting sphere of radius R to a potential φ. The field outside a charged conducting sphere carrying charge Q is just the same as it would be if the entire charge on the sphere were concentrated at its center and is therefore given by Coulomb's law:

$$E(r) = \frac{Q}{r^2}, \qquad r > R \tag{24.2}$$

Inside the conductor there is, of course, no electric field at all. The potential of the sphere is, according to Eq. 16.5, given by

$$\begin{aligned} \varphi(R) &= \frac{Q}{R} = R\frac{Q}{R^2} \\ &= RE(R) \end{aligned} \tag{24.3}$$

This shows that, for a given value of the potential $\varphi(R)$, the larger the sphere, the smaller is the magnitude of the corresponding electric field. For this reason all high voltage devices have smooth, rounded surfaces. Figure 24.1 shows such a device, a "Cockroft-Walton" generator

FIGURE 24.1 This Cockroft-Walton high voltage generator has a potential of 750,000 volts on its topmost terminal. It is used to give protons their initial acceleration at the National Accelerator Laboratory's synchrotron. The final proton energy will be 5×10^{11} eV when the synchrotron is completed (National Accelerator Laboratory.)

that has a potential of 750,000 V on its terminal. If surfaces with small radii of curvature such as corners or points were present, the electric field in their vicinity would be large and the air or other insulating material might break down. (Note that although we speak here of "voltages," all our equations are written in the Gaussian system of units in which the unit of potential is very nearly 300 volts, as explained in Section 16.)

The high voltage terminal of our accelerator must be a large, smooth body, often spherical, supported by insulators. How does one charge such an object to a high potential? Not by connecting it with a wire to a battery, obviously. First, if one had the battery one would not need the terminal. Second, if the wire were small it would have a large electric field around it and could not be maintained at the high potential desired. A method of charging is needed that does not involve the presence of an equal or a still higher potential somewhere else.

An elegant method of charging a conductor even if it is already at a high potential was invented by R. J. Van de Graaff, based on the fact discussed in Section 16 that a charge introduced into a conductor immediately distributes itself over the surface of that conductor, leaving the interior uncharged. The spherical terminal of the accelerator has holes cut in it (with well-rounded edges to avoid causing strong electric fields) through which an endless belt of insulating material passes, as shown in Fig. 24.2. The belt extends to the base of the accelerator, which is at ground potential. Because of the large size of the terminal, at no point along its length is the belt in an electric field strong enough to cause it to break down and conduct charge. If a portion of the belt is charged when it is near the ground, this charge will be carried bodily into the terminal by the moving belt. Inside the terminal a conducting brush gives the charge on the belt a path to get to the outside of the terminal, which it immediately does. The uncharged belt leaves the terminal and returns to ground. In this way charge is built up on the terminal until the potential gets so high that charge leaks off in various

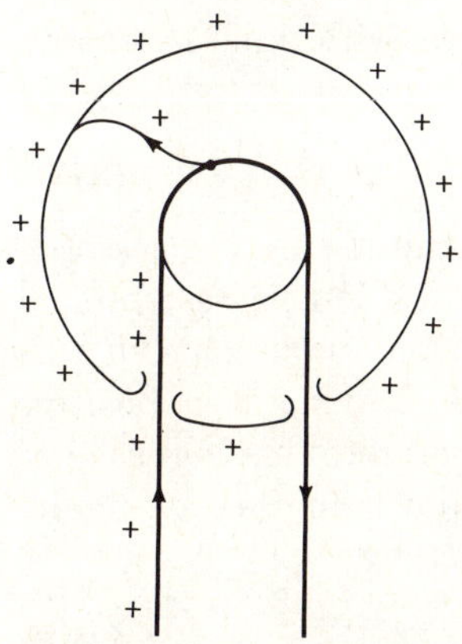

FIGURE 24.2 A Van de Graaff terminal is charged by running a charged belt into the terminal and giving the charge introduced into the terminal in this way a path by which it can move to the outside surface. The belt leaves the terminal uncharged.

ways as fast as it can be brought up. Some modern Van de Graaff generators can produce potentials as high as 14 million volts on the terminal.

The particles to be accelerated are stored in the terminal as atoms or molecules, usually of a gas, and are there converted into positive ions by removing some electrons, often by bombarding them with electrons. These ions are then accelerated along an evacuated accelerating tube connecting the terminal to ground, where they can be used as desired. The accelerating tube itself is an elaborate structure designed to keep the ion beam focused into a narrow pencil and to prevent electrons from being accelerated toward the terminal. With a positive terminal accelerator, any positive ion that can be produced in the terminal can be accelerated. In order to accelerate electrons, the terminal must be negatively charged.

In the tandem Van de Graaff accelerator the potential difference between the terminal and ground is used twice. The particle to be accelerated is made into a negative ion at ground potential by attaching an extra electron to the neutral atom. (More than one extra electron generally will not stick as the two or more extra negative charges repel so strongly that the attraction of the positive nucleus cannot overcome this repulsion.) This negative ion is accelerated along an accelerating tube toward the positive terminal, arriving there with kinetic energy $K^{(1)}$ given by

$$K^{(1)} = e\varphi_t \tag{24.5}$$

if φ_t is the terminal potential and e is the magnitude of the electron charge. Inside the terminal the fast-moving negative ion passes through a thin foil or a short tube filled with gas. Collisions undergone in this passage remove electrons from the negative ions, leaving it a positive ion sometimes lacking several electrons and therefore having several times e as charge. This positive ion is accelerated down another accelerating tube to ground, gaining kinetic energy $K^{(2)}$ given by

$$K^{(2)} = ne\varphi_t \tag{24.6}$$

in the process, ne being the positive charge on the ion. The total kinetic energy of the particle K is now

$$\begin{aligned} K &= K^{(1)} + K^{(2)} \\ &= (n+1)e\varphi_t \end{aligned} \tag{24.7}$$

which is considered greater than the energy obtainable by single-ended

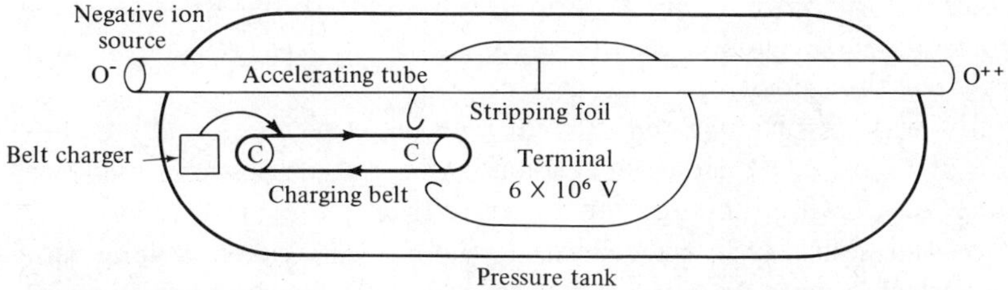

FIGURE 24.3 A tandem Van de Graaff is shown schematically. Negative oxygen ions enter from the left and are accelerated toward the positive terminal. Here they are stripped of some electrons and are accelerated toward ground potential at the other end of the accelerating tube. High pressure gas insulates the terminal from ground.

operation. For example, if oxygen is the ion being accelerated, it can happen that all eight electrons in an oxygen atom are removed in the stripper in the terminal, so that values of n as high as 8 can be achieved. Most of the ions will have less charge than this, so that ions of several discrete energies corresponding to different values of n will occur at the output end of the accelerator. A schematic drawing of such an accelerator is shown in Figure 24.3. The whole thing is enclosed in a pressure tank to provide insulation for the terminal.

The energy of particles accelerated in this direct way by the creation of a potential difference can be increased by forming the negative ions in the negative terminal of one Van de Graaff accelerator and then accelerating them toward the positive terminal of a tandem Van de Graaff where they are stripped as before and then go on to ground as positive ions. It is not possible to go any further, cascading one Van de Graaff after another, because it is not possible to attach extra electrons to fast-moving positive ions so as to make them negative ions. It is easier to knock electrons off than to attach them. Also, for most purposes it is not practical to have the output of the accelerator inside the negative terminal of a Van de Graaff. The accelerated particle beam is to be used for experiments, and these experiments usually cannot be done inside a terminal, so the output is at ground potential. For special purposes the use of a target in a negative terminal has been proposed.

Van de Graaff accelerators are valuable tools for studying atomic nuclei. The energy of the product ions is very precisely controllable, which is a great advantage over natural sources and some other types of accelerators. The maximum output energy is limited because the maximum potential attainable on a terminal is limited by the insulation problem. Not nearly enough energy is available for the creation of antiparticles other than positrons.

The only way to produce really high energy particles is to give them energy a little bit at a time, never having a potential difference corresponding to the final energy or a large fraction of it anywhere. If nuclear particles are to be accelerated, the process takes a long time and the particles must be confined in a small region during the accel-

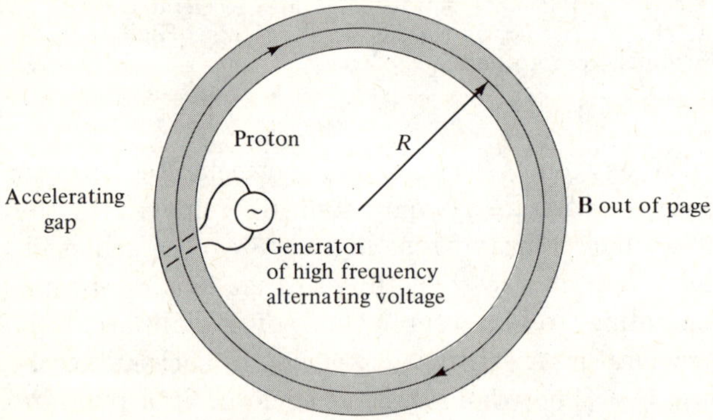

FIGURE 24.4 The magnetic field confining the protons in a proton synchrotron is an almost uniform field in the shaded area. Protons of the proper momentum go round the ring of radius R and can be accelerated by accelerating gaps placed around the ring. The radius R is fixed, so as the proton momentum is increased, the magnetic field also must be increased.

eration, which is accomplished by the use of a magnetic field. If a uniform magnetic field B exists everywhere in an annular region, as shown in Figure 24.4, a charged particle of the correct momentum will move around the ring. The condition for this is that the momentum p of the particle be related to the magnetic field B and the radius R by Eq. 18.7, which is

$$p = \frac{eBR}{c} \tag{24.8}$$

If the particle is not interfered with by being scattered or otherwise, it will continue to move in this circle almost indefinitely. Actually, in practice the field B is not uniform but is made to vary around the ring in such a way that the particles are "focused," so that even if their momentum is not exactly given by Eq. 24.8 or if their direction of motion is not exactly right they will move around the ring without straying outside it. It was the discovery of this method of focusing that made the construction of modern high energy accelerators possible.

The magnetic field confines the particles to the ring. Somewhere on the ring an accelerating gap is inserted. In principle this is a pair of capacitor plates with holes in them for the particles to pass through. These plates are connected to a generator of alternating voltage, and the frequency with which the electric field between the plates oscillates is adjusted so that a particle which is accelerated when it goes through the gap once is accelerated again the next time through. It would not do to connect the plates to a battery producing a steady electric field between the plates. The reason for this is that associated with a steady electric field is an electrostatic potential φ, and the motion of a charged particle in such a field must obey the law of energy conservation

$$K + e\varphi = \text{constant} \tag{24.9}$$

When this equation holds, a steady increase of K as the particle goes round and round the ring is not possible, because whenever the particle comes to a place on the ring where it has been before, it has the same potential energy that it had before and therefore must have the same kinetic energy that it had before. In general, an electric field that varies with the time does not have a simple potential φ associated with it, so that an equation of the form Eq. 24.9 does not hold. Energy is still conserved in the sense that any increase in the kinetic energy K of the particle must occur at the expense of the energy of the system that provides the alternating electric field, but this energy of the system is not describable as a potential energy of the particle depending solely on the position of the particle.

Accepting the fact that an accelerating gap can give a circulating charged particle additional kinetic energy each time it goes through the gap, we can see how to operate an accelerator. As the kinetic energy increases, so does the momentum of the particle. To keep the particle in the ring, the magnetic field must be increased, and to keep the electric field so that it always accelerates the particle when it is in the gap, the frequency with which this field alternates must be changed. The necessity of keeping all these changes in step with each other led to the name *synchrotron* for such accelerators.

Synchrotrons have been built to accelerate protons and electrons. The largest proton synchrotrons at present are those at Serpukhov in the USSR, producing protons of 70 GeV (billion electron volts) energy, at Brookhaven in the United States producing 33 GeV protons, and at CERN in Switzerland producing 28 GeV protons. By increasing the radius of the ring, the final energy of such accelerators can be increased almost (but not quite) indefinitely; but of course the cost becomes large. A proton synchrotron capable of producing protons of 500 GeV energy is under construction at Batavia, Illinois, and should be in operation in 1972.

The rest energy of a proton is 0.938 GeV. The kinetic energy of a 33 GeV proton is 33 GeV. (The energy rating of an accelerator takes account only of the energy it gives the accelerated particle. No credit is given for the rest energy.) Thus the rest energy is only some 3 percent of the total energy of the accelerated proton, and not much of an error is made if we neglect the rest energy. This approximation simplifies the connection between energy and momentum as given in Eq. 22.18, making it the same as for particles with zero rest mass (Eq. 23.2), namely,

$$E = pc \tag{24.10}$$

This enables us to get a simple expression for the radius of the ring needed to accelerate protons to a given energy E when we know the strongest magnetic field that is available. We solve Eq. 23.8 for R and then replace p by E/c to obtain

$$R = \frac{E}{eB} \tag{24.11}$$

In this equation, to get R in centimeters we must insert the value of E as measured in ergs, not GeV; e as measured in statcoulombs; and B as measured in gauss. It is not at present practical to use values of B much greater than 12,000 gauss, so to obtain protons with an energy E = 30 GeV = 4.8×10^{-2} ergs, a ring of radius

$$\frac{4.8 \times 10^{-2}}{(4.8 \times 10^{-10}) \times (1.2 \times 10^{4})} = 0.8 \times 10^{4} \text{ cm}$$

is needed. This is roughly 100 yards or the length of a football field in radius. Unless improved magnets become available, a 500 GeV proton synchrotron has to be about fifteen times as big in radius.

Electron synchrotrons also have been built. The formula Eq. 24.11

also holds for them because when the rest energy was neglected in deriving Eq. 24.11 all reference to the rest mass of the particle being accelerated was eliminated. There is a practical limit to the energy of the accelerated electrons. An accelerated electric charge loses energy in the form of electromagnetic radiation. In the case of electrons of energy somewhere near 1 GeV in a synchrotron, much of this radiation is in the form of visible light, called synchrotron radiation. The rate at which energy is lost in this way depends sensitively on the ratio of the energy of the particle to its rest energy. Because the rest energy of the electron is so much smaller than that of the proton, this energy loss becomes important for electrons at much lower energy than for protons. It seems impractical to use circular machines to accelerate electrons to energies higher than 10 GeV. Protons would have to have an energy almost 2000 times this before they would lose energy so fast by radiation in a machine of the same size.

There are other kinds of particle accelerators. The cyclotron is a well-known type. It operates much like a synchrotron but is simpler. The magnetic field occupies the entire circle, not just the ring at the edge of the circle, and is kept constant in time. As the particle is accelerated, its momentum increases and so does the radius of its orbit in the almost uniform magnetic field. The orbit is a spiral starting at the center and widening out with time. The operation is especially simple when the particle energy is nonrelativistic so that its momentum is given by

$$mv = \frac{eBr}{c}$$

The time it takes the particle to go around once in the field is

$$\begin{aligned} T &= \frac{2\pi r}{v} \\ &= \frac{2\pi mc}{eB} \end{aligned} \qquad (24.12)$$

which is independent of the radius of the orbit, or the particle momentum. There is, therefore, no need to change the frequency at which the accelerating field is applied. This angular frequency is called the *cyclotron frequency* ω_c and is defined by

$$\omega_c = \frac{2\pi}{T} = \frac{eB}{mc} \qquad (24.13)$$

This frequency plays an important role in parts of physics far removed from the production of fast particles.

There are other accelerators in which the particles move in a straight line as they do in a Van de Graaff, although the acceleration is produced by alternating fields, not a static field. These linear accelerators are useful when particles of very different masses are to be accelerated by the same machine, since in a circular machine the radius R of the orbit depends on the mass of the accelerated particle. A linear accelerator for electrons avoids the problem of energy loss by radiation, and at Stanford such a device 2 miles long has been constructed. Eventually it is designed to yield electrons with a maximum energy of 40 GeV.

Problems

1. A lightning rod is supposed to work by preventing the buildup of sufficiently strong electric fields above the protected building by neutralizing the charge in the atmosphere nearby, not by conducting the charge in a lightning bolt to ground. Give an argument based on Eq. 24.3 and the fact that air becomes conducting in very strong electric fields for making lightning rods pointed.
2. Oxygen atoms are accelerated in a Van de Graaff accelerator by knocking off either three or four electrons and accelerating the resulting ion. If the ions are then deflected by passing them through a uniform magnetic field, will the triply charged ions (with three-fourths the kinetic energy of the quadruply charged ions) be deflected more or less than the quadruply charged ones? The larger charge feels a larger magnetic force but is on a particle of higher momentum, so these effects must be compared.
3. What energy protons could be produced by a synchrotron that went around the equator of the earth and had a maximum average magnetic field of 5000 gauss?
4. Some of the light from the Crab Nebula is emitted as "synchrotron radiation" by very energetic electrons moving in a magnetic field. If the magnetic field is about 10^{-4} gauss and if the electron orbits extend over something like 10^{18} cm, what is the approximate energy of these electrons?

25. PARTICLE DETECTORS

In describing various experiments involving the motion of particles that are invisible to the human eye we have said that the particles follow such and such paths and end up at certain points. We devote this section to a brief description of some of the devices used to detect the presence of these particles and to tell something about their motion.

Detectors may be divided into two large categories, point counters and track detectors. The former type detects the presence of a particle in a small volume and may give other information too. A track detector displays the path of the detected particle through the detector with more or less accuracy, depending on the type of detector. All particle detectors make use of the effect a moving charge has on the atoms of the material the detector is made of. Neutral particles, therefore, must be detected in two stages: In the first they produce charged particles in some reaction and in the second the charged particles are then detected. A charged particle moving through matter loses energy to the electrons in the atoms of that matter, generally dissociating them from the atoms and producing *ion pairs* consisting of the negative electrons and the positively charged residual ions. Eventually these ion pairs recombine, preparing the detector to be used again.

Thomson's detector in his e/m experiment was a phosphor. This is a material that emits light when struck by electrons or other charged particles. It is very satisfactory if a large number of charged particles are to be detected collectively. The spot on an oscilloscope tube "detects" the arrival of the electron beam at the location of the spot. The flashes caused by individual electrons are very faint and cannot be seen as separate events. Different phosphors emit light of different colors, and this effect is used in color television tubes.

The detector used in the Rutherford scattering experiment done by Geiger and Marsden also was a phosphor. Here the incident particles were alpha particles instead of electrons, and the resulting flashes or scintillations are brighter. They could be seen individually by an eye that was dark adapted, and the experiment was done by visual counting of these scintillations. This is a tedious and inaccurate method.

A vast improvement on these visual detectors is the scintillation counter. A scintillation counter consists of a scintillator and a photomultiplier tube. The scintillator is a transparent material, such as sodium iodide containing minute amounts of certain impurities, which has the property of giving off a pulse of light when it is traversed by a charged particle. Because the material is transparent, this light can escape from it and be collected at the photomultiplier. This latter is nothing more than a sensitive photoelectric cell, and it produces a pulse of electric current proportional to the amount of light that enters it. Such a counter is sketched in Figure 25.1. The amount of light in the pulse resulting from the passage of a charged particle is proportional to the number of ion pairs formed by the particle. Such detectors can count particles at rates as high at 10^4 per second. When the charged particle stops in the scintillator without producing a nuclear reaction, the number of ion pairs formed is proportional to the initial kinetic energy of the particle, so this can be measured by the size of the pulse put out by the counter. If a reaction is produced, this will cause additional ion pairs to appear and the method does not work, or at least not as accurately. Scintillation counters are excellent for detecting photons. Photons are neutral, but they are likely to transfer most of their energy to electrons, which then cause scintillation. Heavy particles, as we know, cannot

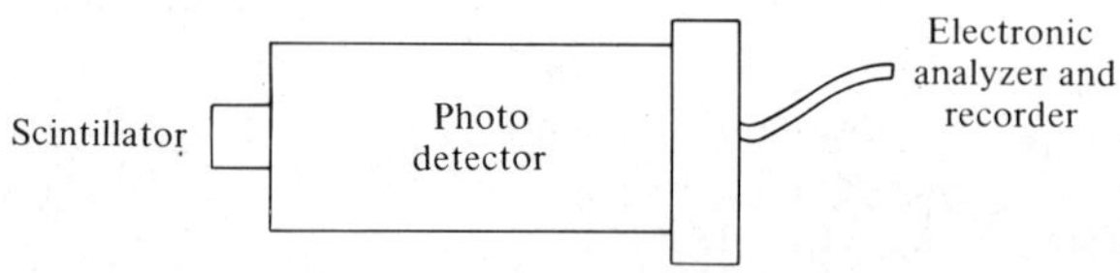

FIGURE 25.1 A scintillation counter consists of a "scintillator" that emits light when a charged particle moves through it and a photodetector that converts the light pulse into an electrical pulse that can be electronically recorded.

transfer much of their energy to a single electron but can lose it in many small transfers to a large number of electrons.

If a heavy charged particle passes through a thin scintillator, it causes a small light pulse to be formed and counted, but the momentum of the particle itself is only slightly altered by the process. It is therefore possible to have several scintillation counters arranged along a line and to see if a light pulse is generated in each of them at essentially the same time. If the counters count "in coincidence," then one knows that a particle went in the direction of the line. Such an arrangement is known as a *counter telescope*. If two counters are far enough apart and if the particle is not going too fast, it is possible to measure the speed of the particle by measuring the time delay between the two counts. This is called a *delayed coincidence*. Delays as small as a few times 10^{-9} sec can be measured.

One may want to know if a charged particle went along a line and was then scattered out of that line in a collision. This can be determined using the arrangement of Figure 25.2. The particle comes from the left,

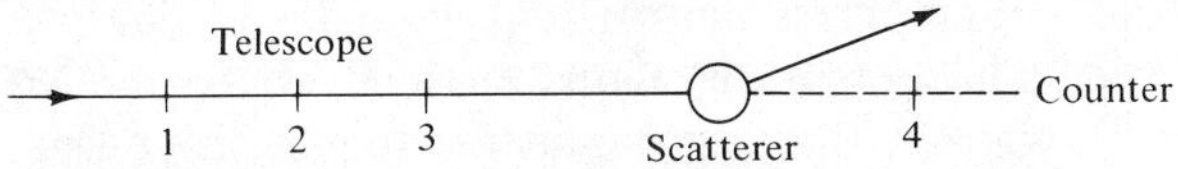

FIGURE 25.2 The occurrence of a scattering event can be detected by having a counter telescope in front of the scatterer and another counter behind the scatterer. If the counter telescope registers the passage of a particle and the other counter does not, the particle must have been scattered out of the original beam.

and its presence and direction of motion are determined by the counter telescope. If the particle is not scattered in the scatterer it will also register in counter 4 behind the scatterer. If it is scattered in the scatterer, counter 4 will not register. Scattering events thus can be identified by operating the counters 1, 2, and 3 of the telescope in coincidence and then counting the telescope output only when counter 4 gives no pulse. The telescope and counter 4 are said to be in *anticoincidence.* This scheme is used when the scatterer is a cloud chamber or a spark chamber, as will be described below.

There are many other kinds of particle counters. The best known one is the *Geiger counter.* This is a useful counter, but it yields pulses of the same size for all particles. The kind of counter best suited to particular situations can be decided only on the basis of very technical considerations that do not interest us here.

A fast charged particle traveling through a gas such as air produces ion pairs in the gas. If the gas is supersaturated with some vapor, droplets of the liquid will form on the ions, producing a cloud. A familiar example of a similar phenomenon is the formation of vapor trails behind high-flying jet airplanes. The *cloud chamber* exploits this phenomenon to make the tracks of charged particles visible. The problem is to provide a volume of gas supersaturated with vapor. This is done by having a volume of gas that is saturated with vapor, which is attained simply by having an excess of the liquid present in the vessel containing the gas. Alcohol is frequently used as the vapor and air as the gas. If the saturated vapor is suddenly cooled, it becomes supersaturated and will condense very rapidly on ions or on dust particles suspended in the air. If neither ions are present nor dust, the vapor will condense on the walls or elsewhere until the normal saturated state is reached. Supersaturation is an unstable condition and does not last long. However, if a charged particle leaves a trail of ions that are there when the supersaturation occurs, a track of cloud will form on the path of the particle.

A cross section of a cloud chamber is shown in Figure 25.3. The top and sides are made of glass. The bottom is a piston that can be moved to expand the gas suddenly, thereby cooling it and producing the necessary supersaturation. After expansion the piston is returned to its original position, and when equilibrium is reached the chamber is ready for another expansion. The cloud tracks are illuminated from the side, the droplets showing up against the bottom of the chamber when viewed

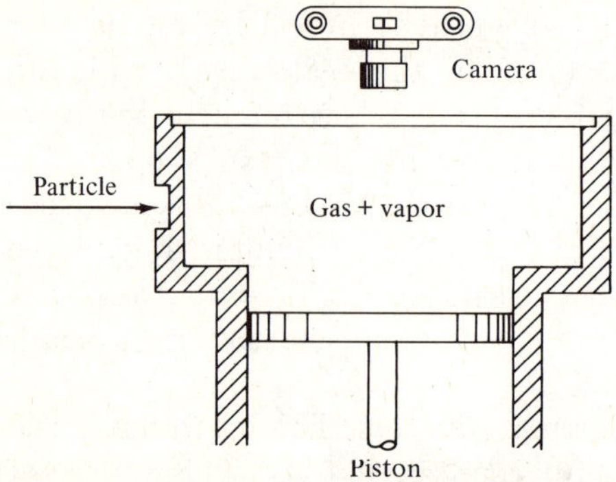

FIGURE 25.3 When the cloud chamber is expanded after the passage of a charged particle through it, the vapor inside becomes supersaturated and condenses on the ion trail left by the particle. This cloud track then can be photographed stereoscopically to record the event.

from the top. Stereoscopic photographs of the tracks are usually taken to provide the record of what took place. A magnetic field may be provided, causing the tracks to curve. Their radius of curvature can be measured, thus determining the momentum of the particles if their charge is known.

A cloud chamber is sensitive only to ions formed just before it is expanded and therefore only for a small fraction of the time. It is possible, however, to trigger the expansion of the cloud chamber by having counters placed before and after it, as in Figure 25.2. A charged particle trips the telescope and does not trip the other counter, so something has happened in the chamber. The chamber is immediately expanded, which can be done quickly enough so that the ions left by the particle have not recombined or drifted too far from their original position, and a record of what happened is obtained. The original positron picture of Anderson shown in Figure 23.3 was taken in this manner, although without the use of the anticoincidence counter 4.

The cloud chamber is a very useful instrument, but it has the serious disadvantage that the matter it contains is of low density, being a gas, so that collisions of the incident particles with nuclei in the chamber are improbable. For some purposes it suffices to put metal plates in the chamber, as in Anderson's positron experiment, but it is then not possible to see individual events, to observe the recoil particles, or to use the simplest of all target nuclei, protons. Filling the chamber with a heavy gas helps, but only in a limited way. This particular disadvantage is overcome by the bubble chamber.

The cloud chamber operates through the temporary production of the unstable supersaturated state of a vapor. The *bubble chamber* operates through the temporary production of the unstable *superheated* state of a liquid. A superheated liquid is one heated above its boiling point. If ions are present, vapor bubbles will form around them and the liquid "boils" locally. If the superheated state is reached when a track of ion pairs exists in the liquid, bubbles form along the track. They are illuminated and stereoscopically photographed. The superheated state is reached by expanding the liquid suddenly; this reduces the pressure and makes it possible for the liquid to vaporize. The degree of superheating must be carefully chosen so that bubbles form on the tracks and not everywhere.

The advantage of the bubble chamber lies in the high density of the liquid as compared to the gas of a cloud chamber. Liquid hydrogen is frequently used, providing a dense target of protons. This requires the bubble chamber to be operated at a temperature of about $-246°C = -411°F$, at which temperature the liquid hydrogen is readily superheated. The bubble chamber has a severe drawback: Its sensitive time is so short and its expansion time is so long that it is not possible to trigger its expansion with counters. By the time the counters have registered and the chamber has expanded, the ions are no longer there to cause a track to form. Bubble chambers are expanded as frequently as possible, and thousands or hundreds of thousands of pictures are taken, on some of which some desired event is recorded. Used in conjunction with high energy accelerators that provide intense beams of particles, bubble chambers have provided us with much of the information we have on fundamental particles. Some bubble chamber pictures are to be seen in Sections 26 and 27.

The last detector to be described here is the *spark chamber*. This consists of many very thin metallic foils separated from each other by a few millimeters, the space between being filled with a gas such as neon. Charged particles are incident in a direction perpendicular to the foils. They go through the chamber leaving a trail of ion pairs behind them. If a high potential difference is now applied to adjacent foils, sparks will jump across where these ion trails are located. The chain of sparks can then be photographed. The construction is shown in Figure 25.4, and a track is shown in Figure 25.5.

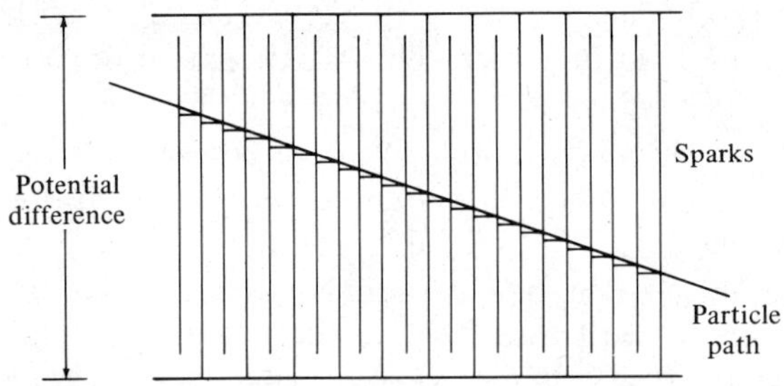

FIGURE 25.4 A spark chamber consists of a series of plates or foils between which a high potential difference can be quickly applied. When a charged particles goes through the chamber leaving a trail of ions, sparks will jump between the plates at the location of the ions when the potential difference appears, and these sparks can then be photographed or otherwise recorded.

The response of a spark chamber is extremely fast compared with that of a bubble chamber or a cloud chamber, as no mechanical motion is needed. The counters can register and the plates can be charged in a time of about 10^{-7} sec. This makes it possible to trigger a spark chamber with counters. Also, the chamber can be made sensitive at short intervals, maybe once every 10^{-4} sec, so that it is an enormous waste of time to photograph the sparks, photography being a slow process. Various other ways of locating the sparks have been developed so that a spark chamber can produce data at very high rates. Its output is now often fed directly into a computer that analyzes the events recorded and presents them untouched by human hands to the experimentalist. The spark chamber does not have the resolving power of the bubble chamber because a spark between two foils or wires is much larger than the bubble formed on an ion. Spark chambers are being rapidly

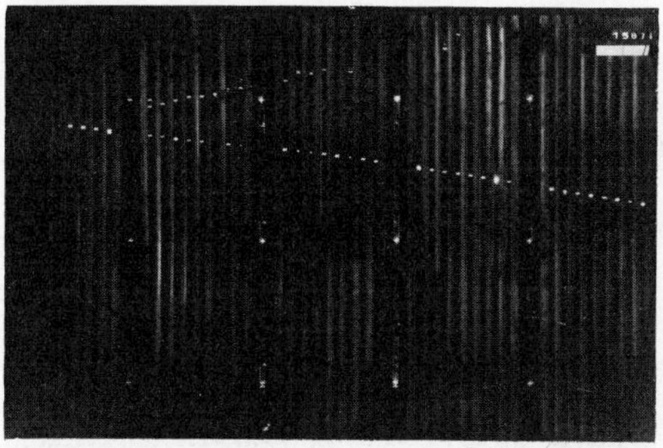

FIGURE 25.5 A neutrino produces a reaction in a spark chamber which was installed at the 33 GeV synchrotron at Brookhaven National Laboratory. The long straight spark track is that of a muon created by the incident neutrino. Another track is also seen entering from the left. This is thought to be due to electron pairs produced by a photon. This spark chamber was used in 1961 and 1962 for the experiment described in Section 28 that proved the existence of two neutrinos. (Brookhaven National Laboratory.)

improved, however, and they are unique in their efficiency in detecting very rare events.

Problems

1. Slow neutrons are readily absorbed by certain nuclei such as boron, giving rise to emission of a photon that can be detected. One way of detecting fast neutrons is to slow them down by passing them through a hydrogen-rich material like water or paraffin and then detecting the slow neutrons. Why does passing fast neutrons through such material slow them down faster than passing them through lead?
2. One might think that neutrinos could be slowed down in a way similar to the slowing down of neutrons as in Problem 1. Give as many reasons as you can for believing that this is not possible even in principle.
3. A bubble chamber 200 cm long (just over 6 feet) has a magnetic field of 20,000 gauss associated with it. What is the radius of curvature of the track of a 25-GeV proton in this chamber? By what angle is the direction of motion of this proton bent in the chamber? (One reason large chambers are needed is to make the curvature accurately measurable.)

26. FUNDAMENTAL PARTICLES: PIONS

All the physics that has been developed so far in this book comes to a focus on the subject of fundamental particles. Here the relativistic mechanics of particles, the action of electric and magnetic fields, particle accelerators and detectors, pair creation and annihilation, some elementary parts of quantum mechanics, and the elementary particles already met with such as the electron, the nucleon, and the photon all become ingredients of a development that could not have been anticipated 20 years ago. A series of new particles has been discovered, a series to which particles are still being added. The events in which these particles are created are simple enough to be analyzed in some detail, and considerable success has been achieved in organizing our knowledge of the phenomena involved, even though no satisfactory underlying theory is known that permits a unified description of the phenomena, as Newtonian mechanics unifies our knowledge of the solar system.

The study of these new particles is called *high energy physics* because the particles appear only in the collision of energetic particles of the kind we already know. The new particles are in general rather heavy, comparable to the proton rather than to the electron, so that they have large rest energies that must be provided as kinetic energy of the colliding particles. High energy physics is one of the most active research areas in science and, after space research, is one of the most expensive. The aim in the next three sections is to explain what kinds of experiments are done and how their results can be interpreted, in this way acquiring some understanding of this newly discovered part of nature.

The new "fundamental" particles are produced in events where the dominant interaction of the colliding particles is the nuclear or *strong* interaction, which is responsible for holding together the nucleons in atomic nuclei. That this interaction is a strong one has been known for a long time. As evidence of its strength, consider the difference in mass between a helium nucleus (alpha particle) and the two protons and two neutrons out of which it is made. This mass difference is equivalent to an energy difference of 28 MeV, the helium nucleus being lighter than the four separate particles. This means that an energy of 28 MeV is needed to take such a nucleus apart. In contrast to this, an energy of only 13 eV, one two-millionth as much, is needed to separate a hydrogen atom into its constituent proton and electron. It is the large energies with which nucleons are bound together in nuclei that make nuclear power sources produce so much energy for so little fuel consumption.

The strong interactions not only are strong but they are also very short ranged. This we already know from experiments like Rutherford's alpha particle scattering from gold. The behavior of alpha particles scattered in thin gold foils can be completely accounted for by assuming only the Coulomb repulsion to act between the gold nucleus and the alpha particle unless the two particles approach within some 10^{-12} cm of each other. When the alpha particles are energetic enough to get closer than this, the scattering deviates from what is expected for Coulomb forces alone and the nuclear force or strong interaction becomes apparent. It is this short range of the strong interaction that makes it inconspicuous in low energy experiments. Strong interactions do not affect electrons or photons at all.

High energy physics had its beginning in the attempt to understand strong interactions. H. Yukawa made the start in 1935 by suggesting that there is a new kind of field, a little like a Coulomb field, generated by a nucleon and that another nucleon in that field experiences a force of attraction. Associated with this field is a potential $V(r)$, as the Coulomb potential is associated with the Coulomb force. This *Yukawa potential,* as it is now called, differs from the Coulomb potential by being much larger in magnitude near its source and by dropping off in magnitude very rapidly with increasing distance from its source. This latter corresponds to the short range of the interaction.

The Coulomb interaction between two charged particles can be described by saying that the two particles exchange photons. The mathematical description of this process is more elaborate than we want to give here, but one can see how this might come about. The emission of a photon must conserve both energy and momentum, so the energy and momentum of the emitting electron are altered by such an emission. The energy and momentum of the electron that absorbs the photon also are altered by the absorption process, both quantities also being conserved here. Thus energy and momentum are exchanged between the two charged particles. The photons taking part in this exchange are not quite like the photons in a beam of light. Photons associated with a light beam are called *real photons.* They can travel long distances, and their energy E is related to their momentum $\mathbf{p}$ by the equation $E = pc$. The photons exchanged between two charged particles are called *virtual photons.* They go no further than between the particles, and their energy E and momentum $\mathbf{p}$ are not related by this equation but can have any values.

To see that a free electron cannot absorb a real photon, imagine that an electron initially at rest absorbs a photon of momentum $\mathbf{p}$ and of energy pc. Then, before the absorption, the energy of the photon plus electron is $pc + m_0c^2$ and the momentum is $\mathbf{p}$. After the absorption both of these must be the same, but the energy is the energy of the electron alone and the momentum is the momentum of the electron alone and these must be related to the rest mass of the electron according to

$$E^2 + p^2c^2 = m_0^2c^4$$

Thus $(pc + m_0c^2)^2 - p^2c^2$ should equal to $m_0^2c^4$, but expanding the square shows that it is equal to $2pm_0c^3 + m_0^2c^4$, instead. A free electron cannot absorb, or emit, a real photon without something else happening at the same time. An electron acted on by an external electric or magnetic field can emit photons, as the operation of all ordinary lamps shows.

The novel feature of Yukawa's idea was that the strong interaction described by the Yukawa potential should be described in terms of the exchange of virtual particles (particles whose energy and momentum are not related as those for a free particle are and which have an ephemeral existence) between nucleons. The strength of the interaction means that many such particles must be exchanged, and the short range of the interaction means that the particles exchanged must have a mass greater than zero, about 200 times the electron mass. The connection between the range of the interaction and the mass of the exchanged particle is not obvious or simple to derive but can be written down in a simple formula. If R is the range and m is the mass of the exchanged particle and $\hbar$ is Planck's constant, the constant characteristic of quantum effects introduced in Section 20, then

$$R = \frac{\hbar}{mc} \tag{26.1}$$

The known range of the strong interaction leads to the mass. The value of this mass is intermediate between that of the electron and that of the nucleon, so the particle was at first called the mesotron—meso meaning middle or intermediate. This name has since been shortened to *meson*.

If the interaction between nucleons is mediated by these mesons, it should be possible to produce real mesons in collisions between energetic nucleons as real photons can be produced in collisions of charged particles. This is observed to happen. A bubble chamber picture showing the reaction

$$\mathrm{p} + \mathrm{p} \rightarrow \mathrm{p} + \mathrm{p} + \pi^+ + \pi^- \tag{26.2}$$

is shown in Figure 26.1. The Yukawa mesons, now called "pi mesons"

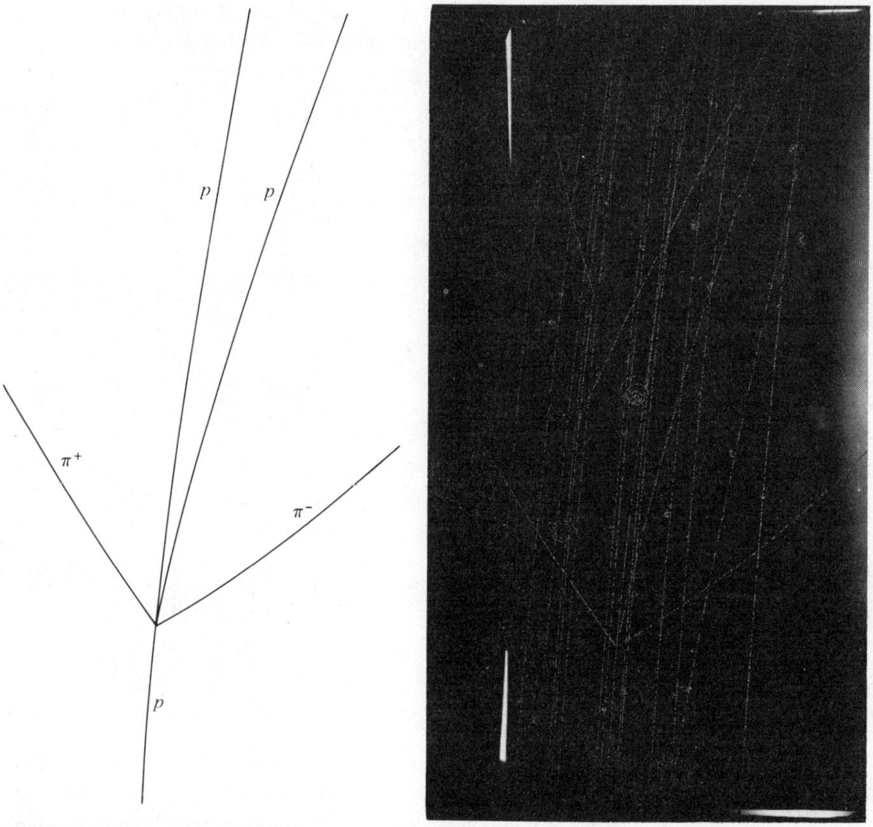

FIGURE 26.1 An energetic proton from the Brookhaven synchrotron strikes a proton at rest in a hydrogen bubble chamber. In the resulting reaction a pion pair is produced and the two protons go on. (Brookhaven National Laboratory.)

or pions to distinguish them from other mesons, can be charged and leave tracks. There is also a neutral pion, the π^0 mentioned in Section 23, which leaves no track in a chamber. The π^+ and π^- constitute a particle-antiparticle pair, while the π^0 is its own antiparticle and can decay into two photons when free, which it does after a mean life of 0.76×10^{-16} sec.

The mass of the charged pion can be determined from the analysis of reactions like Eq. 26.2, although this is not the most accurate method.

The mass of the proton is known, and so is the momentum of the protons in the initial state before the collision, so the total energy initially is

$$E = \sqrt{m_p^2c^4 + p_{p_1}^2c^2} + m_pc^2 \tag{26.3}$$

if p_{p_1} is the momentum of the incident proton. Labeling the other particles as shown in Figure 26.2, the energy in the final state, after the collision, is the sum of four terms, one for each particle.

$$\begin{aligned} E = {} & \sqrt{m_p^2c^4 + p_{p_3}^2c^2} + \sqrt{m_p^2c^4 + p_{p_4}^2c^2} \\ & + \sqrt{m_\pi^2c^4 + p_{\pi_1}^2c^2} + \sqrt{m_\pi^2c^4 + p_{\pi_2}^2c^2} \end{aligned} \tag{26.4}$$

The curvature of the tracks in the magnetic field in the bubble chamber yields the momenta of the four particles, so that equating the initial

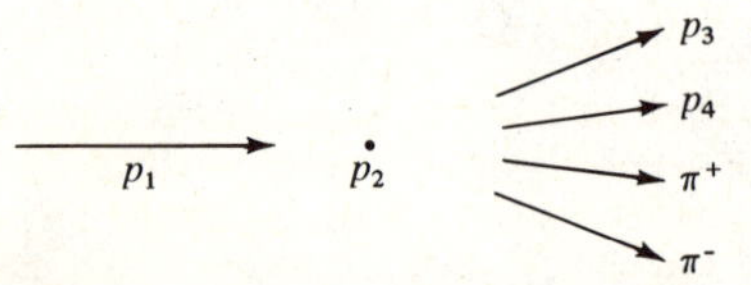

FIGURE 26.2 The energetic proton p_1 strikes the stationary proton p_2 producing a pair of pions. The recoil proton and the scattered proton are also shown.

and final energies (because energy is conserved in collisions) gives an equation in which m_π is the only unknown. This analysis always gives the same answer for m_π, so the assumption that energy is conserved seems justified. There are reactions that look like Eq. 26.2 in the bubble chamber but do not give the pion mass when analyzed. They can be accounted for by assuming the production of one or more π^0's in the same collision.

Pion production reactions are very common in collisions of 1-GeV (10^9-eV) protons with stationary targets, so the probability of such an event taking place is high. This gives another measure of the strength of these strong interactions. A proton with kinetic energy of about 1 GeV moves with a speed that makes its total energy very nearly twice its rest energy, the latter being close to 1 GeV also. Therefore a good estimate of its speed is obtained by setting

$$\frac{1}{\sqrt{1 - (v^2/c^2)}} = 2$$

The result is

$$1 - \frac{v^2}{c^2} = \frac{1}{4}, \qquad \frac{v}{c} = \sqrt{\frac{3}{4}} = 0.866, \qquad v = 2.6 \times 10^{10} \text{ cm/sec}$$

The pions are produced while this proton is passing another proton or a nucleus, therefore, while going a distance not much greater than 10^{-13} cm, and so are produced in an event that lasts only 10^{-23} sec. It takes a strong interaction to produce pions in this short a time interval.

Pions are unstable particles, which means that when they are left alone they decay into lighter particles. The decay of the π^0 according to

$$\pi^0 \rightarrow \gamma + \gamma \tag{26.5}$$

has already been mentioned. On the average a free π^0 lives 0.76×10^{-16} sec before decaying. This is an extremely short time to measure in the laboratory, but it is an extremely long time compared with the production time of 10^{-23} sec, which shows that the *electromagnetic interaction* responsible for this decay into photons is much weaker than the strong interaction. A charged pion cannot decay into photons because of charge conservation, and so it must follow another route and usually decays into a pair of particles, a muon and a neutrino. This decay process can be written

$$\pi^+ \rightarrow \mu^+ + \nu_\mu$$

$$\pi^- \rightarrow \mu^- + \bar{\nu}_\mu \tag{26.6}$$

The mean lifetime of a free charged pion is 2.6×10^{-8} sec, which is 100 million times as long as the mean lifetime of the π^0. The interaction responsible for this decay, appropriately enough, is called the *weak interaction.*

The muons were the first of the particles aside from the nucleons and electrons to be observed experimentally. High energy cosmic ray protons striking the earth's atmosphere from above produce pions in collisions with atmospheric nuclei, and these pions decay into muons and neutrinos on their way down through the atmosphere. The neutrinos are almost impossible to observe because they have only weak interactions with other particles, but the muons are charged and can be detected in the same way that other charged particles are. Unlike pions, muons do not have any strong interactions with nuclei, so the cosmic ray muons are very penetrating and are found in the deepest mines.

L. Alvarez and co-workers have done an experiment using muons to look for unknown chambers in the pyramids in Egypt, taking advantage of their penetrating power. Muon counter telescopes are located in a chamber at the base of the pyramid, and the number of muons coming from different directions is measured. If there is a chamber above the counters, the pyramid should be more transparent to muons coming through it than to those missing it. So far the results indicate an absence of unknown chambers.

The muon itself decays weakly according to the scheme

$$\begin{aligned} \mu^+ &\rightarrow e^+ + \nu_e + \bar{\nu}_\mu \\ \mu^- &\rightarrow e^- + \bar{\nu}_e + \nu_\mu \end{aligned} \tag{26.7}$$

so that the charge of the pion eventually shows up as the charge on an electron or positron. It is only recently (1962) that the existence of two different kinds of neutrinos, electron and muon neutrinos, was recognized. The experimental evidence for this is given later. The mean lifetime of the muons against their decay into electrons and neutrinos is 2.2×10^{-6} sec, a very long lifetime on the scale of times occurring in high energy physics.

The charge of the pion is of magnitude e. This cannot be measured in a Millikan oil drop experiment because this experiment requires a time longer than 10^{-8} sec for its performance, but it can be inferred by tracing through the decay processes. These end up, according to Eqs. 26.7 and 26.6, with an electron or positron and three neutrinos. If the neutrinos had any charge at all they would not be as penetrating as they are and could be detected by some effect produced by their charge, so unless charge conservation is violated the pion charge is $\pm e$.

The spin of the pion is zero. The rules of quantum mechanics require it to be an integer if it is produced in a reaction such as

$$p + p \rightarrow p + n + \pi^+$$

which is observed to take place. The spin of the proton and of the neutron is one-half, so for the spin angular momentum to change by an integer multiple of Planck's constant $\hbar$, as it must according to quantum mechanics, the spin of the π^+ must be an integer. If it were unity or greater, polarized pions could be produced, and the absence of any polarization effects in the scattering or decay of pions indicates the value zero. Particles with integer spin (including spin zero) that can be created singly in collisions are called *bosons,* after S. N. Bose who first studied some of their properties, while particles of half-integer spin that must be created and annihilate in pairs are called *fermions,* after E. Fermi. Mesons and photons are bosons; nucleons and leptons are fermions.

So far we have encountered two different families of strongly interacting particles, the two nucleons and the three pions. How can we summarize neatly the fact that there are just two nucleons and just three pions? A neat way to do this is to adapt to this situation the quantum rule that the possible values of any component of angular momentum of a particle must differ by an integer multiple of $\hbar$. The two nucleons have charges differing by e, and the three pions also have charges differing by integer multiples of e, so one says that there is a new physical quantity I called *isospin.* A component of this isospin, called the 3-component I_3, can have values within a family of particles that differ by integers, and the charge Q of the members of the family is given by the formula

$$\frac{Q}{e} = \frac{B}{2} + I_3 \tag{26.8}$$

The quantity B in this formula is the baryon number of the particle, which was introduced in Section 21. This formula works for the nucleon if we chose $B = 1$ (certainly appropriate and planned for) and $I = \frac{1}{2}$. When $I = \frac{1}{2}$, a component I_3 can have the values $+\frac{1}{2}$ and $-\frac{1}{2}$, which then describe the proton and neutron, respectively. For the pions we must chose $B = 0$, also a desirable choice, and $I = 1$. The possible values of I_3 are then $+1(\pi^+)$, $0(\pi^0)$, and $-1(\pi^-)$. The only use we shall make of isospin is the fact that strong interactions do not change the value of the total I_3 of a system, although electromagnetic and weak interactions can.

The baryon number B is an exactly conserved quantity. This conservation law results from the observed stability of ordinary matter. If a proton could decay into lighter particles, the element hydrogen would not be stable, but hydrogen has never been observed to decay into lighter particles. Experiments that have been done to look for this decay have not detected it and should have detected it if hydrogen decays with any mean life less than 10^{23} years. We shall see that the proton is the lightest of a family of particles called *baryons* and hence cannot decay without violating this baryon conservation law. A neutron is heavier than a proton and can decay into a proton, electron, and neutrino. The neutron also has $B = 1$ for the electron and neutrino $B = 0$, so this conserves B.

Problems

1. The Coulomb field of a charged particle can be represented by lines of force starting on the particle (assuming it to be positive for definiteness) and extending to infinity. The strength of the field is represented by the density of the field lines. The potential introduced by Yukawa can be written

$$V(r) = V_0 \frac{e^{-\mu r}}{r}$$

which decreases much more rapidly with increasing r than the Coulomb potential. Is there any way to represent the strength of this field corresponding to this potential by the density of field lines? Can these lines then begin and end only on the interacting particles as they do in the Coulomb case?

2. In the reaction of Eq. 26.2 illustrated in Figure 26.2, the nature of the incident particle and of the target particle is known, both being protons. The momenta of all particles observed in the reaction can be measured. How does the conservation of momentum and energy enable one to be quite sure that no neutral particle is produced in each of these events?

3. When a charged pion decays while at rest as in the processes of Eq. 26.6, find the fraction of the pion rest energy taken by the neutrino. Is this greater or less than the neutrino energy in the much rarer decay process

$$\pi^+ \rightarrow e^+ + \nu_e$$

(Is there more or less kinetic energy involved in this decay than in that of Eq. 26.6?)

4. When a muon at rest decays as in the processes of Eq. 26.7, two neutral particles that are almost impossible to detect are claimed to be produced. Why does the fact that the electron produced does not always have the same momentum rule out the possibility that only one neutral particle is produced?

27. STRANGE PARTICLES

Among the particles produced when energetic protons collide with stationary target nuclei are the three kinds of pions, positive, neutral, and negative. Of these, the positive and negative pions can be formed into beams of well-defined momentum by the use of deflecting and focusing magnets, so that a beam of π^+ or of π^- of controlled momentum can be sent into a bubble chamber. The reactions produced by these particles in the bubble chamber then can be systematically studied. It is possible to use liquids other than hydrogen in the chamber. If deuterium (heavy hydrogen) is used, one can infer something about the reactions with neutrons rather than protons as targets, as the nucleus of deuterium contains one proton and one neutron rather loosely bound together. If the product particles are charged, they leave tracks so that their momentum can be measured using the curvature of the track in the magnetic field present. If neutral particles are produced, they may decay into charged pairs or may produce secondary reactions that reveal their presence and their characteristics. It is also possible to have beams of other charged particles, but we discuss events produced by pions first.

It might seem that most of the pions produced should decay before they can be used in an experiment. The mean life of a charged pion is only 2.6×10^{-8} sec, so even traveling at a speed near c such a particle should go an average distance of

$$(3 \times 10^{10}) \times (2.6 \times 10^{-8}) = 7.8 \times 10^{2} \text{ cm}$$

or about 25 feet. This is not what happens. There is an effect of relativistic mechanics called *time dilation,* which makes the rate of a moving clock slower than that of a similar clock at rest by the factor $\sqrt{1 - v^2/c^2}$. This could have been derived in Section 22, but it is much more simply derived from the approach to relativity given in Section 30, so we merely state the result here and derive it later. A pion *at rest* has a mean life of 2.6×10^{-8} sec, so a moving pion has an apparently longer life as observed in the laboratory. If τ is its mean life at rest and $\tau(v)$ is its mean life when it is moving with speed v, then

$$\begin{aligned} \tau(v) &= \frac{\tau_0}{\sqrt{1 - (v^2/c^2)}} \\ &= \tau_0 \frac{E}{mc^2} \end{aligned} \tag{27.1}$$

where the relation between total energy and rest energy

$$E = \frac{mc^2}{\sqrt{1 - (v^2/c^2)}}$$

has been used. For a pion of 400 MeV kinetic energy, the factor $E/mc^2 = (K + mc^2)/mc^2$ is $(400 + 139.6)/139.6 = 3.9$, and it grows with increasing energy, so beams of energetic pions can be transported quite far.

The neutral pion has a mean life of only 0.76×10^{-16} sec, so even very energetic ones will not, on the average, get far before decaying. Also, neutral particles cannot be focused into beams, so reactions caused by π^0's must be observed with π's produced in the bubble chamber.

When sufficiently energetic pions enter a hydrogen or deuterium bubble chamber, the following four kinds of reactions are observed to occur frequently, so they must be initiated by a strong interaction:

$$\pi^- + \mathrm{p} \rightarrow \Lambda + \mathrm{K}^0$$
$$\mathrm{K}^0 \rightarrow \pi^+ + \pi^-$$
$$\Lambda \rightarrow \mathrm{p} + \pi^- \qquad (27.2)$$

$$\pi^+ + \mathrm{n} \rightarrow \Lambda + \mathrm{K}^+$$
$$\mathrm{K}^+ \rightarrow \pi^+ + \pi^0 \qquad (27.3)$$
$$\Lambda \rightarrow \mathrm{p} + \pi^-$$

$$\pi^+ + \mathrm{p} \rightarrow \Sigma^+ + \mathrm{K}^+$$
$$\mathrm{K}^+ \rightarrow \pi^+ + \pi^0 \qquad (27.4)$$
$$\Sigma^+ \rightarrow \mathrm{p} + \pi^0$$

$$\pi^- + \mathrm{p} \rightarrow \Sigma^0 + \mathrm{K}^0$$
$$\mathrm{K}^0 \rightarrow \pi^+ + \pi^- \qquad (27.5)$$
$$\Sigma^0 \rightarrow \mathrm{p} + \pi^-$$

Bubble chamber photographs of two of these events are shown in Figures 27.1 and 27.2. The initial particles and the final particles are already known to us. We want to study the unstable intermediate particles.

In Figure 27.1, illustrating the first of these reactions, the track of the incident π^- simply ends, but further along two V-shaped tracks appear signaling the creation of two pairs of charged particles. Measurement of the momenta of these pairs reveals that the momentum of each lies in the direction of the line joining the point of disappearance of the pion to the point of creation of the pair and that the total momentum

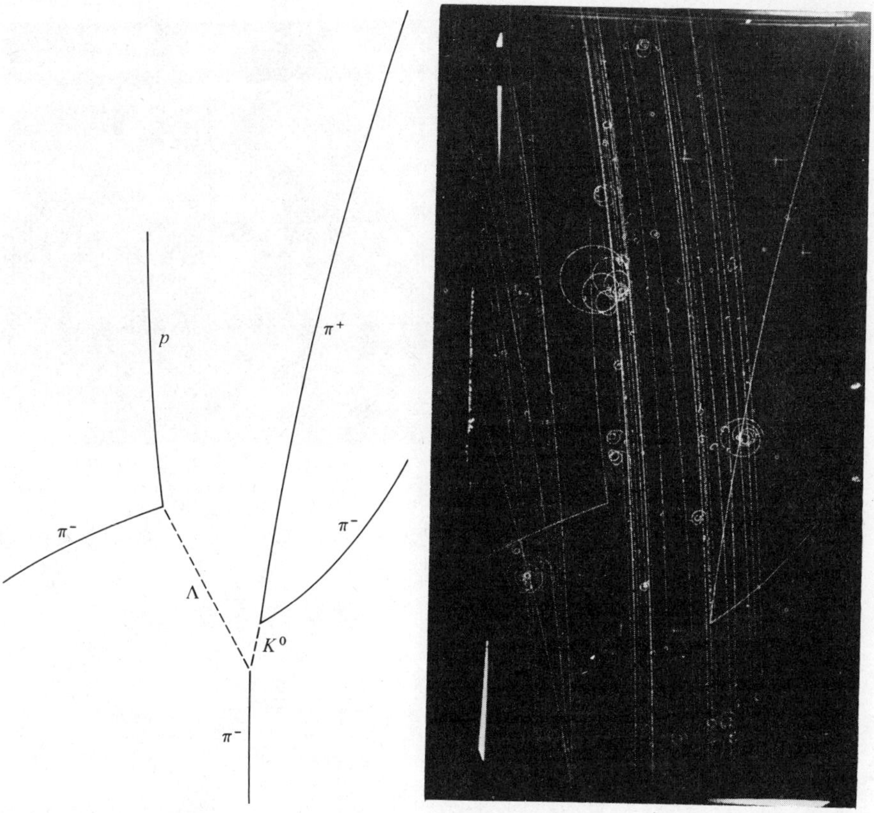

FIGURE 27.1 A negative pion strikes a proton at rest in a hydrogen bubble chamber producing two neutral particles, a K^0 and a Λ, both of which subsequently decaying into pairs of charged particles. (Brookhaven National Laboratory.)

of the two pairs is the momentum of the incident pion. This justifies the form assumed in Eq. 27.2. Two neutral particles are created in the π^-p collision, and these particles then decay. The V-tracks are not caused by another collision, because if they were the rest energy of the second target would have to appear somewhere; the total energy of the two pairs is just the total energy of the intial pion and the first target proton.

Let us look at these reactions in the light of the various conservation

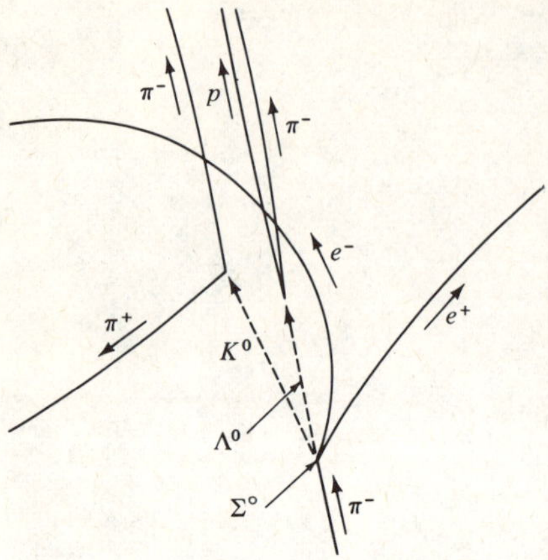

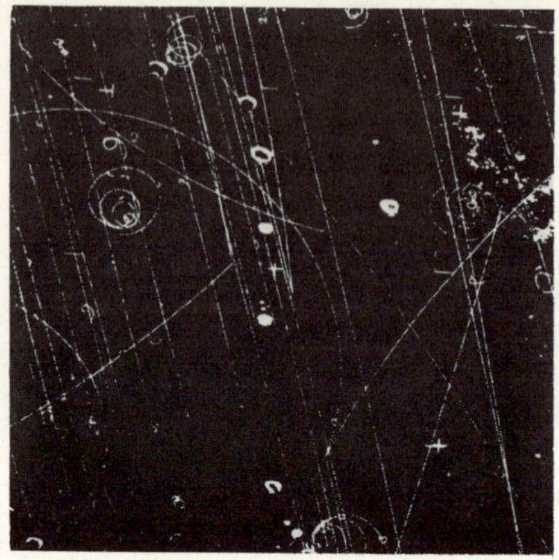

FIGURE 27.2 A π^- collides with a stationary proton in a hydrogen bubble chamber, producing the following sequence of events:

$$\pi^- + \mathrm{p} \rightarrow \Sigma^0 + \mathrm{K}^0$$
$$\mathrm{K}^0 \rightarrow \pi^- + \pi^+$$
$$\Sigma^0 \rightarrow \Lambda + \mathrm{e}^+ + \mathrm{e}^-$$
$$\Lambda \rightarrow \pi^- + \mathrm{p}$$

The decay of the Σ^0 into a Λ and an electron-positron pair is rather rare, occuring only about 5 times per thousand decays. Usually a photon is produced instead of the pair. (Ealing Corporation, Cambridge, Mass.)

principles that should apply to them. First, energy and momentum are conserved if the reactions proceed as written, though this is not obvious from casual inspection of the tracks. Next, charge is conserved if all the charged particles seen are assumed to have charges of magnitude e. Although this cannot be checked accurately in all cases, there is no reason to doubt that this is so. If the Λ and Σ's are baryons with $B = 1$, then baryon number is conserved in these reactions. They therefore are assigned baryon number $+1$, and the K's are assigned baryon number zero like the pions. The K's are mesons called either K-mesons or kaons.

Angular momentum conservation is harder to verify, but it can be done. According to the quantum rule that any component of an angular momentum can change in value only by integral multiples of Planck's constant $\hbar$, the Λ and the Σ's must have half-integer spin and the K's must have integer spin; this is seen by looking at the decays of these particles, remembering the nucleon has spin one-half and the pion has spin zero. The actual value of the spin is determined by the study of polarization effects in the decays. The analysis is complicated, but the result is simple: The Λ and the Σ's have spin one-half like the nucleons, and the K's have spin zero like the pions.

Even though all the conservation laws are valid in these processes, there is something queer about them. Why can a Λ be produced in the length of time it takes a pion to cross a proton, a time of something like 10^{-23} sec, and yet take so long to decay that during its mean life it can travel several centimeters? If it can be produced by a strong interaction, why can it not decay by means of that same strong interaction? The long mean life indicates that the decay is by the weak interaction, the same type that eventually causes a charged pion to decay into a muon and neutrino. No good explanation of this is known, but there is a relatively simple way to describe what happens: A new physical quantity is defined and is assumed to be conserved in all strong and electromagnetic processes, but not necessarily conserved in weak processes. The above production processes conserve this new quantity while the decay processes do not.

The new physical quantity introduced for the purpose of systematizing these reactions is called *strangeness* and is denoted by S. It was introduced into particle physics independently by M. Gell-Mann and by K. Nishijima in 1953. Like charge and baryon number it is assumed to be additive, so that the strangeness of a two-particle system is the sum of the strangenesses of its parts. The strangeness of the nucleons and the pions is taken to be zero, there being no need to introduce this concept for reactions in which only these particles appear. The Λ is arbitrarily assigned strangeness -1, and therefore the K^0 is assigned strangeness $+1$ to satisfy the conservation of strangeness in the production of the Λ and K^0 in Eq. 27.2. We now must see how to assign strangeness quantum numbers to other particles in a way that is consistent with experiment and with the new conservation principle.

Looking at the reaction in Eq. 27.3, we see that the total value of S initially is zero; the pion and the neutron both have $S = 0$. The Λ here has the same mass, charge, and mean life as the one met before, so it is the same kind of particle that is produced in the reaction Eq. 27.2 and has $S = -1$, which makes us assign to the positive kaon the value $S = +1$ like the K^0. None of the decays in these first two reactions conserves strangeness, so none of them can proceed by a strong interaction. Now we look at reaction Eq. 27.4, where again the initial value of S is zero. The Σ^+ is produced in association with the K^+, whose strangeness is already known to be $+1$, so the Σ^+ must have $S = -1$, like the Λ. Finally, we look at reaction Eq. 27.5 and recognize the neutral kaon produced, so we assign the Σ^0 the value $S = 1$ also. Another reaction that takes place frequently is seen in Figure 27.3. It is

$$\pi^- + p \rightarrow \Sigma^- + K^+ \tag{27.6}$$

which introduces another new particle, the Σ^-, which must have $S = -1$ because K^+ has strangeness $+1$.

There are now five types of particles that have been identified, the nucleons, the pions, the kaons, the Λ, and the Σ's. These are grouped in this way because it seems appropriate to do so. The masses of the two nucleons are not the same, but are very close, much closer to each other than to the mass of any other baryon. Similarly the pion masses

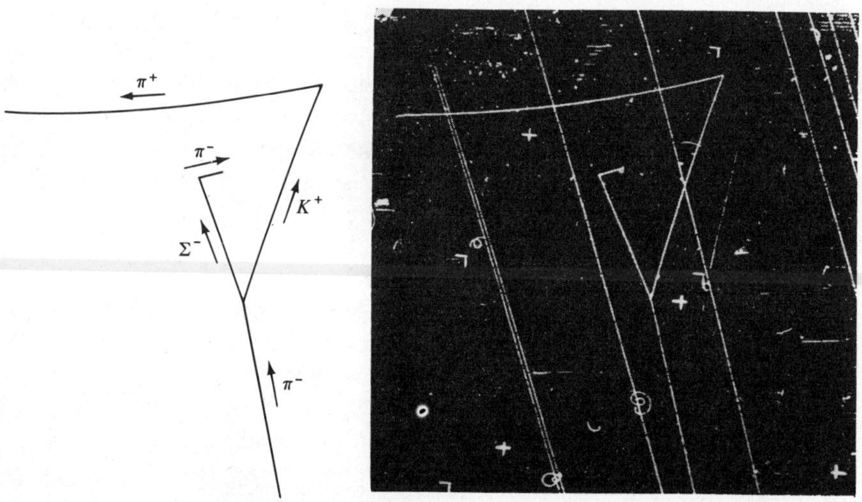

FIGURE 27.3 A Σ^- is produced by a π^- colliding with a proton in a hydrogen bubble chamber. The sequence of events is

$$\pi^- + p \rightarrow \Sigma^- + K^+$$
$K^+ \rightarrow \pi^+ + \pi^0$
$\Sigma^- \rightarrow \pi^- + n$

While neither the π^0 nor the neutron is observed, the decays are consistent with emission of particles with the π^0 mass and the neutron mass, and are inconsistent with emission of particles of different masses. (Ealing Corporation, Cambridge, Mass.)

are not all equal, but the two charged pions have the same mass and one not very different from that of the π^0. The masses of the three Σ's are nearly equal too, while the Λ stands alone. The positive and neutral kaons have nearly equal masses. All of these statements can be verified by looking at Table 27.1, where the properties of these particles are given.

TABLE 27.1
Properties and Quantum Numbers of Particles

Particle	Rest Mass MeV	Spin	Baryon No.	Isospin I, I_3	Strangeness	Mean Life sec	Principal Decay
γ	0	1	0	—	0	∞	—
ν	0	$\frac{1}{2}$	0	—	0	∞	—
e	0.511	$\frac{1}{2}$	0	—	0	∞	—
μ	105.7	$\frac{1}{2}$	0	—	0	2.20×10^{-6}	$e\nu\nu$
$\Pi^{\pm}$	139.6	0	0	$1, \pm 1$	0	2.60×10^{-8}	$\mu\nu$
Π^0	135.0	0	0	1, 0	0	0.76×10^{-16}	$\gamma\gamma$
$K^{\pm}$	493.8	0	0	$\frac{1}{2}, \pm\frac{1}{2}$	± 1	1.24×10^{-8}	$\mu\nu$
K^0	497.8	0	0	$\frac{1}{2}, -\frac{1}{2}$	1	50% 0.86×10^{-10} 50% 5.38×10^{-8}	$\Pi^+\Pi^-$ $\Pi e\nu$
p	938.3	$\frac{1}{2}$	1	$\frac{1}{2}, +\frac{1}{2}$	0	∞	—
n	939.6	$\frac{1}{2}$	1	$\frac{1}{2}, -\frac{1}{2}$	0	0.93×10^3	$pe^-\nu$
Λ	1115.6	$\frac{1}{2}$	1	0, 0	-1	2.51×10^{-10}	$p\Pi^-$
Σ^+	1189.4	$\frac{1}{2}$	1	1, 1	-1	0.80×10^{-10}	$p\Pi^0$
Σ^0	1192.5	$\frac{1}{2}$	1	1, 0	-1	$<10^{-14}$	$\Lambda\gamma$
Σ^-	1197.3	$\frac{1}{2}$	1	$1, -1$	-1	1.5×10^{-10}	$n\Pi^-$
Ξ^0	1314.7	$\frac{1}{2}$	1	$\frac{1}{2}, +\frac{1}{2}$	-2	3×10^{-10}	$\Lambda\Pi^0$
Ξ^-	1321.3	$\frac{1}{2}$	1	$\frac{1}{2}, -\frac{1}{2}$	-2	1.66×10^{-10}	$\Lambda\Pi^-$
Ω^-	1672.5	$\frac{1}{2}$	1	0, 0	-3	1.3×10^{-10}	$\Xi^0\Pi^-$

(Data from Particle Data Group, Lawrence Radiation Laboratory, Berkeley, Calif., July 1, 1970. By permission.)

The two nucleons and the three pions are distinguished among themselves by the value of I_3, the 3-component of the isospin. The nucleons have $I = \frac{1}{2}$ and possible values of I_3 of $+\frac{1}{2}$ and $-\frac{1}{2}$ corresponding to proton and neutron. The pions have $I = 1$ and possible values of I_3 of $+1$, 0, and -1 corresponding to π^+, π^0, and π^-. For these particles the value of the charge quantum number is given by the formula Eq. 26.8, which is

$$Q = e\left(\frac{B}{2} + I_3\right) \tag{27.7}$$

If the new particles are to fit into this scheme, it is natural to assign the Λ an isospin 0 since it has no charged counterparts, to assign the Σ's the value $I = 1$ since there are three charge states possible, and to assign the kaons $I = \frac{1}{2}$. This, however, does not satisfy Eq. 27.7 as it stands. The trouble can be fixed by modifying this equation, writing instead

$$Q = e\left(\frac{B + S}{2} + I_3\right) \tag{27.8}$$

This new form for the equation is the same as the old as far as pions and nucleons are concerned because for them $S = 0$. For the new particles it is easy to see that Eq. 27.8 is satisfied.

Each of the reactions given above serves to determine the strangeness of one of the particles appearing, so there has been no chance for a contradiction between "theory" and experiment. There are possible checks, among which is the nonreaction

$$\Lambda + p \nrightarrow n + p$$

which could happen in a collision between a Λ and a proton in the bubble chamber except for conservation of strangeness in strong and electromagnetic interactions. Another nonreaction forbidden only by strangeness conservation is

$$p + p \nrightarrow \Sigma^+ + \Sigma^+$$

This conserves charge, baryon number, and isospin but not strangeness, and has never been seen to happen.

Two more particles produced in proton-nucleon collisions belong in the group being discussed, the K^- and the $\overline{K^0}$. The K^- is a negative kaon, having the same mass as the K^+ and being the antiparticle of the K^+. It is produced with the K^+ in proton-nucleon collisions and therefore must have strangeness -1 in order to conserve strangeness in the pair production process. A confirmation of this assignment comes from the nonreaction

$$\pi^0 + n \nrightarrow \Sigma^+ + K^-$$

which would conserve strangeness if the K^- had the same value of S as the K^+. The $\overline{K^0}$ is the isospin mate of the K^- and is the antiparticle of the K^0 met before.

The production of negative kaons with $S = -1$ is very interesting, especially since enough of them can be made in a high energy synchrotron to form into a beam that can then produce reactions. Here the initial state has strangeness -1, and one might hope to get some new particles. The hope is fulfilled. For example,

$$\begin{array}{l} K^- + p \to \Xi^0 + K^0 \\ \qquad\qquad\quad \hookrightarrow \pi^+ + \pi^- \\ \qquad\quad \hookrightarrow \Lambda + \pi^0 \end{array} \tag{27.9}$$

The K^0 produced has strangeness $+1$, so the Ξ^0 has $S = -2$. This then decays with unit change of strangeness into a Λ and a pion. If the K^0 were a $\overline{K^0}$ with $S = -1$, the Ξ^0 would have $S = 0$ and need not decay into a Λ but could go directly into a nucleon and a pion, which it never does, so this assignment is excluded. It also appears from this that even weak decays involving a change of two units in S are forbidden.

According to the charge formula Eq. 27.8 applied to the Ξ^0, we have

$$0 = e\left(\frac{1-2}{2} + I_3\right)$$

so that $I_3 = \frac{1}{2}$. There should then be another Ξ particle, with $I_3 = -\frac{1}{2}$ and consequently with $Q = -e$, the Ξ^-. Such a particle has been observed in the reaction shown in Figure 27.4:

$$\begin{array}{l} K^- + p \to \Xi^- + K^0 + \pi^+ \\ \qquad\qquad\quad \hookrightarrow \pi^+ + \pi^- \\ \qquad\quad \hookrightarrow \Lambda + \pi^- \end{array} \tag{27.10}$$

which fits in very nicely. The nonreaction

$$\pi^- + n \not\to \Xi^- + K^+ + K^-$$

confirms the different strangeness of the K^+ and K^-.

There is one further particle that must be mentioned here, the Ω^-. This is a particle with $S = -3$ whose existence was hoped for on theoretical grounds. It was first seen in the spectacular event shown in Figure 27.5. The chain of reactions visible in the chamber is unusually long.

$$\begin{array}{l} K^- + p \to \Omega^- + K^0 + K^+ \\ \qquad\quad \hookrightarrow \Xi^0 + \pi^- \\ \qquad\qquad\quad \hookrightarrow \Lambda + \gamma^- + \gamma^- \\ \qquad\qquad\qquad\qquad\qquad\quad \hookrightarrow e^+ + e^- \\ \qquad\qquad\qquad\qquad\quad \hookrightarrow e^+ + e^- \\ \qquad\qquad\quad\quad \hookrightarrow p + \pi^- \end{array} \tag{27.11}$$

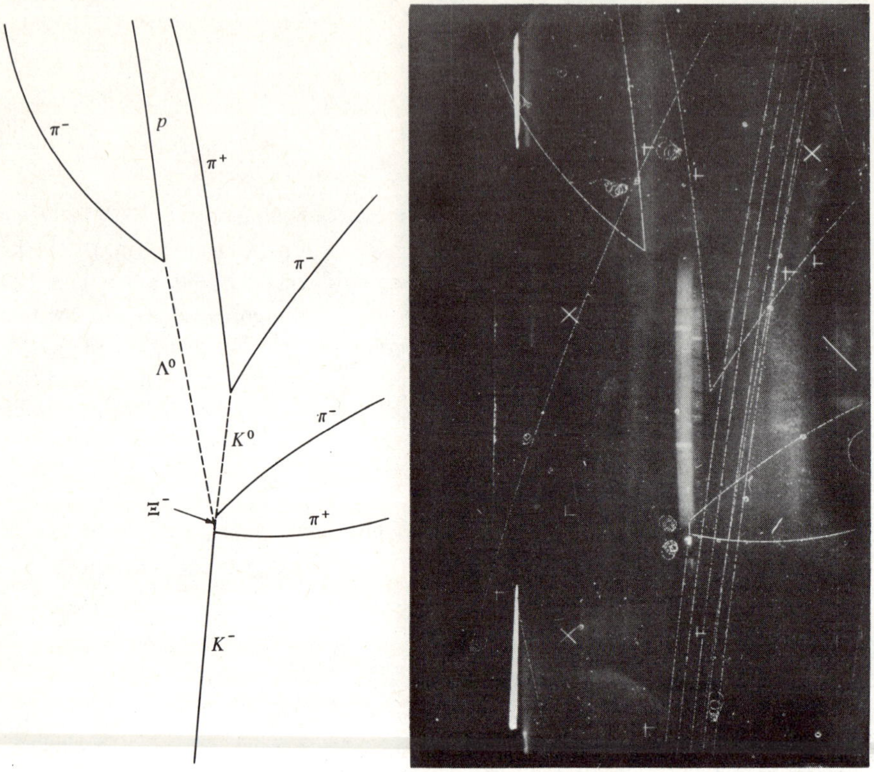

FIGURE 27.4 A negative kaon strikes a proton at rest in a hydrogen bubble chamber producing two strange particles, a Ξ^0 and a K^0, both of which then decay. (Brookhaven National Laboratory.)

The reader should check that in the production S is conserved if Ω^- has $S = -3$ and that in the weak decays occurring there is always a strangeness change of one unit. The pair production by the photons is not a decay process but involves a collision of the photon with a target.

The same theoretical ideas that led one to expect an Ω^- to exist also lead one to believe that there are no other baryons that can decay only weakly except, perhaps, some with masses considerably higher than that of the Ω^- and hence not producible with present-day accelerators.

The antiparticles corresponding to all the baryons and mesons met with here except the $\overline{\Omega^-}$ have been observed. The π^+ and π^- are a particle-

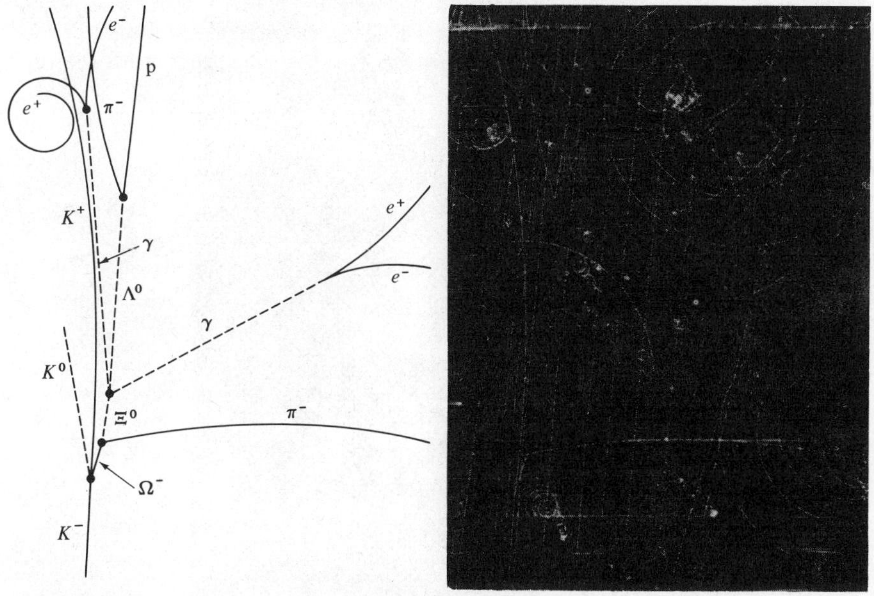

FIGURE 27.5 A negative kaon strikes a proton at rest in a hydrogen bubble chamber producing three strange particles including the particle Ω^- with strangeness —3. Several subsequent decays make the identification of the particles involved quite unambiguous. (Brookhaven National Laboratory.)

antiparticle pair, as are the K^+ and the K^- and the K^0 and $\overline{K^0}$. The π^0 is its own antiparticle. The antibaryons are produced in reactions initiated by antiprotons. These are produced in reactions such as

$$\mathrm{p} + \mathrm{p} \rightarrow \mathrm{p} + \mathrm{p} + \mathrm{p} + \bar{\mathrm{p}}$$

and can be made into beams. Figure 27.6 shows the reaction

$$\bar{\mathrm{p}} + \mathrm{p} \rightarrow \Xi^- + \overline{\Xi^-}$$

$$\overline{\Xi^-} \rightarrow \bar{\Lambda} + \pi^+, \quad \bar{\Lambda} \rightarrow \bar{\mathrm{p}} + \pi^+ \tag{27.12}$$

$$\Xi^- \rightarrow \Lambda + \pi^-, \quad \Lambda \rightarrow \mathrm{p} + \pi^-$$

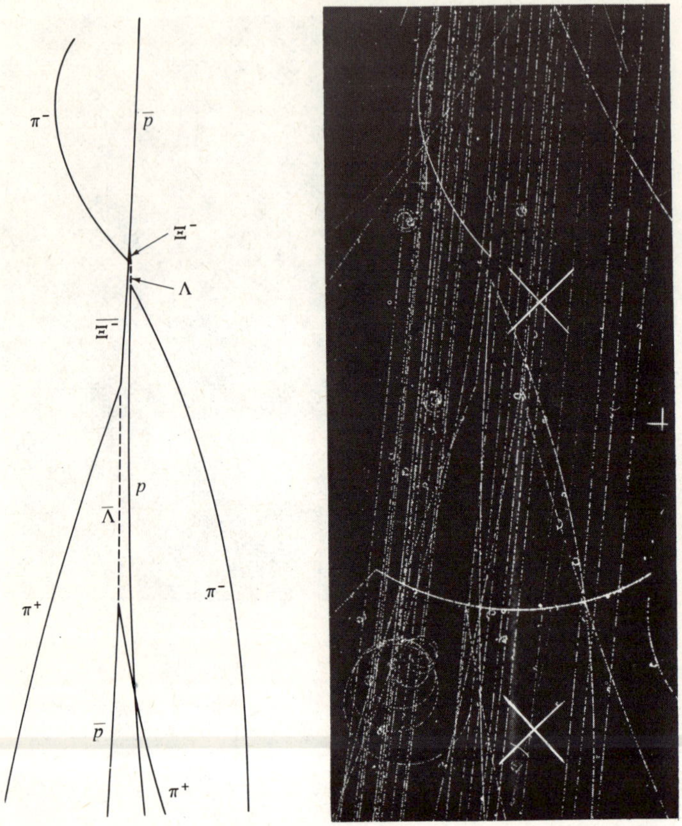

FIGURE 27.6 An antiproton strikes a proton at rest in a hydrogen bubble chamber producing a baryon-antibaryon pair, these being a Ξ^- and a $\overline{\Xi^-}$, each of which then decays into a Λ or $\bar{\Lambda}$ and then back into a p or $\bar{p}$. The net result of the whole event is the production of two pairs of pions. (Brookhaven National Laboratory.)

This confirms the idea that each particle has an antiparticle of equal mass and of opposite charge, baryon number, strangeness, and 3-component of isospin. The $\overline{\Omega^-}$ is expected to be very hard to produce, and its failure to appear so far is not an indication that the idea has broken down.

The main ideas that have emerged from the experiments described here can be rather briefly summarized. There are baryons and mesons that can be produced in association with one another by the strong interaction and that cannot decay separately by the strong or electromagnetic interaction because these two interactions conserve the new physical quantity, strangeness. Strangeness quantum numbers can be assigned to all the observed particles in such a way that all observed strong and electromagnetic interaction do conserve strangeness and that all observed weak interactions change it by unity at most. There are many relations between the particles into which we have not gone, mainly because the discussion would get too technical and detailed and partly because there is no good overall picture of what happens—no theory that encompasses a large range of phenomena. This is one reason why the area of high energy physics is one of intense activity both from the experimental and the theoretical sides.

Problems

1. Below is a list of reactions, some of which are possible and others not. Make a check list of conservation laws and see which of these reactions are possible via the strong interactions. Assume that reaching threshold energy is always possible.

   (a) $p + p \to p + p + p + \bar{p} + \pi^+ + \pi^-$
   (b) $p + p \to \pi^+ + \pi^-$
   (c) $p + p \to \Sigma^+ + \Sigma^-$
   (d) $p + p \to \Sigma^+ + K^+ + \pi^+ + \pi^-$
   (e) $K^- + p \to n + K^0$
   (f) $K^- + p \to \Lambda + K^+$
   (g) $K^- + p \to \Sigma^- + K^+$
   (h) $K^- + p \to \Xi^- + K^- + p$

2. Below is a list of decay processes. Indicate whether they can proceed via a weak interaction or not at all and what physical principles determine your answers.

   (a) $\pi^+ \to e^+ + \nu_e$
   (b) $n \to p + e^- + \bar{\nu}_e$
   (c) $p \to n + e^+ + \nu_e$
   (d) $\Sigma^+ \to p + \pi^0$
   (e) $\Lambda \to \Xi^- + e^-$
   (f) $\Lambda \to \Xi^- + e^+$
   (g) $\Xi^+ (= \overline{\Xi^-}) \to p + K^- + \pi^+$
   (h) $\Xi^- \to n + \pi^-$

3. In the reaction of Eq. 27.11 describing the production and decay of the Ω^-, verify that in each part of the process the rules outlined in this section and in earlier ones concerning the conservation of charge, baryon number, lepton number, strangeness, and isospin are obeyed where they are alleged to be.

28. THE WEAK INTERACTION

The existence of the weak interaction has been noted earlier in connection with the slow decay of particles such as the charged pions, kaons, muons, and others. Historically this interaction was first observed as the cause of the emission of beta-particles (electrons) by naturally occurring radioactive elements such as radium E, an isotope of bismuth. Even this early observation led to apparent paradoxes, and more recently new paradoxes have appeared, most of which have been resolved in a quite satisfactory manner. This section gives an account of these important recent developments.

When a naturally radioactive nucleus emits an electron, it gains one unit of charge and becomes an isotope of the next higher element in the periodic table. The mass, and therefore the energy, of both the initial and final nuclei seem to have definite values, but the kinetic energy of the emitted electron is not always the same. The momentum of this electron can be measured by observing the curvature of its path in a magnetic field in the usual way, and the distribution of the momenta of such electrons emitted from Rb^{86} (an isotope of rubidium) is shown in Figure 28.1. These momenta range from zero to a maximum value, so the energy carried away by the electron varies from $m_e c^2$ to a maximum. It is hard to see how energy can be conserved under these circumstances, since the energy carried away by the electron matches the energy difference between the initial and final nuclei only when the electron kinetic energy has its maximum value.

There is another difficulty involved in the process: If an electron is created in the act of beta-emission, while the number of nucleons in the nucleus remains fixed by baryon conservation, the spin angular momentum of the system changes by an amount $\frac{1}{2}\hbar$, while quantum mechanics allows only changes by an integral multiple of Planck's constant $\hbar$. Both of these difficulties are overcome if it is assumed that the creation of the electron is accompanied by the creation of another very light neutral particle of spin one-half. This way out of the energy paradox was proposed by Pauli, and a theory for the process was devel-

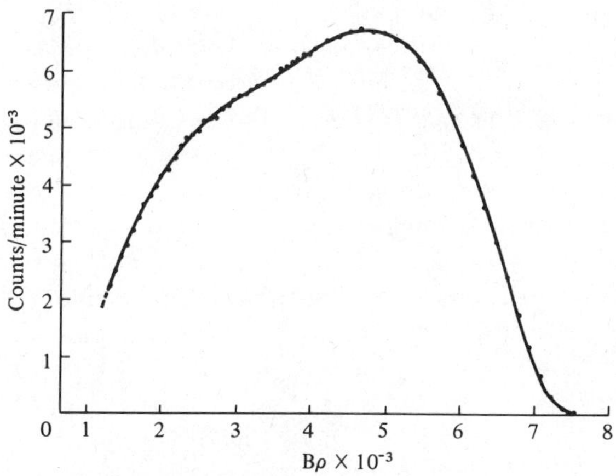

FIGURE 28.1 The distribution of momenta of electrons emitted by Rb^{86} is shown. The momentum is given in units of rB, the product of the radius of the path, r, in the magnetic field B. The kinetic energy corresponding to the maximum momentum plus the electron rest energy is the energy lost by the rubidium nucleus in the decay process. (After Zaffarano, Kern, and Mitchell. By permission.)

oped by Fermi on this basis, a theory that fits experimental observations very well.

The most fundamental example of beta-decay is that of the neutron. The decay process is written

$$n \rightarrow p + e^- + \overline{\nu}_e \tag{28.1}$$

where the neutron, proton, and electron are all well-known particles. The new, neutral particle emitted is called an *antineutrino*. Its mass is too small to be measured, and it is assumed to have no interaction with other particles except the weak interaction, which makes it extraordinarily difficult to detect and which, therefore, accounts for its not being detected in any usual experiment. One could have called the particle occurring in the neutron decay a "neutrino" rather than an "antineutrino," but calling it an antineutrino makes the formulation of a new conservation law to be described later in this section much simpler.

The decay Eq. 28.1 is that of a free neutron; the emission of electrons by radioactive nuclei involves one of the neutrons in the nucleus being transformed into a proton by this same process.

One is inclined to believe in the existence of a particle when its existence enables one to maintain the laws of energy conservation and of angular momentum conservation and to make a theory that describes the momentum distribution of the emitted electrons accurately, but this belief is greatly strengthened if one can observe reactions initiated by these particles. In order to do this it is necessary to have an enormous number of antineutrinos because the probability of any one of them producing a reaction is so small. A nuclear reactor provides a large number because in it the fuel such as U^{235} is converted into lighter elements in which the ratio of the number of neutrons to the number of protons is smaller. For U^{235} whose atomic number is 92 the ratio is

$$\frac{235 - 92}{92} = 1.56$$

and for a product nucleus such as tin, atomic number 50, it is

$$\frac{118 - 50}{50} = 1.36$$

This shows that in a nuclear reactor neutrons must be converted into protons, and antineutrinos are a product of this transformation.

C. L. Cowan and F. Reines were the first to observe reactions produced by these particles from a reactor. They looked for the reaction

$$\bar{\nu}_e + p \rightarrow n + e^+ \tag{28.2}$$

which should take place. The target protons were provided by the hydrogen in the water in a large tank near the reactor, and the occurrence of the reaction was detected in the following way: The positron formed annihilates with an atomic electron and produces two photons very promptly. The neutron is slowed down by collisions with protons in the water and is eventually absorbed in some cadmium, where it causes a nuclear reaction and the emission of more photons. The occurrence of two sets of photons with a short time interval between them is characteristic of the reaction Eq. 28.2 under the experimental circumstances, and just about the expected number of events were seen. This gives a very convincing proof that these particles really exist.

The antineutrinos detected by Cowan and Reines were of relatively low energy. The Fermi theory of weak interaction processes predicts that very energetic neutrinos are much more likely to produce events than are medium energy ones, so the best way to study neutrinos is to produce them at high energy and then try to detect events caused by them. There is available a rather good source of high energy neutrinos and antineutrinos, the decay of energetic pions. If the pion in the decay processes

$$\pi^+ \rightarrow \mu^+ + \nu_\mu$$

$$\pi^+ \rightarrow e^+ + \nu_e$$

is one of high momentum, the decay products will move in a direction almost parallel to the direction of motion of the original pion, so that a beam of charged pions is gradually transformed into a beam of muons and neutrinos, the conversion being more complete the farther the beam travels. This transformation takes a considerable distance to occur because of the time dilation effect mentioned earlier. The undecayed pions and the charged decay products μ^+ and e^+ can be filtered out of the beam by passing it through a sufficient thickness of iron, and then one is left with an essentially pure neutrino beam.

An experiment using such a neutrino beam was done at Brookhaven and at CERN in Geneva and has been repeated several times. It is known that when π^+'s decay, 10,000 μ^+ are created for every e^+ that appears; this means that in the neutrino beam there should be 10^4 ν_μ for every ν_e if, indeed, there is any difference between these two neutrinos. The object of the first experiment was to see whether they are alike or are distinct. If they are alike, when very energetic neutrinos strike nucleons they should produce muons and electrons in roughly equal numbers. If they are distinct, one would expect the ν_μ to produce only muons and the ν_e to produce only electrons in such collisions.

In the Brookhaven experiment the neutrino beam was filtered through 13.5 m (44 feet) of iron to remove the penetrating muons and then went into a spark chamber. A plan view of the Brookhaven neutrino experiment is shown in Figure 28.2. Any neutrino event producing charged particles would trigger the chamber, and the resulting tracks were recorded. A bubble chamber would not do here because it cannot be triggered in time to show tracks, and random pictures would almost certainly not reveal any neutrino event; these were detected at a rate of about 4 per 24-hour day. The beam consisted, as has been said, almost entirely of neutrinos associated with muons, and in the spark chamber only the creation of muons was seen, so the conclusion was reached that the two kinds of neutrinos really are distinct. There are, then, two types of leptons: the muon type and the electron type.

There is a law of baryon conservation according to which the total baryon number of a system is conserved under all kinds of interactions. There are two laws of lepton conservation, one for muon type leptons

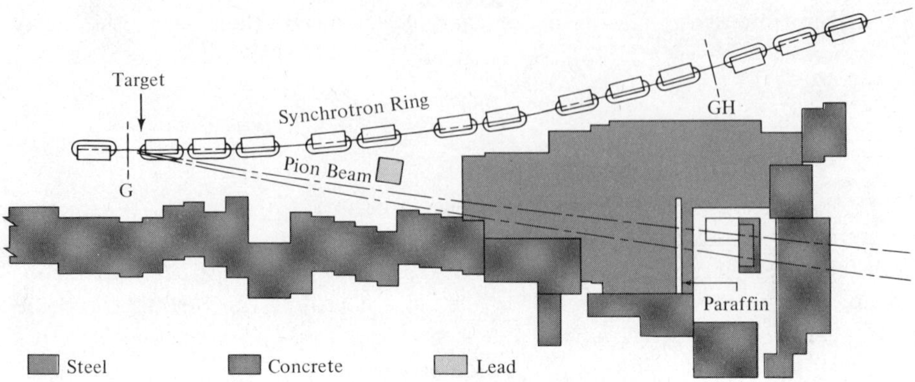

FIGURE 28.2 A plan view of the Brookhaven neutrino experiment is seen. The synchrotron ring is at the top of the figure. The pions produced by protons striking the target have 21 meters in which to decay into muons and neutrinos before striking the 13.5 meter iron wall. The neutrinos, which easily penetrate the wall, then enter the spark chamber-counter array. (Brookhaven National Laboratory)

and one for electron type leptons. These are less interesting than the baryon conservation law because there are only two known muon type leptons, the μ^- and ν_μ, and their antiparticles, μ^+ and $\overline{\nu_\mu}$, while there are many different baryons. The electron type leptons are also restricted to the e^- and ν_e together with their antiparticles the e^+ and $\overline{\nu_e}$. Speculations have been made about other leptons, but none have been observed up to now. All the reactions involving the creation or transformation of leptons written down are consistent with the conservation of these two kinds of leptons separately, as the reader may verify.

Study of the weak interaction has produced one major surprise: It used to be believed that nature was completely symmetrical in its fundamental laws between right and left. This is not in contradiction with the existence of more right-handed than left-handed people on earth; as far as we know this latter is a purely accidental circumstance. The evidence for right-left symmetry of the fundamental laws of physics is extremely good if only the strong and electromagnetic interactions are considered. It is the electromagnetic interaction that is responsible for nearly all chemical properties of matter, so the laws of chemistry are symmetric between right and left. This symmetry or its absence shows up whenever it is necessary to introduce a right-hand rule in a physical statement, as in defining the direction of an angular momentum vector or in defining the direction of the magnetic field around a current. Whenever the symmetry exists, the rule has to be applied twice to find a physically observable effect such as the force between two currents or the direction of tilt of a top acted on by a force, so that the replacement of all right-hand rules by left-hand rules would make no difference as far as descriptions of observable effects are concerned. Only intermediate stages of the discussions would be altered. Discovery of an observable physical effect that requires applying a right-hand rule only once in its description demonstrates the absence of symmetry between right and left in that system.

The first experiment to show that there is not complete symmetry between left and right in processes caused by the weak interaction was done by C. S. Wu and her co-workers at the National Bureau of Standards. The idea of the experiment was to line up the spins of the nuclei of an electron emitting radioactive substance and then to see if the electron is more probably emitted in the direction of the spins or in the opposite direction. The spin direction is defined by a right-hand rule, but the direction of travel of an electron requires no such rule for its definition, so the statement that the direction of emission is more probably parallel to the spin than antiparallel to it involves the right-hand rule only once. A suitable substance to use is an isotope of cobalt, Co^{60}, which is used in medicine as a source of gamma rays for cancer therapy. The medically important gamma ray is emitted very shortly after the beta particle of interest to us here. The cobalt spins can be polarized at very low temperature with the help of a magnetic field. The experimental arrangement is shown schematically in Fig. 28.3. The

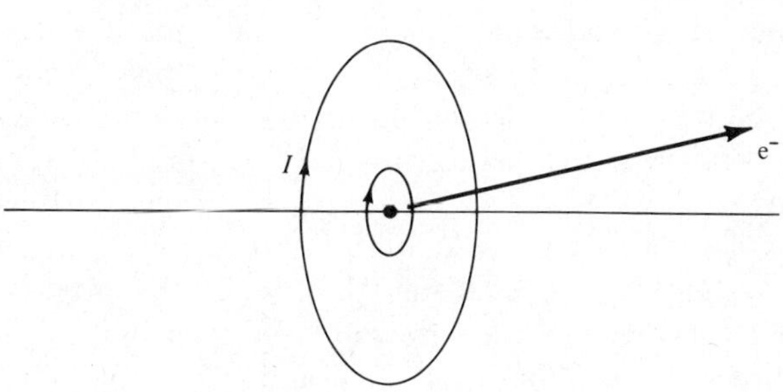

FIGURE 28.3 A $cobalt^{60}$ nucleus at the center of a loop of wire tends to align itself so that its rotation is in the direction of the current in the loop. This situation is left-right symmetric. The nucleus emits electrons which come out more frequently on one side of the loop than on the other, which is not symmetric between left and right.

Co^{60} nucleus is shown at the center of a current-carrying loop of wire; this current provides the magnetic field that causes the nuclear "compass needle" to line up. As illustrated, everything in the setup has left-right symmetry: The two sides of the wire loop are the same, and the nucleus is "spinning" around an axis perpendicular to the plane of the loop. Despite this symmetry, the electron emitted by the nucleus is observed to come out more often to the right in the figure than to the left. This can be described by saying that the electrons come out preferentially in a direction opposite to the direction of the nucleon spin. Because this spin is defined by a right-hand rule, and because the electron's direction is independent of a right-hand rule, one can state a physically observable property of this system using the right-hand rule only once, and there is no symmetry between right and left in this system. Nothing like this arises in events caused by the strong or electromagnetic interactions. It has been looked for intensively and never found.

The operation of interchanging right and left is called reflection, or the parity operation, and the symmetry of nature between right and left can be formulated as a conservation principle, the conservation of parity. The weak interaction does not conserve parity, which is merely a technically convenient way of saying that it distinguishes between right and left.

A consequence of the nonconservation of parity observed is that neutrinos are emitted polarized; experiments somewhat too elaborate to be described here show that the neutrino always has its spin pointing in the direction opposite to its momentum and that the antineutrino has its spin in the direction of its momentum. One says that the neutrino is left-handed; the antineutrino is right-handed. Such a statement makes sense only for particles of zero rest mass. Suppose that someone said, "All electrons are left-handed." Then one could take an electron moving with speed v along the positive x-axis of an inertial coordinate system and would see its spin vector pointing in the negative x-direction, as shown in Fig. 28.4(a). One could then look at this same electron from an inertial coordinate system traveling faster than v in the original system, as in (b). From this system the electron would appear to be spinning in the same direction as before, but it would now be moving in the negative x-direction, so its spin and momentum would be parallel and the electron would be right-handed. A particle of zero rest mass always moves with speed c, and it is impossible to get a coordinate system moving fast enough to pass such a particle, as the coordinate system introduced above passed the electron, so it is impossible to reverse the handedness of a massless particle in this way.

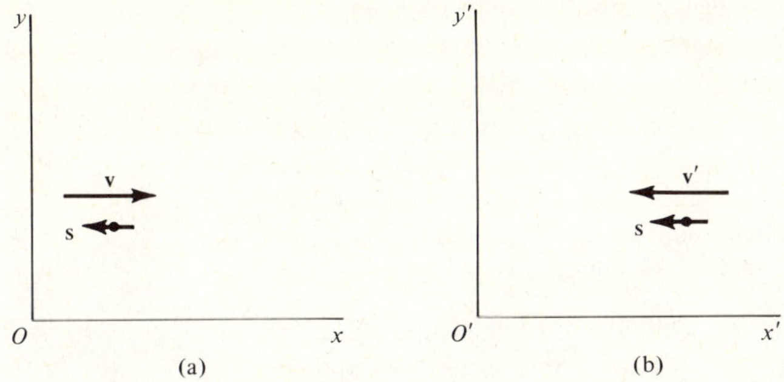

FIGURE 28.4 In (a) an electron with its spin s pointing in the negative x-direction is traveling to the right in the positive x-direction. If this same electron is looked at from a reference frame moving to the right faster than the electron as in (b), it will be seen to be moving to the left. Thus a particle which looks "left-handed" in one reference frame can look "right-handed" in another. Only when the particle's mass is zero so its speed is c is this not possible.

The nonconservation of parity first made its appearance in the decay of the neutral kaons, but was not recognized as such because parity conservation was believed in thoroughly. The basis for this belief was very firm, but it was restricted to effects produced by the strong and electromagnetic interactions, which even yet are observed to conserve parity. T. D. Lee and C. N. Yang were the first to point to the lack of experimental evidence for parity conservation in weak interaction events, and they suggested the experiment showing the violation. It now appears that more than a violation of parity conservation occurs in the decay of the neutral kaons, but this situation is still too obscure for us to try to describe it here.

Problems

1. From the mass difference of the neutron and proton find the range of possible neutrino energies in the neutron beta decay. (The center of mass of the system after decay may be taken to be the proton for the purpose of this problem, so while the proton may have some momentum in the rest frame of the neutron, its kinetic energy is negligible, since $K = p^2/2m$, and p is small while m is large.)
2. Find the threshold energy for the antineutrino in reaction Eq. 28.2, including of course the neutron-proton mass difference.

3. Suppose a pion at rest decays into a muon and a neutrino that move along the y-axis of a reference frame. Find the values of p_y for the two decay products. Find the x-components of the momenta of these two particles in a reference frame in which the pion is moving in the $+x$-direction with an energy $E = 20\ m_\pi c^2$. Sketch the directions of these particles in this reference frame. (This illustrates the statement made earlier that when fast pions decay the decay products move almost parallel to the pion's motion.)
4. Sketch the mirror image of Figure 28.3 including the spin vector of the cobalt nucleus. Note that if in the original figure the electron momentum has a negative component in the direction of the nuclear spin, in the mirror image it has a positive component in this direction. Nonconservation of parity occurs because electrons are observed to be more likely to have a negative than a positive component in this direction.

CHAPTER 6 Waves and Relativity

29. LIGHT WAVES AND OTHER WAVES

Light has played an important part more than once in our discussion of the physics of particles. First it appeared in an indirect way, its speed being a limiting speed for any massive particle. Actually in Section 22, where this aspect entered, all that was done was to say that the limiting speed of a particle is the speed of light. One could conceive of a limiting speed for particles that had nothing whatever to do with the speed of light, but experimentally the two are equal. It was in the next section, Section 23, that some connection between the limiting speed for particles and the speed of light was established. It was shown there that particles of zero rest mass must move with speed c, and if light is thought to consist (in some sense at least) of a stream of massless particles, then naturally the speed of these particles is the speed of light. The particle aspect of light also was shown to make a great deal of sense in the processes where the members of a particle-antiparticle pair annihilate each other to generate photons and in the process of electron-position pair creation used to detect energetic photons.

The study of light, however, did not begin with the study of processes involving energetic photons, or indeed with the processes of the emission and absorption of light by anything. Rather it began with the study of the way light travels from one place to another, and this study of the propagation of light leads to a very different view of it than does the study of its emission and absorption. The view arrived at through a study of the propagation of light is that light is a wave phenomenon. In order to understand this wave aspect of light we must say something about waves in general, using very simple mechanical examples to make the ideas clear. First we describe waves that travel along a stretched string. Then we say something about surface waves on water, where the phenomena of interference can be clearly illustrated. Finally the experiment is described that

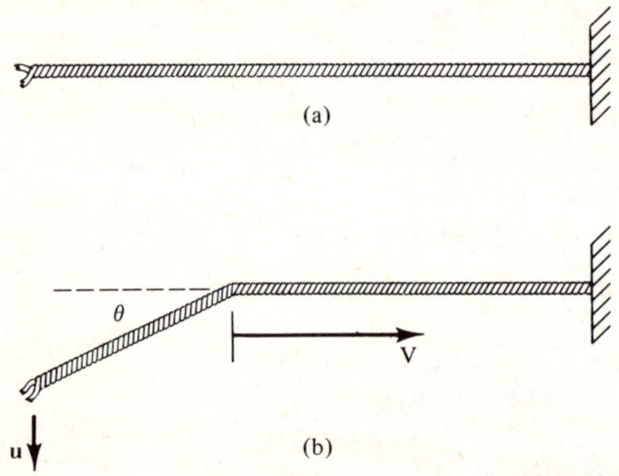

FIGURE 29.1 (a) A string under tension and with both ends fixed is straight, and remains straight. (b) If one end of the string in (a) is started downward with velocity u, the string is no longer straight but forms an angle. The vertex of this angle travels with speed V along the string.

showed that light exhibits interference effects and therefore must involve waves. This crucial experiment was done by Thomas Young in 1801.

The simplest type of wave is that which occurs when a tightly stretched string is pulled away from its equilibrium form of a straight line. Let us imagine that we have a very long, uniform string with one end fastened to a rigid support and the other end held in one hand as in Figure 29.1(a). The string is pulled so that there is a tension T in it and the string is initially at rest.

Let us now imagine that the end of the string is moved downward, perpendicular to the length of the string, with a constant speed u starting at time $t = 0$. The string no longer is straight, but at a time $t > 0$ it has the form shown in Figure 29.1(b). It consists of two segments separated by a corner where the direction of the string changes. The segment away from the moving hand is at rest as it was before; the segment near the moving hand is straight but makes an angle θ with the undisturbed segment and is moving down with speed u. The junction point of the two segments moves to the right with speed V.

It is not hard to understand what is going on. A string is flexible, so it can exert a force only along itself. In order to accelerate any short segment of the string sideways, the string must undergo a change of direction at that segment so that there will be an unbalanced sidewise force on the segment. No part of the string will start to move until it forms the corner between the two nonparallel segments. This corner travels along the string with a definite speed V. What travels with speed V is the disturbance in the string; no part of the string itself need

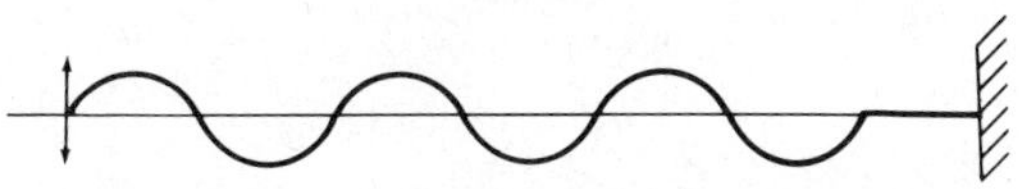

FIGURE 29.2 The form of a stretched string whose end has executed three cycles of harmonic motion is shown. There are three cycles of a sine curve present, the rest of the string not yet being disturbed and therefore straight.

move with this speed. A traveling disturbance is called a *wave*, so here we have a wave traveling down the string with speed V.

By applying Newton's second law of motion to each short segment of the string, thereby regarding each such segment as a particle to which Newton's laws apply, it is not hard to show that the wave speed V depends only on the tension T in the string and on the mass per unit length of the string μ. The derivation is not given here because it is not of sufficient importance for our purpose; the result is

$$V = \sqrt{\frac{T}{\mu}} \tag{29.1}$$

The lighter and tighter the string, the faster a wave travels along it. This wave speed is quite independent of how the end of the string is moved. It can be moved up and down with harmonic motion, for example. If it is, after three cycles of the motion the string will have the shape shown in Figure 29.2, assuming that the string is so long that the disturbance will not reach the end in this time.

The kind of wave illustrated in Figure 29.2 is an especially important one. It is a periodic wave because every point on the string moves up and down with harmonic motion of the same frequency ν as the end where the wave originates. During one period of the motion, $\tau(= 1/\nu)$, the disturbance represented by a crest of the wave travels a distance λ given by

$$\lambda = V\tau = \frac{V}{\nu} \tag{29.2}$$

and during this time the motion of any point on the string will have gone through one complete cycle of its harmonic motion, so λ is the distance between two points on the string that are moving in the same way at the same time. λ is called the *wavelength* of the wave. Equation 29.2 can be written as

$$\lambda\nu = V \tag{29.3}$$

Another example of a wave that is often seen is a surface wave on water. In equilibrium the surface of water is flat because of the force of gravity, but if such a flat surface is disturbed at a point, ripples spread out from that point as center. These ripples are due to a predominantly up and down motion of the water, but the disturbance itself spreads out horizontally over the surface of the water. As in a string, the wave speed V is independent of the amplitude of waves. The simplest form of surface waves is that of the waves coming from a point where a harmonic disturbance is created, as by having a pointed object move up and down with harmonic motion while just dipping into the water. Periodic waves move out in circles, the radius of a circle where the water level is raised (a crest) increasing uniformly with the time according to

$$r = Vt \tag{29.4}$$

A photograph of such a wave is shown in Figure 29.3. Here again the wavelength of the wave is the distance between adjacent crests or troughs.

A very important wave effect, called *interference,* occurs when two sources of waves are operating on the same water surface at the same time. Figure 29.4 illustrates the situation being considered. S_1 is a source of harmonic waves of wavelength λ, and S_2 is another source of waves of the same wavelength. The two sources are arranged so that they

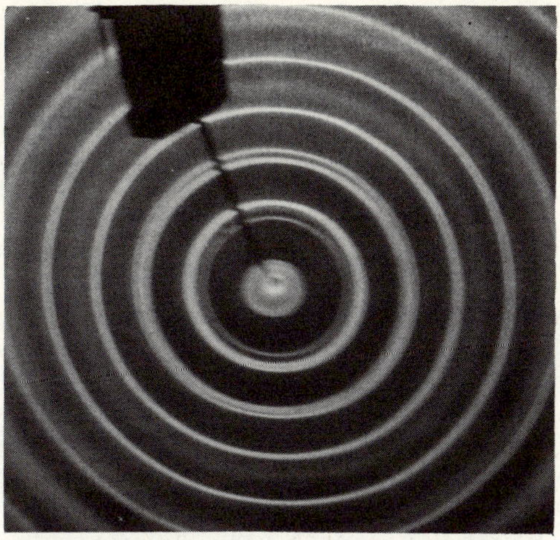

FIGURE 29.3 Surface waves on water are generated by a vibrating point touching the surface. They travel out from the source in expanding circles. (From *PSSC Physics*, D. C. Heath, Lexington, Mass., 1965.)

move up and down together, crests leaving the two of them simultaneously. Such sources are said to be *in phase* with each other. We ask how the water surface moves at the point P.

If only the source S_1 were operating, the water at P would move up and down with the frequency ν of the source, with an amplitude

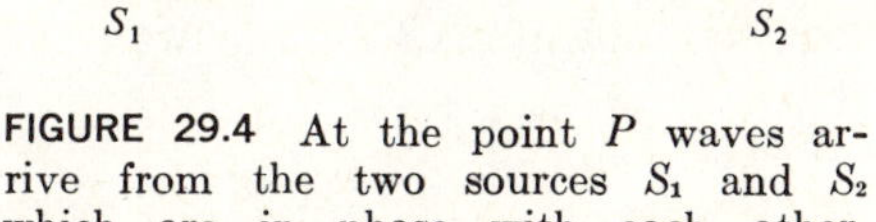

FIGURE 29.4 At the point P waves arrive from the two sources S_1 and S_2 which are in phase with each other. Whether the interference at P is constructive or destructive depends on the difference in the number of wavelengths between P and S_1 and between P and S_2.

that depends on the strength of the source S_1 and on the distance of P from S_1. If this distance r_1 is a whole number of wavelengths, then there is a crest at P at the same time there is a crest at S_1, or the wave at P is in phase with the source S_1. If this distance is a whole number of wavelength plus an extra half wavelength, then there is a trough at P when there is a crest S_1 and the wave at P is *out of phase* with S_1.

If only the source S_2 is operating, exactly similar statements about the wave at P can be made as were made about the wave from S_1. If r_2 is an integral number of wavelengths, the wave at P is in phase with S_2; and so on.

If both S_1 and S_2 are operating, then the disturbance at P is the sum of the disturbances coming from S_1 and S_2 separately. This is a fundamental characteristic of wave motion known as the *principle of superposition*. When two waves are present simultaneously, the net disturbance is the sum of the disturbances coming from the waves separately. This has an interesting consequence when P is a whole number of wavelengths from S_1 and a whole number plus an extra half wavelength from S_2. Recalling that we arranged the sources to vibrate in phase so that crests leave the two at the same time, at P a crest from S_1 will arrive at the same time that a trough arrives from S_2, and the disturbance will be *smaller* than either separate wave acting alone would produce. If the two amplitudes are of equal magnitude at P, the disturbance there will be zero, the rise of the surface due to the crest of one wave being canceled by the fall of the surface due to the trough of the second wave. This mutual annuling of two waves is called *destructive interference*. If the amplitudes are not equal, the interference is not completely destructive and some disturbance is present at P. If P is a whole number of wavelengths from both sources, the two disturbances will be in phase there and will add, so we have *constructive interference*. Figure 29.5 is a photograph of a water surface with two sources acting. The places where constructive interference and destructive interference are located can be easily distinguished.

When the two sources are in phase, the condition for constructive or destructive interference may be written down in a simple way:

Constructive interference $\quad r_1 - r_2 = n\lambda$

Destructive interference $\quad r_1 - r_2 = (n + \frac{1}{2})\lambda \qquad (29.5)$

with n an integer. The examples given above are special cases of these conditions where r_1 itself is an integer multiple of λ.

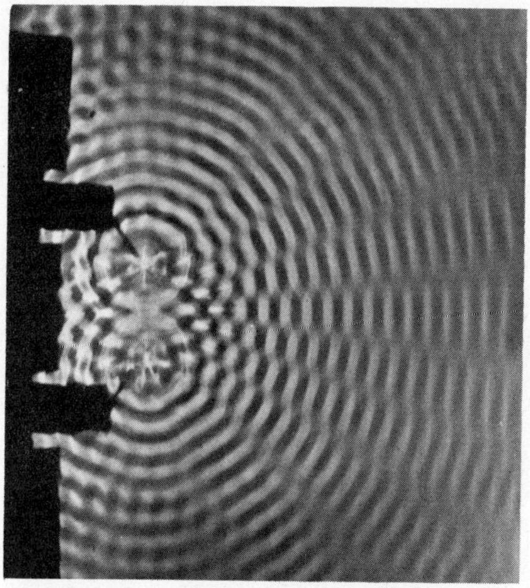

FIGURE 29.5 Surface waves on water generated by two sources vibrating in phase interfere with each other. The interference is constructive where the light and dark bands are most sharply defined, and is destructive where definition is lacking. (From *PSSC Physics*, D. C. Heath, Lexington, Mass., 1965.)

The wave nature of light was clearly demonstrated by Young in 1801 by the use of an optical version of the two-source experiment described above. This experiment consisted of illuminating two very small holes in an opaque screen with light from a single source and then looking at the pattern of illumination on a screen further from the light. The apparatus is sketched in Figure 29.6. What he saw on the second screen was a series of light and dark bands that can be interpreted only as results of interference, alternately constructive and destructive. A photograph of the illumination is shown in Figure 29.7. In this photograph Young's lamp is replaced with a laser, resulting in many more visible fringes than Young could obtain, but this does not alter the idea or the interpretation.

Young's experiment not only demonstrated the wave nature of light through making the interference effects visible but also gave an estimate of the wavelength of light. The separation of the light and dark fringes depends on the wavelength of the light λ, on the separation of the holes

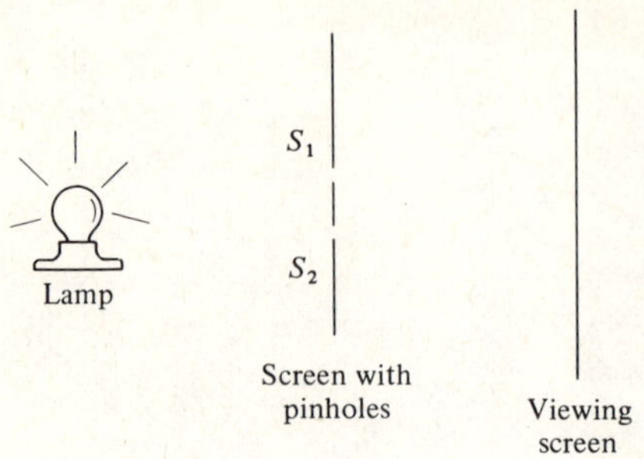

FIGURE 29.6 Young's experiment is done with a lamp illuminating two pinholes or slits in an opaque screen. The pinholes act like the two sources in Figure 29.4, and an interference pattern is seen on the viewing screen.

d, and on the distance between the screens *D*. The connection between these quantities can be seen in Figure 29.8. The two angles marked θ in the figure are equal because their arms are perpendicular in pairs. At *P*, equally distant from S_1 and S_2, the interference is constructive and the screen is bright. At *P′*, where the distance from S_2 is a half wavelength longer than the distance from S_1, the interference is destructive and the screen is dark. If *x* is the separation of *P′* and *P*, then

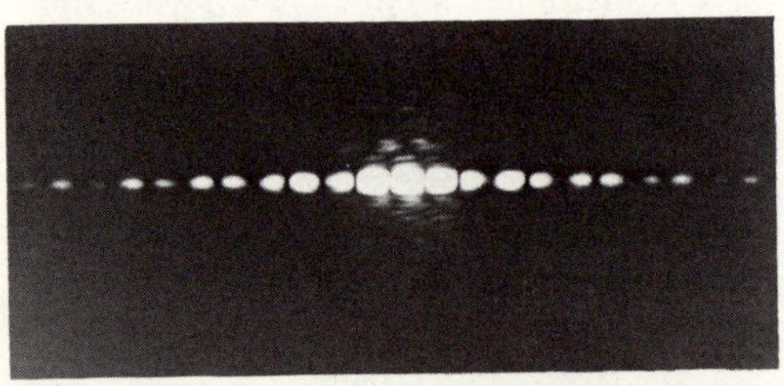

FIGURE 29.7 The interference pattern produced on a screen by two parallel slits illuminated from the other side by a laser is shown. The center of the photograph is overexposed in order to make the bright spots far to the sides visible. Constructive interference with a pathlength difference of ten wavelengths is clearly visible. A longer exposure would make many more bright spots visible. (J. N. McGruer and P. Stehle.)

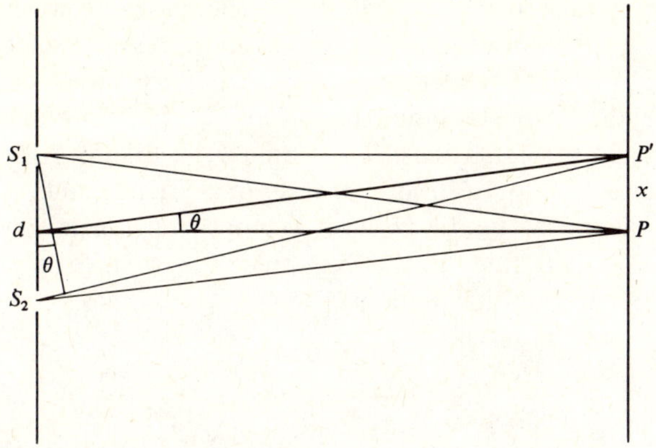

FIGURE 29.8 If light from the two sources starts out in phase, at the symmetric point P there will be constructive interference. There will be constructive interference at the point P' provided that the path-length difference from P' to the two sources is an integral number of wavelengths. This difference is seen to be $d \sin \theta$ where d is the source separation and θ is the angle shown.

one has

$$\theta = \frac{x}{D} = \frac{\lambda/2}{d} \tag{29.6}$$

to a very good approximation when d is much smaller than D. Thus

$$\lambda = \frac{2 \times d}{D} \tag{29.7}$$

and all quantities on the right can be measured. Young did not use the light of a single wavelength, so his interference fringes were not clear, light of different colors having different wavelengths and, therefore, producing fringes at different separations x.

A more accurate determination of the wavelength of light can be made with a *diffraction grating.* This is a device very much like Young's screen with two pinholes; it consists of a series of equally spaced slits in a screen. The advantage in using many slits rather than just two is that constructive interference takes place over a much smaller range of angles θ, so that the angle at which the interference is completely constructive can be measured more accurately with a grating than with two holes or two slits.

The values of λ corresponding to visible light lie in the range between about 3.5×10^{-5} and 7×10^{-5} cm, the short wavelength being violet and the longer wavelength red. These wavelengths are very short on the length scale of everyday objects, and this accounts for the fact that light seems to travel in straight lines. The angle θ in Figure 29.8, the angle between the direction of a bright interference fringe and a dark one, can be taken as an estimate of how much the propagation of light deviates from a straight line. If the two holes are 1 cm apart, for light of wavelength 5×10^{-5} cm this angle is

$$\begin{aligned} \theta &= \frac{\lambda}{2d} = \frac{5 \times 10^{-5}}{2} \\ &= 2.5 \times 10^{-5} \text{ rad} \qquad (29.8) \\ &= 1.43 \times 10^{-3} \text{ degree} \\ &= 5.2 \text{ sec} \end{aligned}$$

which is a very small angle indeed—hardly noticeable without careful observation. Young had to use a considerably smaller separation of his pinholes than this to observe interference.

The speed of light can be measured by measuring the time it takes light to cover a known distance, usually over a path from the source to a mirror and back to a point near the source. A pulse of light is started out at a known time, and the time of its arrival back at the starting point is noted. In this way the speed in vacuum is found to be 2.9979×10^{10} cm/sec. An experiment like this does not reveal what kind of thing moves with this speed. It could be a wave or it could be a stream of photons. In fact what moves with this speed is light, which has both particle and wave aspects but is not entirely either.

Problems

1. Two sources of circular waves of wavelength λ are located at a distance of 5λ from each other. If the sources are in phase with one another, in how many different directions can there be constructive interference? In how many different directions can there be destructive interference?
2. One way of discovering that there are two holes in a screen is to find an interference pattern characteristic of two-hole interference on one side of the screen when the other side is illuminated with light of wavelength λ. Discuss what happens to the interference patterns as the two holes get closer together, and show that it washes out completely when they get to within about a wavelength of each other.

(This is connected with the general fact that to distinguish objects by use of light, their separation must be greater than the wavelength of the light used.)

3. Sodium emits yellow light of wavelength 5.890×10^{-5} cm and of wavelength 5.896×10^{-15} cm. If the first of these is to be diffracted through an angle of 30° by a pair of holes as in Young's experiment, what must the hole separation be? Will light of the second wavelength be diffracted through a larger or a smaller angle?

30. LIGHT WAVES AND RELATIVITY

Waves in a string, surface waves on water, and sound waves in air are disturbances in material media whose laws of motion are Newton's laws, so these waves can be studied by applying the principles of particle mechanics. The concept "wave" arose in connection with such waves, and for a long time waves in a mechanical medium were the only ones considered to be possible. By 1815 the wave nature of light was firmly established both experimentally and theoretically, so a material medium, the ether, was postulated as the medium in which light waves propagate. One naturally wanted to find out as much as possible about this medium, which has to have some rather remarkable properties. For example, interstellar space must be filled with it because we receive light from the stars, and yet the planets must be able to move through it without resistance because they have been moving around the sun throughout the long history of the solar system. Aside from these difficulties, there were others of a more detailed sort; no mechanical model of the ether has ever been made that allows for the propagation of waves like light waves without at the same time allowing for the propagation of other kinds of waves not observed in nature.

Perhaps the discovery that made the ether theory useless for the description of optical phenomena was that motion of the earth through the ether cannot be detected by optical experiments. We saw in Section 22 that the motion of charged particles with speeds approaching that of light involves new effects that we have successfully described in a way not involving a preferred reference frame in which the limiting speed c has to be measured. If such a reference frame had been found, one would not be surprised if the ether of optics were at rest in it. Historically the physics of fast particles developed much later than the physics of optical experiments. Light always travels with speed c relative to something even when emitted by a candle, while making particles move with speeds near that of light requires elaborate equipment. Here some of the optical evidence for the nonexistence of an ether will be described. The picture one is led to is the same theory of relativity that appeared in Section 22, but a rather different view of the theory emerges from this approach.

Suppose that an ether exists and that light is a disturbance in it, moving through it with speed c. If the earth moves through the ether with velocity $\mathbf{v}$, the speed of light as measured on the earth should

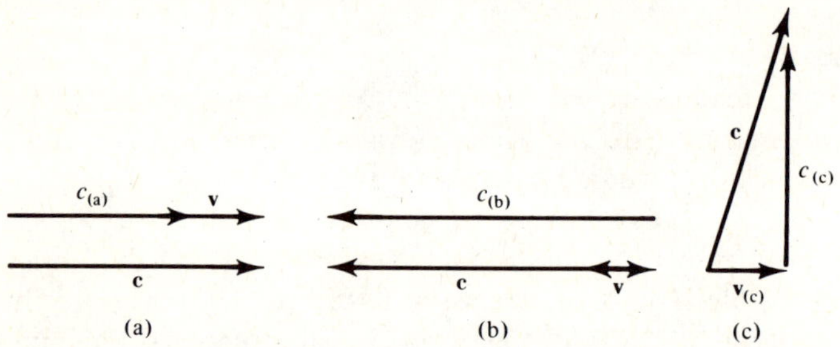

FIGURE 30.1 The earth moves through the "ether" with velocity $\mathbf{v}$ and light moves through it with velocity $\mathbf{c}$. The velocity of light relative to the earth then depends on the relative directions of the two velocities. In (a) they are parallel, so the relative speed is the difference of the two speeds. In (b) they are antiparallel, so the relative speed is the sum of the two speeds. In (c) the relative velocity is perpendicular to the velocity of the earth through the "ether."

depend on the direction the light travels. Three possible situations are shown in Figure 30.1: In (a) the earth and the light are traveling parallel to each other through the ether; in (b) they are traveling antiparallel; and in (c) they are traveling so that the velocity of light relative to the earth is at right angles to the earth's velocity $\mathbf{v}$. If a measurement of the speed of light is made in a laboratory on the earth, the results to be expected in these three situations are, since the vector velocities are expected to add,

$$
\begin{aligned}
&\text{(a)} \quad c_a = c - v \\
&\text{(b)} \quad c_b = c + v \qquad (30.1) \\
&\text{(c)} \quad c_c = \sqrt{c^2 - v^2}
\end{aligned}
$$

This is a prediction that can be submitted to experimental test.

The speed of the earth through the ether is not known, of course, so what value should be used for v? A reasonable lower limit to put on it is the speed of the earth in its motion around the sun, about 3×10^6 cm/sec. If this is an overestimate at one time, it should be an underestimate 6 months later when the direction of the earth's velocity relative to the sun has reversed. An experiment that can detect a speed such that $v/c = 10^{-4}$ should give a result if the fundamental ideas used in designing the experiment are right.

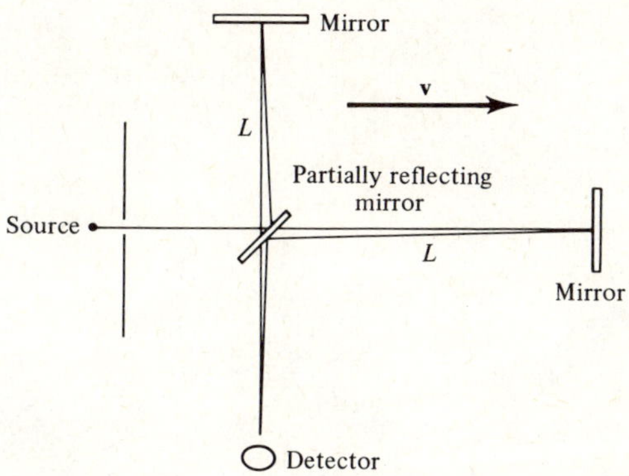

FIGURE 30.2 The Michelson interferometer consists of a partially reflecting mirror that splits a beam of light from the source, sending one part out and back along each of the two arms. Both beams illuminate the detector, and they combine with constructive interference or destructive interference, depending on the difference in the lengths of the two paths followed.

The most famous and direct check on this prediction was made by A. A. Michelson and E. Morley using a Michelson interferometer. In this instrument a beam of light from a source S is split in two by a partially reflecting and partially transmitting mirror, as shown in Figure 30.2. The two resulting beams travel at right angles to each other and are reflected back on themselves by two other, completely reflecting mirrors. These two reflected beams combine to give an interference pattern at D. There is an interference maximum if crests from the two beams arrive simultaneously at D and a minimum if a crest meets a trough. Whether a crest of one wave meets a crest or a trough of the other wave depends on how long it takes for light to follow the two possible paths.

In the experiment of Michelson and Morley, the two paths were made of equal lengths L. If the earth's velocity is in the direction shown in the figure, one beam travels first parallel and then antiparallel to **v** and the other travels perpendicular to **v**. The time intervals for the two beams to travel from the partially reflecting mirror to the other mirrors and back to the partially reflecting mirror are different by an amount that depends on the ratio of the speed v to the speed of light c.

1. Parallel-antiparallel beam

$$\begin{aligned} t_{\text{par}} &= \frac{L}{c_a} + \frac{L}{c_b} \\ &= \frac{L}{c - v} + \frac{L}{c + v} \qquad (30.2) \\ &= \frac{2Lc}{c^2 - v^2} \\ &= \frac{2L}{c} \cdot \frac{1}{1 - (v^2/c^2)} \end{aligned}$$

2. Perpendicular beam

$$\begin{aligned} t_{\text{perp}} &= \frac{2L}{c_c} \\ &= \frac{2L}{\sqrt{c^2 - v^2}} \qquad (30.3) \\ &= \frac{2L}{c} \frac{1}{\sqrt{1 - (v^2/c^2)}} \end{aligned}$$

The difference lies in the factors containing $1 - (v^2/c^2)$, this appearing once without and once with the square root sign.

The time difference for the two beams is therefore

$$\begin{aligned} \Delta t &= t_{\text{par}} - t_{\text{perp}} \\ &= \frac{2L}{c}\left[\frac{1}{1 - (v^2/c^2)} - \frac{1}{\sqrt{1 - (v^2/c^2)}}\right] \end{aligned}$$

which can be evaluated for any value of v/c. To see how big this is and how it depends on the value of v/c it is useful to use the binomial expansion introduced in Section 22, which is very good when v/c is much less than unity. This approximation can be written

$$(1 + x)^n \approx 1 + nx$$

for any value of n as long as x is small. If x is not very small, the approximation can be improved by adding more terms on the right. This is applied here with $x = v^2/c^2$ and $n = -1$ or $-\frac{1}{2}$. Thus

$$\frac{1}{1 - (v^2/c^2)} = \left(1 - \frac{v^2}{c^2}\right)^{-1} \approx 1 + \frac{v^2}{c}$$

$$\frac{1}{\sqrt{1 - (v^2/c^2)}} = \left(1 - \frac{v^2}{c^2}\right)^{-\frac{1}{2}} \approx 1 + \frac{1}{2}\frac{v^2}{c^2}$$

Using this in the formula for Δt simplifies it to

$$\begin{aligned} \Delta t &= \frac{2L}{c} \cdot \frac{1}{2}\frac{v^2}{c^2} \\ &= \frac{L}{c} \cdot \frac{v^2}{c^2} \end{aligned} \tag{30.4}$$

For v/c about 10^{-4} as expected here, $v^2/c^2 \approx 10^{-8}$ or one in 100 million, so the assumption that v^2/c^2 is small is surely justified.

When the time difference Δt is an integral number of periods of vibration of the light used, the interference at D should be constructive. If Δt changes, the interference changes from destructive to constructive and back to destructive as Δt changes by one period τ of the light. This shows why the interferometer is so well suited to detect minute changes in Δt; the period τ of light with $\lambda = 6 \times 10^{-5}$ cm (orange light) is only 2×10^{-15} sec, while the time needed for light to travel the distance $2L = 1100$ cm used by Michelson and Morley is 3.7×10^{-8} sec. A change of one whole period is, therefore, a change of about 1 part in 2×10^7, not much larger than the change 1 part in 10^8 being looked for. Δt can be changed simply by rotating the whole apparatus slowly about an axis perpendicular to $\mathbf{v}$. When the apparatus has turned through 90°, the arm for which the light transit time was shorter becomes that for which the transit time is longer and vice versa, effectively changing Δt into $-\Delta t$ and so producing a change $2\,\Delta t$, so the interference should slowly change from completely destructive to less than completely destructive as the turning progresses. If a change of one period occurred so that $2\,\Delta t = \tau$, the interference would change from destructive to constructive and back to destructive, and we would have

$$\lambda = c\tau = 2c\,\Delta t$$

Using the expression Eq. 30.4 for Δt and replacing τ by λ/c, where λ is the wavelength of the light, this becomes

$$\lambda = 2c \times \frac{L}{c} \times \frac{v^2}{c^2} \tag{30.5}$$

$$\frac{v^2}{c^2} = \frac{\lambda}{2L}$$

Now the length L of the arm of the interferometer can be made very many wavelengths. If L is 11 meters, as used by Michelson and Morley, and λ is 6×10^{-5} cm, one cycle of change of the interference would correspond to a value

$$\frac{v^2}{c^2} = \frac{\lambda}{2L}$$

$$= 2.7 \times 10^{-8}$$

or

$$\frac{v}{c} = 1.6 \times 10^{-4}$$

Actually a change $2\,\Delta t$ as small as $0.04\,\tau$ could have been detected in the exeriment, so that the smallest expected value of v/c should have led to an easily visible change in the interference pattern. The result of the experiment was that no shift could be observed and thus no value of v/c distinguishable from zero could be found. This was a surprising and disturbing outcome.

We already know something about how this is to be described. The null result of the ether-drift experiments is the same as the observation made in Section 22, namely, that the limiting speed of a particle c, which is the speed of light, is the same in all inertial reference systems. Velocities must be added in accordance with the Einstein addition law rather than simply as ordinary vectors. Looking at these matters from the point of view of waves rather than that of particle motion is enlightening, and we do this here briefly.

If a pulse of light is emitted from the origin of an inertial reference system S at time zero, at a later time t the disturbance will be located on a spherical shell with O as center and with radius ct. The cartesian coordinates x, y, z, of a point on the wave front will be related to the time by

$$x^2 + y^2 + z^2 = c^2t^2 \tag{30.6}$$

which is nothing but the statement that the square of the radius of the sphere is the square of ct.

If this same pulse of light is viewed from a different inertial coordinate system S' whose axes are parallel to those of S and whose origin O' coincides with O at time zero, but not before or after this time, the result of the Michelson-Morley experiment tells us that this same pulse is seen to travel in all directions with the speed c, so in this coordinate system at time t' the light pulse will be located on the surface of a sphere of radius ct'. The coordinates x', y', z' of a point on this sphere are related to t′ by

$$x'^2 + y'^2 + z'^2 = c^2t'^2 \tag{30.7}$$

How is it possible that two observers who are at the same place when a light pulse is emitted and who then travel apart at some constamt speed v later both observe the light wave front to be a sphere with themselves at the center, as seen in Figure 30.3? This would be geo-

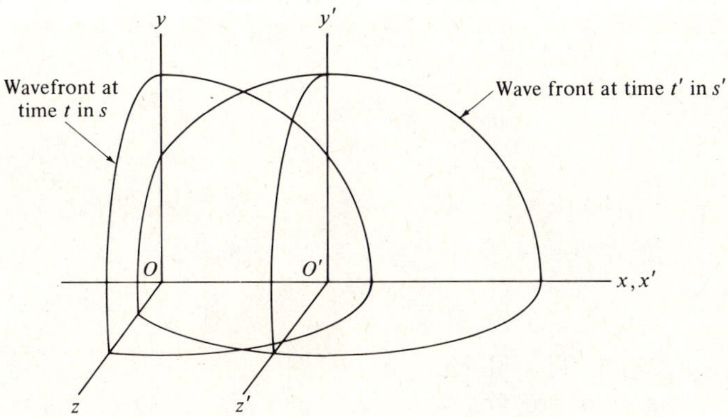

FIGURE 30.3 An observer in reference frame S sees the wave front of a light pulse spread out spherically, and an observer in frame S' sees the same light pulse also spreading out spherically. The wavefront cannot be two spheres at the same time, so the two observers must measure time in their frames with clocks that differ from each other.

metrically impossible if the two observers meant the same thing when they specify the time at which they locate the wave front. If the two observers' ideas of when two events are simultaneous do not coincide, this paradoxical result can be accounted for.

The condition Eqs. 31.8 and 31.9 can be written in the form

$$c^2t^2 - (x^2 + y^2 + z^2) = 0$$
$$c^2t'^2 - (x'^2 + y'^2 + z'^2) = 0 \tag{30.8}$$

which implies that the value of this combination of time and coordinates has the same value, zero, in all inertial coordinate systems whose origins coincide at time zero. In relativistic particle mechanics we saw that the quantity

$$E^2 - p^2c^2 = m_0^2c^4$$

has the same value in all inertial coordinate systems and that this could be accounted for by saying that the energy E and momentum $\mathbf{p}$ of a particle in one such system are related to the energy E' and momentum $\mathbf{p}'$ of that particle seen from another such system by a Lorentz transformation, Eq. 22.15 or 22.17. If we assume that the time t and the coordinates $\mathbf{r}$ corresponding to a point on the trajectory of a particle are related to the time t' and coordinates $\mathbf{r}'$ of that same point seen from another coordinate system S' by a Lorentz transformation, then things work out in agreement with observation. We therefore write

$$\begin{aligned} t' &= \frac{t - (v/c^2)x}{\sqrt{1 - (v^2/c^2)}} \\ x' &= \frac{x - vt}{\sqrt{1 - (v^2/c^2)}} \\ y' &= y \\ z' &= z \end{aligned} \tag{30.9}$$

in analogy with Eq. 22.15. The placing of factors of c differs because the dimensions of E and $\mathbf{p}$ are not the same, nor are those of t and r. If in Eq. 22.15 E were replaced by E/c, which has the same dimensions as p, then E/c, $\mathbf{p}$ transform in exactly the same way that ct, $\mathbf{r}$ do. The transformation Eq. 31.9 makes the origin of S', $\mathbf{r}' = 0$ have the coordinates $x = vt$, $y = 0$, and $z = 0$ in S, so S' is moving with speed v along the x-axis of S.

It is now easy to see that if $c^2t^2 - r^2 = 0$, so is $c^2t'^2 - r'^2$. It is also easy to check that when v/c is negligible compared to unity these equations give the transformation one would expect from everyday low speed experience where no question of different rates of clocks appears so that $t = t'$ and where the usual vector addition of velocities is applicable. Under these circumstances we know that if S' is moving uniformly to the right along the x-axis of S with speed v, the coordinates of a point P at rest in S are related to its coordinates in S' by

$$x' = x - vt$$

$$y' = y$$

$$z' = z \tag{30.10}$$

The application of the Lorentz transformation to the time and the coordinates of a point on the trajectory of a particle or on a wave front of a light wave does not contradict everyday experience, but it does cause a disruption of some firmly held ideas about things. Two of these are shown here, the Lorentz contraction and the time dilation.

If a rod of length L is at rest along the x-axis in the reference system S, what is its apparent length as seen by an observer at rest in S'? First one must be clear about the meaning of the question. To measure the length of a moving object, one must record the position of the two ends at the same time. In S' the rod appears to be moving to the left with speed v, as shown in Figure 30.4. At time $t' = 0$ the two origins coincide, so at this time the left end of the rod is at the origin in both S' and S. At $t' = 0$, where is the right end? For the right end, $x = L$, so the first of Eqs. 31.9 says that when the right is located by the observer in S', $t = vL/c^2$, since $t' = 0$. Inserting $x = L$ and this value of t in the second equation of Eqs. 31.9 then gives

$$x' = \frac{L - (v^2/c^2)L}{\sqrt{1 - (v^2/c^2)}}$$

$$= L\sqrt{1 - (v^2/c^2)} \tag{30.11}$$

which is less than L by the factor

$$\sqrt{1 - (v^2/c^2)}$$

This is known as the Lorentz contraction.

The reader should verify that if the rod lies along the y-axis of S, its length as seen from S' is again L and is not contracted. The Lorentz

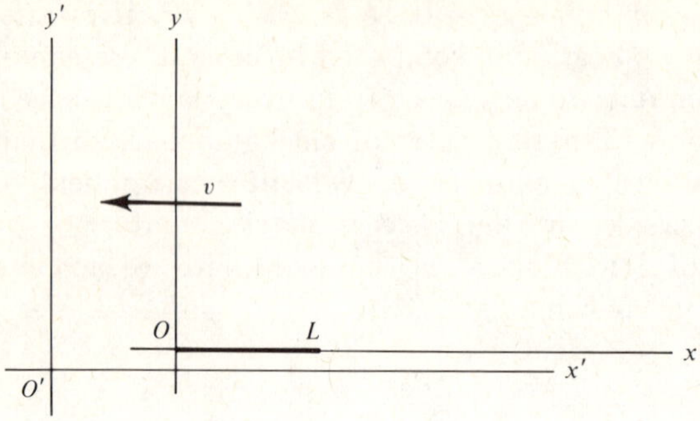

FIGURE 30.4 A rod of length L as measured in reference frame S in which the rod is at rest, appears to an observer in S' to be moving to the left with speed v. To this observer the rod appears shorter than L. This happens because according to the observer in S, the observer in S' looks at the position of the right end of the rod later than he looks at the left end, and in the interval between these times the rod moves.

contraction occurs only in the direction of motion, and there is no contraction perpendicular to the direction of motion.

The "cause" of the Lorentz contraction is the fact that according to an observer in S, the measurer in S' has not looked at the two ends of the rod at the same time but has looked at the left end at time $t = 0$ and the right end at time $t = vL/c^2$. The observer in S' maintains, however, that according to *his* time both locations were made at $t' = 0$. It was Einstein who gave the first penetrating analysis of the measuring of time and length measurements in moving coordinate systems and who showed the profound meaning of the Lorentz transformations for our ideas of space and time.

Another worthwhile exercise for the reader is to see that if the rod were at rest along the x'-axis of S', an observer in S would see it contracted by the same Lorentz factor. The relation between S and S' is completely reciprocal.

Now let us consider two clocks constructed in exactly the same way. One clock is at rest at the origin of S and the other at the origin of S' and both are set to zero when the two origins coincide. When the clock in S is seen by an observer in S to read T, what does the clock in S' read? Things have been arranged so that Eqs. 31.9 hold; when t, $\mathbf{r}$ vanish so do t', $\mathbf{r}'$, as Eqs. 31.9 require. Now when $t = T$, the position of the origin O' in S is at $x = vT$, so we put $t = T$, $x = vT$ into the first of Eqs. 31.9 and obtain

$$t' = \frac{T - (v^2/c^2)T}{\sqrt{1 - (v^2/c^2)}}$$

$$= T\sqrt{1 - v^2/c^2} \tag{30.12}$$

which is less than T by the factor

$$\sqrt{1 - (v^2/c^2)}$$

The moving clock seems to run slower than a similarly constructed stationary clock. Thus the mean life of a moving radioactive atom is greater than that of a stationary atom as measured by a stationary clock.

This result is again symmetrical between S and S'. A clock stationary in S is moving in S' and appears to an observer in S' to run slower than a clock stationary in S'.

The time dilation effect leads to what is called the "twin paradox." A popular form of this is the following: One of a pair of identical twins remains on earth while the other makes an interplanetary flight at a speed approaching that of light and returns to earth. He will appear to be younger than the twin who remained at home because the clock by which he ages runs more slowly than the clock of the nontraveler. The "paradox" arises from the apparent reciprocity between the twins; as seen from the traveler the nontraveler's clock runs more slowly, so the nontraveler should be younger. This reciprocity is not present, however, because the traveler is not at rest in an inertial frame all the time he is traveling because he must turn around and come back. There is a vast literature on this subject extending from the original statement of the paradox to the present.

The relative youth of traveling particles compared to stationary particles has been established experimentally. High energy muons are kept in a storage ring, consisting of a magnetic field and a vacuum chamber in which the muons move in circles. The number of muons passing a fixed point on the ring decreases with time as the muons decay, the mean life of such a muon being

$$\frac{1}{\sqrt{1 - (v^2/c^2)}}$$

times as great as that of a muon at rest outside the storage ring. The high energy muon corresponds exactly to the traveling twin and the muon at rest to the nontraveling twin. This experiment was done by F. J. M. Farley and co-workers at CERN in Geneva. Little ground for further controversy on this point seems to be left, but no doubt there will continue to be resistance to the conclusion that one can stay younger by traveling because it violates a preconception most of us have or had.

Problems

1. Someone standing on a bridge drops stones at the rate of two per second into the water below. Suppose the resulting circular waves spread out with a speed of 200 cm/sec (about $7\frac{1}{2}$ mph) if the water underneath the bridge is still. Sketch the form of the waves that would result if the water were flowing at (a) 100 cm/sec and (b) 300 cm/sec. The latter case is similar to that of a boat moving through the water at a speed greater than the wave velocity and leaving a V-shaped bow wave. Sonic booms are related phenomena occurring when airplanes move through the air faster than sound waves.
2. By making the water correspond to the ether and the person on the bridge correspond to the experimenter on the earth, show that the situation in (a) of Problem 1 is similar to that predicted by the ether theory. In what way would the situation (b) contradict the theory of relativity?
3. Two particles move along the x-axis of a coordinate system, each with speed $\frac{3}{4}c$, toward each other. What is the *relative* speed of these two particles as seen by an observer in this reference frame? That is, at what rate is the distance between them as measured by this observer changing with his time? Is this result in conflict with the general principles of relativity theory?
4. What is the relative speed of the two particles described in the previous problem as determined by an observer moving along with one of the particles?
5. If Michelson and Morley had been able to detect a change in transit time Δt in their interferometer of nothing smaller than a half period, how long would they have had to make the instrument in order to detect a motion with the orbital speed of the earth around the sun?
6. Suppose there is a truck of length L and height h that can park under an overpass and just be hidden from view from far above when parked in the right place.
   (a) Show that when the truck is moving, one might expect the truck to be hidden from the far-up observer for a

time Δt given by

$$\Delta t = \frac{L - L\sqrt{1 - (v^2/c^2)}}{v} \approx \frac{Lv}{2c^2}$$

The first numerator is just the amount by which the moving truck is shorter than the overpass.

(b) Neglecting the Lorentz contraction for the moment, show that light coming from the bottom of the back of the truck as it disappears under the overpass arrives at the far-up observer later than light from the top of the front of the truck when it emerges, which is when the back disappears if the contraction is neglected, by the amount $\Delta t'$ given by

$$\Delta t' = \frac{h}{c}$$

(c) For small enough values of v it is clear that $\Delta t' > \Delta t$. Show this means that the truck, when moving but not going too fast, will never be completely hidden from the observer by the overpass even though the Lorentz contraction takes place. Note that the *top* of the truck will be hidden for the time Δt but not the entire truck.

31. ELECTROMAGNETIC WAVES

Light waves have been shown not to be mechanical waves in any ordinary medium; the experimental fact that the state of motion of the medium cannot be detected rules this out. This leaves the question: What is the nature of light waves? Of what do they consist and how can their generation be described and how do they affect ordinary matter? The answer to these questions is given by the theory of electromagnetic fields as developed especially by J. C. Maxwell in the middle of one. One of the things the theory of relativity changed was the concept of electric and magnetic fields. Maxwell imagined these as states of stress in the ether, while now they are regarded as basic quantities that do not have to be described in still more elementary terms. The reason for this change of attitude is that these fields can be described very accurately and quite simply by themselves, while no mechanical ether has ever given either an accurate or a simple description of these fields nor has an ether theory suggested new phenomena to look for that have been found. Electric and magnetic fields are assumed to be able to exist in a vacuum, a region free of all matter.

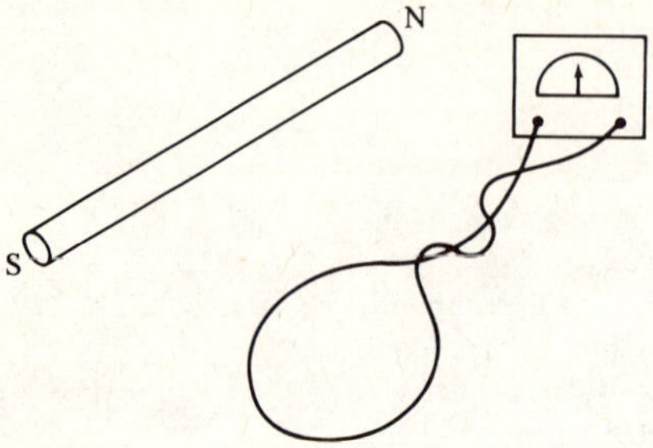

FIGURE 31.1 While either the magnet or the wire loop is in motion the galvanometer shows a deflection, indicating the flow of current through it. If both are moving, but in a way that involves no relative motion of one with respect to the other, there is no deflection.

In earlier sections electric fields produced by stationary charges and magnetic fields produced by steady electric currents were studied. No wave phenomena are involved in these static fields, but when the fields are varying in time, wave phenomena do appear. A changing electric field generates a magnetic field, and a changing magnetic field generates an electric field. It is this interrelation between the two kinds of fields that causes electromagnetic waves to exist.

The production of an electric field by a changing magnetic field was discovered by Michael Faraday. The effect is best studied by looking at a simple example of it. Consider a loop of wire connected to a galvanometer, an instrument that shows a needle deflection when any charge passes through it, and a magnet as shown in Figure 31.1. When the loop and magnet are stationary with respect to each other there is no deflection of the galvanometer needle. If there is relative motion of the magnet and loop there is generally a deflection whose size and direction depend on the circumstances.

If the magnet is kept stationary in the laboratory and the loop of wire is moved, it is easy to understand why the galvanometer shows a deflection. There are charges in the wire that are free to move along the wire but not to leave it, and when these are dragged through the magnetic field by the motion of the wire they experience a magnetic force. To analyze the situation carefully, let us consider a rectangular loop of wire in a nonuniform magnetic field perpendicular to the plane of the loop as in Figure 31.2. Along the top side the magnetic field is stronger than along the bottom side. If the loop is moving upward with velocity **u**, positive charges in the upper side will feel a force to the right, that on a positive charge q being of magnitude

$$F_{\text{upper}} = q\frac{u}{c}B_{\text{strong}} \tag{31.1}$$

Similarly a charge q in the lower side will feel a force, also to the right, of magnitude

$$F_{\text{lower}} = q\frac{u}{c}B_{\text{weak}} \tag{31.2}$$

Charges in both upper and lower sides cannot move to the right because if they did there would be a growing accumulation of positive charge on the right, leading eventually to a repulsion strong enough to prevent any further such accumulation; any part of the wire must stay almost neutral. Because the force in the upper side is greater than that in the lower side, charge will tend to move clockwise around the loop. This charge must go through the galvanometer, so the latter shows a deflection. There is no magnetic force along the wire on the other two sides; the force on a positive charge there, too, is to the right and is thus across the wire. The reader may verify that the galvanometer deflection is the same if the movable charges are negative instead of positive,

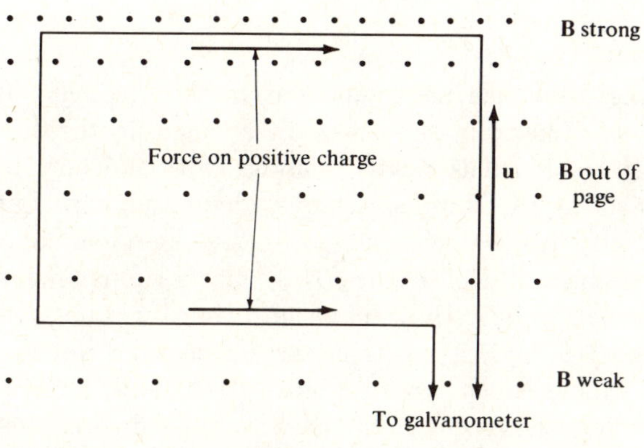

FIGURE 31.2 With the wire loop moving upward into the stronger field, charges in the upper side feel a stronger magnetic force than charges in the lower side, and a current tends to flow in the clockwise direction.

granted that the instrument reverses its response if either the sign of the charge or the direction of motion of charge through the instrument is reversed.

The convenient measure of the tendency of charge to move around the loop is the work that would be done on a particle with unit charge on making one circuit of the loop. It must be noted right away that this work is not done by the magnetic field; no work is ever done by a magnetic field acting on a charged particle because the force exerted by the field is always perpendicular to the direction of motion of the

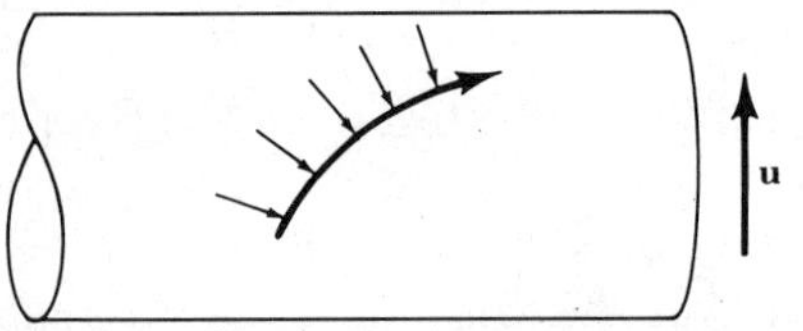

FIGURE 31.3 A positive charge in a wire dragged up through a magnetic field out of the page will experience a force that tends to move it to the right. An electron in the wire will be pushed to the left. In either case the upward component of velocity is reduced, and work must be done to keep the charge moving upward with velocity component **u**.

particle, and therefore this force has no component in the direction of the particle's displacement. What happens is a little complicated but quite understandable. When a wire is made to move with velocity **u**, a charged particle confined to the wire can move along the wire but must have the same velocity as the wire in a direction perpendicular to the wire. Thus in the upper wire of Figure 31.2, charges must move upward with speed u no matter how they move horizontally along the wire. Because of the magnetic field, a positive particle moving upward tends to veer off to the right, as shown in Figure 31.3. This reduces its upward speed, and the forces keeping it in the wire must do work on it to give it the needed upward speed. It is the forces that confine the charged particle to the wire that do the work, not the magnetic force. This effect is closely related to the Hall effect, which is the subject of Section 43.

The work done on a unit charge during one trip around the loop is called the *electromotive force* around the loop, usually abbreviated emf. This is an unfortunate use of the word *force,* but it is traditional.

From the discussion just given, we see that emf about the loop of Figure 31.2 is proportional to the magnitude of **u** and to the difference in the magnitudes of **B** on the top and bottom. This can be summed up very neatly if we introduce the flux of magnetic field through the loop. The magnetic flux through the loop is defined as the total number of lines of magnetic field linking the circuit or going through a surface with the loop as boundary. If lines going through this surface in one direction are counted as giving a positive flux, lines going through in the other direction give a negative flux. Flux of magnetic field is usually denoted by Φ (capital phi). Then the result of this experiment is that the emf in the loop is proportional to the rate at which the flux Φ is changing:

$$\text{emf} = \frac{1}{c}\frac{\Delta\Phi}{\Delta t} \tag{31.3}$$

The proportionality factor is $1/c$ because of the way we have chosen units of emf and magnetic field. The units of emf are ergs per statcoulomb or statvolts and are therefore the same as those of electric potential; the units of B are gauss; and c is the speed of light. Equation 31.3 is *Faraday's law of induction.* We tell how to determine the direction of the induced emf a little further on. In Figure 31.2 the flux Φ out through the loop is obviously increasing as the loop moves into a region of stronger field.

No electric field is applied to the loop in the above process, nor does the magnetic field at any fixed point in space change with time. The emf is ultimately due to the stronger magnetic force on the charges in one arm of the circuit than in the others, not to an external electric field. Now, however, we can repeat the experiment with only one change: The loop of wire is held fixed, and the magnet or whatever else produces the inhomogeneous magnetic field is moved down the page with speed u. The relative motion of the magnet and the loop is the same as before, and the deflection of the galvanometer also is found to be the same as before. Experimentally it makes no difference if the magnet is moved and the coil is held stationary in the laboratory or if the magnet is held fixed and the coil is moved or if both are moved. All that matters is the *relative* motion of the two. With a stationary loop, however, there is no magnetic force to act on the charges in the wire, urging them around and through the galvanometer, because the charges are not moving. The only other kind of force that acts on charges is an electric force, so there must be an electric field present that has a stronger component to the right in the upper arm than in the lower arm. Now, too, the magnetic field at a fixed point in space is changing with time. In this way we see that a changing magnetic field produces an electric field. The exact equation describing this relation between electric and magnetic fields is one of the famous Maxwell equations governing the electromagnetic field. We do not write it down; it is a more detailed version of Eq. 31.3. The electric field is produced by a changing magnetic

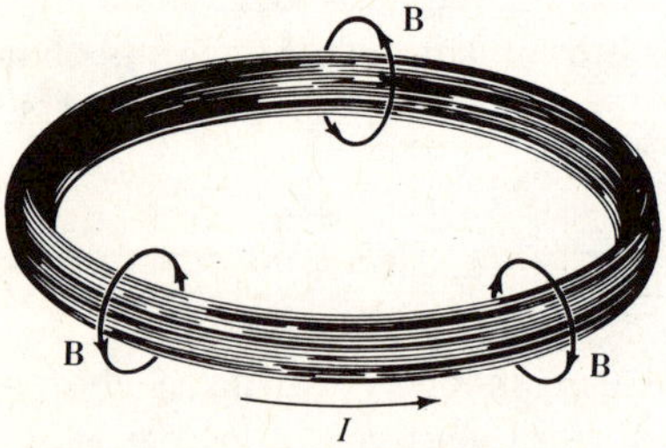

FIGURE 31.4 The current I flowing in the wire loop produces a magnetic field with a flux upward through the loop.

field whether or not a conductor is present in which a current can be induced. The circuit serves to show the existence of the field.

If a changing magnetic flux through a loop of wire induces an emf and consequently a current in that wire, one wants to know in which direction the current flows. A current flowing through the loop itself produces a magnetic field and a flux through the loop. This can be seen in Figure 31.4, the field lines due to the current I shown yielding an upward flux through the loop. The rule for determining the direction of the emf and current induced by a changing flux is this: The induced current flows in such a direction that the magnetic field it produces tends to reduce the change in the magnetic flux. Thus if the original flux is increasing, the induced current produces a flux of the opposite sign; if the original flux is decreasing, the induced current produces a flux of the same sign as the original. The current shown in Figure 31.4 could be induced by a downward magnetic field whose strength is increasing with time or by an upward magnetic field whose strength is decreasing with time, as both of these yield a flux change opposite in direction to the flux produced by the current I. This rule is called *Lenz's law*.

The production of an electric field by a changing magnetic field is readily observed with the use of magnets and loops of wire and galvanometers, once one knows what to look for. The production of a magnetic field by a changing electric field is much harder to observe directly because there are no magnetic currents that can be induced and detected as electric currents can be. The existence of the effect was suggested by Maxwell primarily to make the roles of electric and magnetic fields as symmetrical as possible in his theory. The existence of this effect in addition to Faraday's law of induction makes the existence of electromagnetic waves possible. Once started, by an antenna perhaps, the changing magnetic field produces an electric field whose change produces a magnetic field, and so on. When analyzed mathematically, this wave motion is found to travel with speed c, this quantity being exactly the one introduced in Eq. 31.3 and earlier in the expression Eq. 17.3 for the magnetic field produced by an electric current. In an electromagnetic wave the peak value of the electric field in dynes per statcoulomb is equal to the peak value of the magnetic field measured in gauss.

The connection between electromagnetic waves and light was suspected by Maxwell because the speed of these waves c could be measured by measuring the force between current-carrying wires, and the value so determined was very nearly the measured value of the speed of light.

An electromagnetic wave traveling along the z-axis involves electric and magnetic fields in the x- and y-directions. A possible wave is shown in Figure 31.5. The electric and magnetic fields are not confined to the region of the z-axis but have constant values on planes parallel to the xy-plane, so the picture represents a very wide beam traveling parallel to the z-axis. The wave illustrated is polarized; the electric field of the wave lies in the xz-plane, so the wave is a transverse wave like the one-dimensional waves in a string studied earlier. One could have another electromagnetic wave traveling parallel to the z-axis, this one with its electric field parallel or antiparallel to the y axis, as shown in Figure 31.6. Any other direction of the electric field perpendicular to the z-direction can be obtained by combining fields along the x- and y-axes, so there are only two independent kinds of polarization for a wave going in any given direction.

The existence of two polarizations of an electromagnetic wave is another reason for identifying such waves with light waves. It had been observed as early as 1669 by E. Bartholin that there are two ways

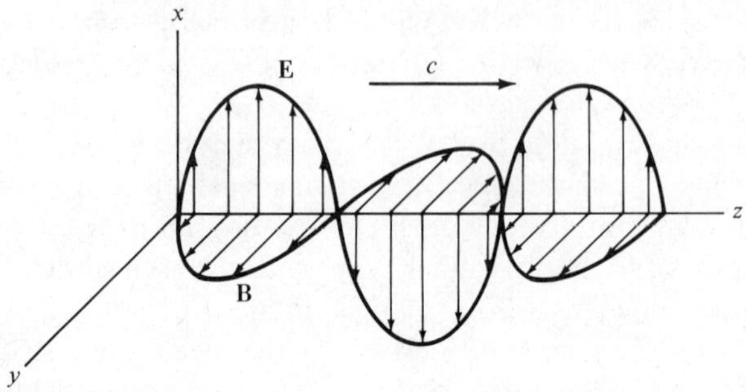

FIGURE 31.5 One kind of electromagnetic wave traveling parallel to the z-axis consists of an electric field in the x-direction and a magnetic field in the y-direction. Such a wave is polarized, the xz-plane being the plane of polarization.

that light can travel through certain transparent crystals. If one looks at a spot of light through a crystal of iceland spar (calcite), it appears double. The light waves of two different polarizations travel in different directions through the crystal, as shown in Figure 31.7. The distinction between the two waves as being a matter of polarization was not made until much later. In 1817 Young proposed the transverse nature of light waves to explain this and other polarization effects. In those days the ether was considered essential, and one difficulty with mechanical models of the ether was that they always permitted compressional waves (longitudinal waves) along with the transverse waves. No longitudinal light waves are known, all light being capable of being polarized and therefore

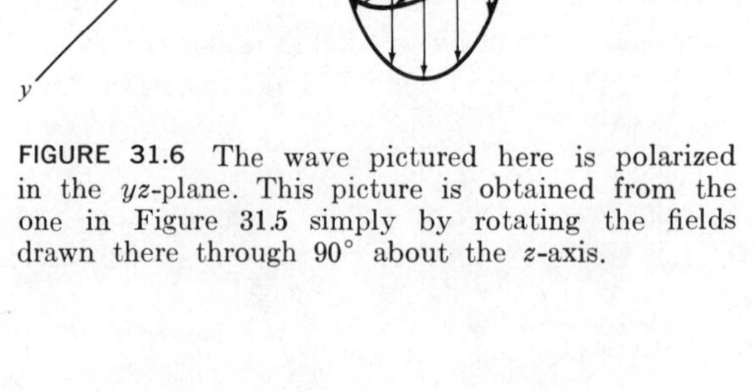

FIGURE 31.6 The wave pictured here is polarized in the yz-plane. This picture is obtained from the one in Figure 31.5 simply by rotating the fields drawn there through 90° about the z-axis.

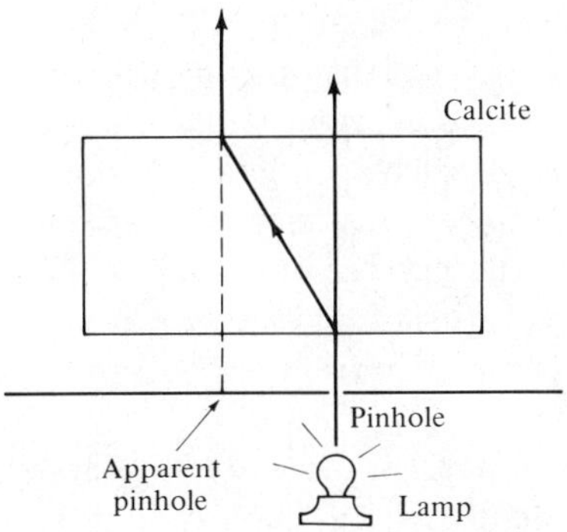

FIGURE 31.7 The way light travels through a crystal of calcite depends on the plane of polarization of the light. A point source of light appears double when seen through the crystal, and the light in each of the transmitted rays is polarized.

necessarily transverse. Electromagnetic waves are always transverse, so this difficulty is completely avoided by the electromagnetic theory of light.

Electromagnetic waves of long wavelength are familiar. Radio waves broadcast by a station at 100 on a radio dial have a frequency of 100 kilohertz or 10^5 hertz (cycles per second). Their wavelength is given by the general formula for wave motion:

$$\begin{aligned}\lambda &= \frac{c}{\nu} \\ &= \frac{3 \times 10^{10}}{10^5} \\ &= 3 \times 10^5 \text{ cm} \qquad (31.4)\end{aligned}$$

This is 3 km or about 2 miles. Such waves bend easily around large objects such as buildings and hills. Television stations broadcast at frequencies near 100 megahertz = 10^8 hertz, so the wavelength of these waves is only 300 cm or 10 feet. Waves of this wavelength do not bend around buildings and hills very well and are likely to be reflected by such objects as light is from a mirror. Radar waves are in the range between 1 and 10 cm wavelength.

In the region of wavelengths shorter than those of visible light, there is first the ultraviolet light just above the visible on the frequency scale.

At still higher frequencies there are then X-rays and γ-rays. For X-rays, the individual atoms in many crystals have the right spacing to form a kind of diffraction grating, a fact that makes X-rays the usual tool for the study of the structure of crystals. Observation of the places where X-rays of a certain wavelength interfere constructively enables the crystallographer to locate the atoms in the crystal and thus to find its structure. For γ-rays, whose wavelengths are still shorter, no diffraction grating exists and wavelength determinations must be made indirectly.

The existence of light waves and of electromagnetic waves in general makes one feel that the electromagnetic field has an independent existence. Static fields can be regarded as just a convenient way of talking about Coulomb's law or Ampere's law. The static fields of force around a charge have no existence without the charge, but a light wave does have existence independent of its source in the way that a meson has an existence independent of the particle that decayed and created the meson. Light from distant galaxies has been traveling for up to 4×10^9 years to reach us, and anything with so long an existence exists. Light carries energy and momentum with it, properties we have used in earlier sections where the photon was introduced, though there the wave nature of the radiation was not in evidence. It requires quantum theory to give an account of light that contains both the wave aspects and the particle aspects in exactly the way needed to describe the results of observations.

Problems

1. Two circular loops of wire are side by side, as shown in the figure. If a current is started in one loop by means of a battery being connected, use Lenz's law to find the direction of the current induced momentarily in the second loop. Sketch the magnetic fields produced by the current in the first loop and in the second loop. Where do these tend to cancel and where to reinforce each other?

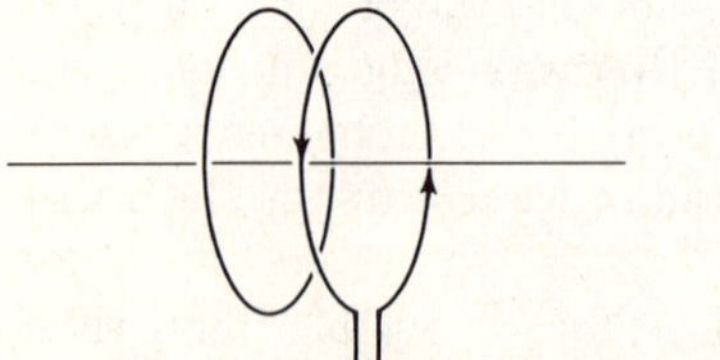

2. Two circular loops of wire are in the same plane, as shown in the figure. If a current is started in one loop by means of a battery being connected, sketch the field produced by the current. Use Lenz's law to determine the direction of the current in the second loop.

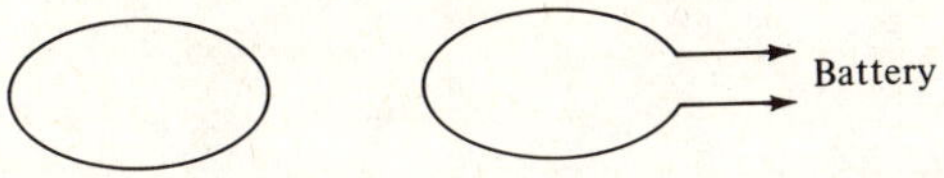

3. Make a sketch similar to Figure 31.5 showing a wave with the electric field parallel to the x-axis but traveling in the opposite direction along the z-axis. (Imagine that you are looking at the figure from the far side, so that the wave shown is going to your left.)
4. An eclipsing binary star is a pair of stars, one being much brighter than the other, which revolve about their center of mass in such a way that the earth is in the plane of their motion. Thus once each revolution the bright star is eclipsed by the darker one. If red light travels through space faster than blue light, what color effects would appear as the bright star disappears behind the dark one and then as it reappears on the other side? The absence of such color effects is good evidence that light of different colors travels at the same speed.

CHAPTER 7 Systems of Many Particles; Kinetic Theory

32. IDEAL GASES

The states in which matter appears are often classified as solid, liquid, and gas. A solid material has a definite volume and shape that it tends to maintain when external conditions are changed. A liquid has a definite volume but no definite shape. A liquid in a container on the earth adapts itself to the shape of the container, filling it until a level is reached corresponding to the volume of liquid present. A liquid in a region free of a gravitational field would not do this but would tend to assume a spherical form like a raindrop because of an effect called surface tension, an effect that causes a given volume of liquid to have as small a surface area as possible. A gas has neither definite volume nor definite shape. A gas will fill any container in which it is confined. If the container is small so that the change in velocity of a molecule of the gas due to the acceleration of gravity in going from the bottom to the top is unimportant, the gas will fill the container uniformly. If the container is large, the density of the gas is less high up than low down, as in the atmosphere. We restrict ourselves to containers small enough to be considered as uniformly filled.

The volume of the gas V is the volume of the container in which it is confined. If the container is a cylinder fitted with a piston as in Figure 32.1, the gas is found to exert a force upon the piston. If there are two pistons the forces on them are observed to be proportional to their areas. This experimental fact makes it useful to introduce the concept of force per unit area, or pressure, denoted by p. The force on any flat area A of wall of the container is then pA and is perpendicular to the surface.

A third variable is needed to describe the state of a given mass of gas of a given kind, the temperature. We have a subjective feeling for the meaning of temperature and for whether one object is at a higher

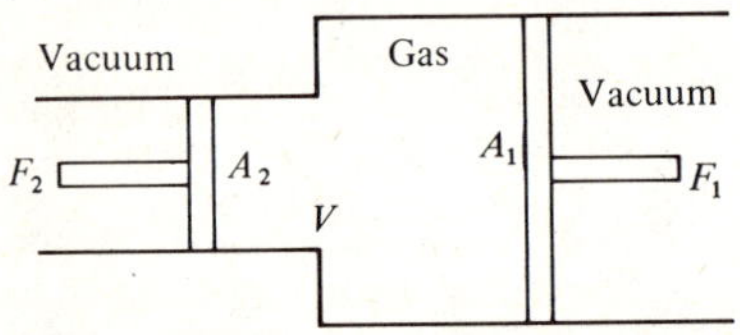

FIGURE 32.1 A gas is confined in a volume V between two pistons of different areas. If there is no gas on the outside of the pistons, the force on them is proportional to their area. The force per unit area is then the same for all surfaces on which the gas presses.

temperature than another. A precise definition of temperature is conveniently made in terms of gases, but without knowing how to determine the size of temperature differences we can observe that objects in contact with one another approach the same temperature. This makes is possible to measure the temperature of the air by having a thermometer in contact with it; eventually the thermometer will reach the same temperature as the surrounding air. The temperature of a cylinder of gas can be kept constant by putting it in contact with a very large object, much larger than the cylinder, whose temperature is supposed to remain the same at all times. Such an object is called a *heat bath* or *heat reservoir*.

If the gas in a cylinder is kept at a constant temperature, a simple relation between the pressure and volume is found to exist, namely, that

$$pV = \text{constant} \tag{32.1}$$

This is known as *Boyle's law*. The value of the constant on the right is proportional to the mass of gas in the cylinder. Boyle's law is not completely accurate; it becomes more accurate as the volume increases so that the density of the gas is reduced. For very dilute gases it is very accurately fulfilled. When the pressure gets high, in some cases the gas becomes a liquid or a solid, and when this happens Boyle's law breaks down completely. It is quite inaccurate at pressures near the liquefaction pressure. We confine our attention to gases of low enough density to make Boyle's law accurately fulfilled.

Boyle's law gives a description of the bulk behavior of a given mass of a dilute gas at constant temperature. Now we consider what happens when the temperature of the gas is raised. The fact that it is higher than before can be verified by simply feeling the cylinder if it is not too hot or too cold. When the gas is warmed, the value of the product pV increases, but if the temperature is then kept constant at its new value by means of a warmer heat bath, pV remains constant as p and V are varied, so that Boyle's law is also valid at the new temperature. This property of a dilute gas, that pV increases with increasing temperature, may be used to give a quantitative meaning to the concept of temperature.

A constant volume gas thermometer consists of a bulb containing the gas to be used as the thermometric substance, usually hydrogen or helium, and a means of measuring the pressure of the gas when it occupies a predetermined volume. By definition the temperature T is given by

$$pV = AT + B$$

where the value of the constant A depends on the size of the temperature unit chosen and the value of B depends on the choice of the zero of the temperature scale. In scientific work, in medicine, and in most countries except the United States and Canada, the celsius (centigrade) temperature scale is used. In this scale the temperature of water freezing at atmospheric pressure is chosen to be zero and the temperature of boiling water at atmospheric pressure is chosen to be 100°C. All other temperatures are then measured by evaluating pV in a gas thermometer in contact with the system whose temperature is to be measured and by inserting this value in the expression

$$T = 100 \frac{(pV)_T - (pV)_f}{(pV)_b - (pV)_f} \tag{32.2}$$

Here $(pV)_b$ means the value of pV for this thermometer in contact with boiling water, and analogously for $(pV)_f$. It is easily seen that inserting

$(pV)_f$ for $(pV)_T$ yields $T = 0$ and $(pV)_b$ yields $T = 100$. The gas thermometer provides a means for establishing a temperature scale. When temperatures too low or too high for the use of a gas thermometer are to be measured, some other property of a suitable material must be used to replace the quantity pV.

The fahrenheit scale, popularly used in the United States and Canada, differs from the celsius in choosing the temperature of freezing water to be 32°F and that of boiling water to be 212°F. We shall not use this scale at all.

As the temperature is lowered the value of pV decreases, and it can be imagined to decrease to zero. According to Eq. 32.2 this value of pV will be reached when

$$T = T_0 = 100 \frac{-(pV)_f}{(pV)_b - (pV)_f} \tag{32.3}$$

This is called "absolute zero." No gas can have a negative volume or a negative pressure, so no temperature lower than this can be defined by a gas thermometer. Absolute zero is defined, therefore, by a process of extrapolation since all gases condense or freeze at temperatures higher than this.

One might expect the value of T_0 given by Eq. 32.3 to depend upon the gas being used, but experimentally it does not. T_0 is a universal constant independent of the gas used whose value is found to be —273.16°C. The fact that this constant is the same for all gases leads one to suspect that all gases are constituted of matter in a way very nearly independent of their chemical constitution.

The mathematical form of Eq. 32.2 becomes simpler if the temperature at which pV should vanish is chosen to be zero. The absolute or kelvin temperature is measured from here, and it is defined by

$$T_K = 100 \frac{(pV)_T}{(pV)_b - (pV)_f} \tag{32.4}$$

The relation of kelvin to celsius temperature is

$$T_K = T_C + 273.16 \tag{32.5}$$

Absolute temperature or kelvin temperature is measured in kelvins, the kelvin being one celsius degree in magnitude (abbreviated to K). Measuring the temperature in kelvins, Eq. 43.2 can be written

$$pV = AT_K$$

From now on all temperatures will be measured in kelvins (often called "degrees kelvin"), so the subscript K will be dropped. The temperature $T = 0$ is absolute zero.

The constant A above is proportional to the mass of gas used in the thermometer. It is advantageous to measure the amount of gas in moles rather than in grams. One mole of a gas has a mass in grams equal to the molecular weight of the gas. If the thermometer contains n moles of gas, then we write

$$pV = nRT \tag{32.6}$$

where R is a new constant. The advantage of this way of writing is

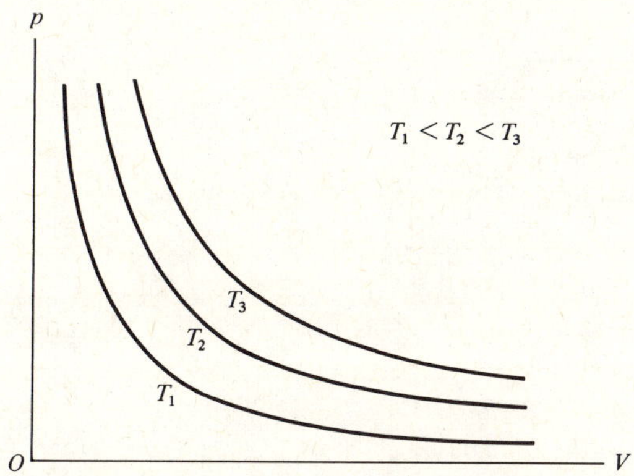

FIGURE 32.2 The variation of pressure with volume of an ideal gas is shown for three different temperatures. Each curve corresponds to Boyle's law for one particular temperature.

that the R introduced in this manner is found to be the same for all gases, as was T_0. R is called the "universal gas constant." It has the value

$$\begin{aligned} R &= 8.314 \times 10^7 \text{ erg/mole K} \\ &= 5.193 \times 10^{19} \text{ eV/mole K} \end{aligned} \tag{32.7}$$

The fact that R, like T_0, is the same for all gases is another indication that all gases have very similar structures. This is not true for liquids or solids. Matter in these forms has properties that depend strongly on the material involved.

The form of the pressure-volume relation for a gas according to the gas law, Eq. 32.6, is shown in Figure 32.2 for several temperatures.

A gas that acts in this way is called an *ideal gas,* and real gases at low density and high temperature are very nearly ideal in their behavior. This is shown in Figure 32.3 where the experimental value of pV for some real gases is plotted as a function of p. For an ideal gas this is constant; for real gases over large ranges of temperatures and pressures it is very nearly constant.

According to the gas law, Eq. 32.6, the volume of a sample of gas at constant pressure is proportional to the temperature, a result known

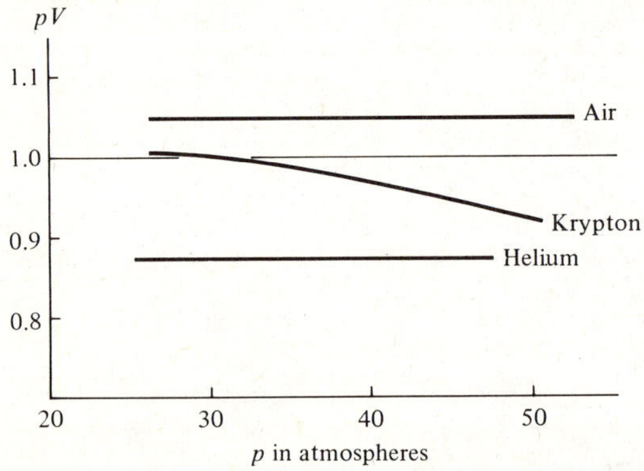

FIGURE 32.3 The value of pV for three real gases is shown as it depends on the pressure p. This product is strictly constant for an ideal gas and varies rather little for real gases even at the pressures shown, which are high.

as Charles' law. When the temperature is defined by a gas thermometer, Charles' law is true by definition. There is another, more fundamental way to define temperature that is independent of the notion of an ideal gas. If this more basic definition of temperature is used, then Charles' law is an observed law of nature and not just a definition. We shall not give this other definition here; it would lead us too far into the theory of heat.

Problems

1. (a) Find the fahrenheit temperature corresponding to 0 K.
   (b) At what temperature are the readings of a fahrenheit and a celsius thermometer the same?
   (c) What is normal body temperature on the celsius scale? (It is 98.6°F.)

2. The gas in a cylinder is expanded slowly from volume V to volume $1.2V$ at constant pressure.
   (a) What change in temperature must take place during this expansion?
   (b) How much work is done by the gas on the piston during the process?
   (c) Does it take more or less energy to warm a gas from temperature T_1 to T_2 when the volume is kept constant or when the pressure is kept constant? Why is there any difference?

33. THE STATISTICAL DESCRIPTION OF MATTER

We know that matter in bulk is made up of atoms and molecules; some of the experimental evidence for this was treated in detail in the first part of this book. The methods of measuring the size and mass of these particles is known, and hence the number of them in a given piece of bulk matter can be found. The number of molecules in 1 mole of a substance is Avogadro's number $N_0 = 6.02 \times 10^{23}$. This is a very large number. The distance from the earth to the sun is only 1.5×10^{14} millimeters, so if the molecules in 1 mole were strung in a line between the earth and sun, there would have to be 4 billion of them per millimeter. Four billion is not much less than the population of the earth. When such large numbers are involved, it is clear that the particles cannot be considered individually but must be studied statistically instead. The assumption that Newton's laws of motion apply to these particles has many observable consequences even when only statistical effects can be measured, and we shall encounter situations where the behavior of large masses of matter shows that the classical laws of Newton do not always apply.

The first problem that must be discussed is how to describe an assemblage of a great many particles. What can be said about the velocity of the assemblage $\mathbf{v}$ or its x-component v_x? It is obviously hopeless to enumerate the values of v_x for the individual particles, but a great deal of information can be given in the following way: The range of possible values of v_x for any one particle extends from $-\infty$ to $+\infty$. Actually no one particle can have a value of v_x greater than c, the speed of light, but a few particles of a large assemblage may still have very high speeds, so we put no upper limit on the speeds that may be present. The range from $-\infty$ to $+\infty$ is then divided up into equal small intervals of size Δv_x, as in Figure 33.1. One now can specify how many particles have an x-component of velocity in the range Δv_x between

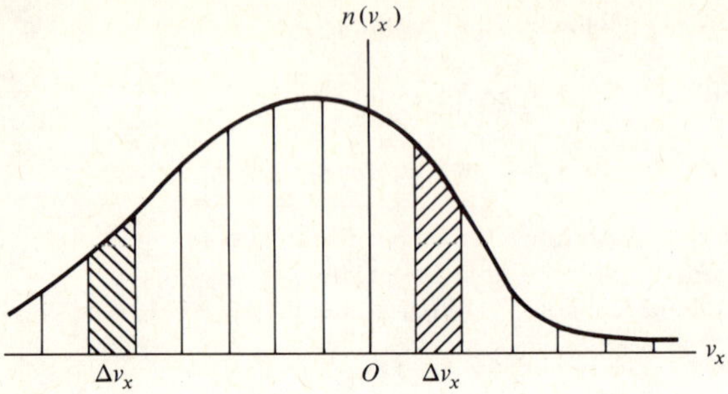

FIGURE 33.1 The range of possible values of v_x is divided into equal intervals Δv_x. The number of molecules with x-component of velocity in any one interval, such as either of the shaded ones, is the area under the graph of the distribution function of the strip with that interval as base.

v_x and $v_x + \Delta v_x$. This is written

$$n(v_x)\ \Delta v_x$$

because the larger the range, the larger is the number of particles with values of v_x in that range. The intervals Δv_x are chosen large enough so that there are many particles in all the intervals corresponding to small values of v_x and yet small enough so that the variation of v_x within one interval is not significant. The intervals corresponding to very large values of v_x will contain few if any particles.

The function $n(v_x)$ is called a *distribution function;* it specifies how the actual values of v_x for the particles are distributed over the possible values. It does not identify the particles that lie in any specified interval, it gives only the number of particles that have values of v_x in that interval. Distribution functions can be given for other quantities, such as y-components of velocity or kinetic energy or angular momentum. The most important distribution functions are those for components of velocity and for energy.

The chief use of a distribution function is to calculate averages. As an example, take the distribution in v_x, the x-component of velocity. Denoting the average value of any quantity by putting a bar over the symbol for that quantity, then by definition of the word *average* we write

$$\bar{v}_x = \frac{1}{N} \sum_{\text{all } v_x} v_x n(v_x)\ \Delta v_x \tag{33.1}$$

Here N is the total number of particles in the assemblage,

$$N = \sum_{\text{all } v_x} n(v_x)\,\Delta v_x \tag{33.2}$$

The symbol Σ stands for "sum," and in Eq. 33.1 it means the sum of contributions of the form $v_x n(v_x)\,\Delta v_x$, one from each interval Δv_x into which the range of possible values of v_x has been divided. If Δv_x is chosen to be very small, the number of intervals is correspondingly large and the sum consists of many small terms. If Δv_x is chosen to be larger, the number of terms in the sum is smaller but each term is larger and the value of the sum is the same. The intervals must be small enough so that the variation of v_x and of $n(v_x)$ between adjacent intervals is unimportant. If all intervals are to be included in the sum, it is said to be "over all v_x." If only intervals in which v_x is positive are to be included, it is said to be a sum "over all positive v_x," and similarly for other ranges of summation.

The process symbolized in Eq. 33.1 is just the ordinary one of finding an average. The average age of a group of people can be found by finding the number n_1 who are one, the number n_2 who are two, up to whatever age is necessary, say 75. The average age $\bar{A}$ is given by

$$\bar{A} = \frac{(1 \times n_1) + (2 \times n_2) + (3 \times n_3) + \cdots + (75 \times n_{75})}{n_1 + n_2 + n_3 + \cdots + n_{75}}$$

which is exactly Eq. 33.1 applied to ages, the age interval being 1 year.

The average of v_x^2 is one that often occurs. It is defined in a way similar to the one used for v_x, namely,

$$\overline{v_x^2} = \frac{1}{N} \sum_{\text{all } v_x} v_x^2 n(v_x)\,\Delta v_x \tag{33.3}$$

The bar is drawn over the entire symbol v_x^2. One also can evaluate the square of the average v_x. This is written $\bar{v}_x^2$ and is never greater than $\overline{v_x^2}$. For example, if there are two particles, one with $v_x = 1$ cm/sec and the other with $v_x = 7$ cm/sec, then

$$\bar{v}_x = \frac{1 + 7}{2} = 4 \text{ cm/sec}$$

$$\overline{v_x^2} = \frac{1 + 49}{2} = 25 \text{ cm}^2/\text{sec}^2$$

which is consistent with the previous statement since $25 > 4^2 = 16$.

If $\overline{v_x}$ is positive, it means that the assemblage of particles is moving as a whole in the positive x-direction, so there is a flow in this direction. If there is no net flow in any direction, then

$$\bar{v}_x = \bar{v}_y = \bar{v}_z = 0, \qquad \bar{\mathbf{v}} = 0$$

Even when this is so, it does not follow that $\overline{v_x^2}$ is zero. Both positive and negative values of v_x yield positive values of v_x^2, so there is no cancellation between different terms in the sum defining v_x^2. $\overline{v_x^2} = 0$ means that no particle is moving along the x-axis in either direction; each molecule individually must have $v_x = 0$ for $\overline{v_x^2}$ to vanish.

In the sections that follow we shall see how some of the measurable properties of bulk matter can be accounted for as statistical averages over the distribution of atoms or molecules, especially over the distribution in velocity.

Problems

1. A system has two indistinguishable particles that may be in any one of the three segments a, b, c of the x-axis as shown in the figure. Make a table showing all the possible distributions of the particles over these segments, the three columns of the table showing the number of molecules in each segment for each distribution.

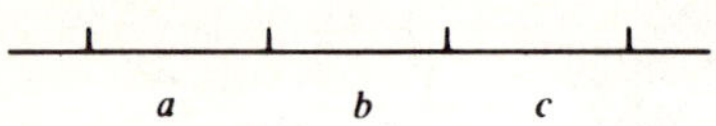

2. Assigning the segment a in Problem 1 the value $x = 0$, b the value $x = 1$, and c the value $x = 2$, for the distribution with only a empty find the value of (a) $\bar{x}$, (b) $\overline{x^2}$, and (c) $\overline{(x - \bar{x})^2}$. Repeat this assigning segment a the value $x = -1$, b the value $x = 0$, and c the value $x = +1$. Explain why the value of $\overline{(x - \bar{x})^2}$ is unaffected by this change of value assignments.
3. The quantity $\overline{(x - \bar{x})^2}$ is a measure of how wide a distribution is; for it to vanish all values of x occurring in the distribution must be equal, and equal naturally in that case to $\bar{x}$. Evaluate the mean square deviation from the mean for each of the distributions in Problem 1, and thereby rank them in order of width.

34. MOLECULAR MODEL OF AN IDEAL GAS

The observed fact that all gases behave very much alike when their density is low suggests that this behavior is a consequence of matter's being made up of small, discrete particles or molecules without reference to any detailed properties of these molecules themselves. The properties of the separate molecules determine the chemical behavior of the gas, of course. At low density the molecules are far apart and can be expected not to have much influence on one another; whatever mutual influence there is should decrease with decreasing density as then the intermolecular spacing increases. The starting point for a theory of gases is, therefore, to make a model of a gas in which the gas molecules do not interact with each other at all. After this is done, one can ask what new effects might be expected in the presence of intermolecular forces, but this is a very complicated matter and is not well understood.

We take as a model of an ideal gas a collection of N identical molecules, each of mass m, which move at random inside the container that defines the volume V of the gas. For the sake of simplicity the effect of gravity on the molecular motion is ignored so that between collisions the molecules are considered to move in straight lines. Only in very large containers does gravity have an appreciable effect. At first we assume that the only collisions are with the walls of the container, but later we shall see that allowing some intermolecular collisions to occur does not change the results. We assume that the gas is in equilibrium so that the number of molecules per unit volume can be treated as a constant in time and so that the distribution of velocity over the molecules is unchanging, although the velocities of individual molecules will change abruptly as they strike a wall. The number of molecules with a component of velocity v_x in the range between v_x and $v_x + \Delta_x$ is denoted by $n(v_x)\ \Delta v_x$; nothing is implied by this about the y- and z-components of $\mathbf{v}$, which may have any values. Only v_x concerns us here. The total number of molecules N is given by

$$N = \sum_{\text{all } v_x} n(v_x)\ \Delta v_x \tag{34.1}$$

in which the sum is over all values of v_x, that is, over all the intervals Δv_x into which the range of possible values of v_x has been divided.

To derive some of the consequences of our model it is useful to consider the container to be divided into two parts, R and L, by an imaginary plane surface. This plane may be taken as the yz-plane of a cartesian coordinate system, as shown in Figure 34.1. This dividing surface of area A is introduced only for the purpose of finding what crosses it, and it has no influence on the gas molecules at all.

First we calculate the number of molecules crossing the surface in the time interval Δt. For a molecule to cross the surface it must have

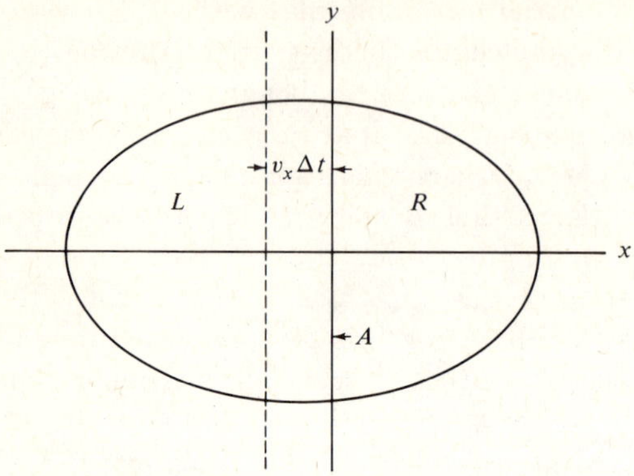

FIGURE 34.1 A volume containing a gas is divided into two halves by the yz-plane of a coordinate system. For a molecule to cross this plane from left to right in the time interval Δt, it must be within the distance $v_x \Delta t$ of the plane with v_x having a positive value.

a component of velocity toward the surface, so a molecule to the left, in L, must have a positive value of v_x and one to the right, in R, must have a negative value of this component to cross A. In order to cross the surface in the time interval Δt, at the start of that interval a molecule must be no farther from the surface than it can go in time Δt, namely, than a distance $v_x \,\Delta t$. All the molecules with v_x in the range between v_x and $v_x + \Delta v_x$ that are within a distance $v_x \,\Delta t$ of the surface will cross it during this time interval. The number of such molecules in the container is

$$[n(v_x)\,\Delta v_x] \times \left(\frac{Av_x\,\Delta t}{V}\right) \tag{34.2}$$

The first factor is the number of molecules in this range of v_x, anywhere in the container. The second factor is the fractional volume occupied by those molecules close enough to the surface to cross it in time Δt; $Av_x\,\Delta t$ is the volume of a layer $v_x\,\Delta t$ thick beside the surface of area A, and V is the volume of the entire container. This layer is shown in Figure 34.1 for a positive value of v_x.

The total number of molecules crossing from left to right per unit time is now seen to be

$$N^{L\to R} = \frac{A}{V} \sum_{v_x>0} v_x n(v_x)\,\Delta v_x \tag{34.3}$$

where, as indicated, only positive values of v_x are included in the sum. Similarly the number crossing from right to left per unit time is

$$N^{R\to L} = \frac{A}{V}\sum_{v_x<0} (-v_x)n(v_x)\,\Delta v_x \tag{34.4}$$

where only negative values of v_x are included. The minus sign enters because v_x itself is here always negative, but the number of molecules crossing per unit time must be positive.

If the gas is in equilibrium, the number of molecules on one side of the dividing surface does not change with time, so the number entering this side must be equal to the number leaving it. The number entering the side R per unit time is $N^{L\to R}$, and that leaving is $N^{R\to L}$, so in equilibrium

$$N^{L\to R} - N^{R\to L} = 0 \tag{34.5}$$

The left side of this equation is the net number of molecules entering R, being the difference between the totals of those entering and those leaving. Using the expressions just derived for these quantities, we get

$$\begin{aligned} 0 &= \frac{A}{V}\sum_{v_x>0} v_x n(v_x)\,\Delta v_x + \frac{A}{V}\sum_{v_x<0} v_x n(v_x)\,\Delta v_x \\ &= \frac{A}{V}\sum_{\text{all } v_x} v_x n(v_x)\,\Delta v_x \qquad (34.6) \\ &= \frac{AN}{V}\,\bar{v}_x \end{aligned}$$

Thus in equilibrium the average value $\bar{v}_x$ must vanish. This is an intuitively obvious statement, because if the molecules had an average velocity different from zero, the gas as a whole would be moving with this velocity. We have tacitly assumed that the container is at rest, so the gas is going nowhere. The average value of v_y and v_z also vanish for the same reason.

Next let us find the amount of x-component of momentum carried across this surface per unit time by the molecules crossing from left to right. We already know that the number of molecules in the range from v_x to $v_x + \Delta v_x$ crossing per unit time is $(A/V)v_x n(v_x)\,\Delta v_x$, as given in Eq. 34.2. Each of these molecules brings with it an x-component of momentum whose value is mv_x. The total x-component of momentum carried across the surface from left to right per unit time is then found by adding up the contributions from all molecules crossing the surface from left to right. It is

$$
\begin{aligned}
P_x^{L\to R} &= \frac{A}{V}\sum_{v_x>0}(mv_x)v_x n(v_x)\,\Delta v_x \\
&= \frac{A}{V}\sum_{v_x>0} mv_x^2 n(v_x)\,\Delta v_x
\end{aligned}
\tag{34.7}
$$

In the same way, each molecule crossing the surface from right to left takes with it the amount mv_x of x-component of momentum, which is negative because molecules moving in this direction have negative values of v_x. The total x-component of momentum carried across from right to left is

$$
\begin{aligned}
P_x^{R\to L} &= \frac{A}{V}\sum_{v_x<0} mv_x(-v_x)n(v_x)\,\Delta v_x \\
&= -\frac{A}{V}\sum_{v_x<0} mv_x^2 n(v_x)\,\Delta v_x
\end{aligned}
\tag{34.8}
$$

Here the negative sign in the final expression follows from the negative value of the x-component of momentum possessed by molecules crossing from right to left. Both $P_x^{L\to R}$ and $P_x^{R\to L}$ correspond to a gain of x-component of momentum by R because R is importing positive amounts from L and is exporting negative amounts of this commodity to L. The net transport per unit time of x-component of momentum from L to R is the imports minus the exports. This is not zero but is, in fact,

$$
\begin{aligned}
P_x^{L\to R} - P_x^{R\to L} &= \frac{A}{V}\sum_{v_x>0} mv_x^2 n(v_x)\,\Delta v_x + \frac{A}{V}\sum_{v_x<0} mv_x^2 n(v_x)\,\Delta v_x \\
&= \frac{A}{V}\sum_{\text{all } v_x} mv_x^2 n(v_x)\,\Delta v_x \\
&= \frac{AN}{V}\,m\overline{v_x^2}
\end{aligned}
\tag{34.9}
$$

It is proportional to the mean value of the square of v_x.

What does such a flow of momentum mean? It does not mean that there is a flow of particles from left to right or from right to left; that is zero as we have seen. The mass of gas on either side of the dividing surface is not changing with time. We recall, however, that the force on an object can be defined as the time rate of change of the momentum of that object:

$$\mathbf{F} = m\mathbf{a} = \frac{\Delta(m\mathbf{v})}{\Delta t}$$

If the momentum of the gas on the right is increasing at the rate given by Eq. 34.9 due to the molecules crossing the dividing surface in both directions, there is a force on that gas given by

$$F_x = \frac{AN}{V} m\overline{v_x^2} \tag{34.10}$$

or there is a pressure exerted on the gas in R by that in L of magnitude

$$p = \frac{F_x}{A} = \frac{N}{V} m\overline{v_x^2} \tag{34.11}$$

The gas in R is not accelerating as it would if this force were the only one acting. It is prevented from doing so by the container. The momentum that the gas in R receives from L is passed on to the wall of the container, so the gas exerts a force outward on the wall of its container. If the container wall ruptures, then the gas indeed accelerates under the force from L and an explosion takes place whose violence depends on the value of p, among other things. By looking at the momentum change in L instead of that in R as was done above, it can be seen that the gas in R exerts a force to the left on the gas in L, which in turn is prevented from accelerating to the left by the left side of the container. Newton's third law of motion is obeyed by these pressure forces.

Notice that our derivation of the expression Eq. 34.11 for the pressure shows that the pressure exists in the gas itself and not just on the container walls. Nothing needs to be said about the details of a collision between a molecule and the wall, the pressure being in fact quite independent of whether the wall is smooth and hard or rough and soft, as long as the gas is in equilibrium.

Equation 34.11 is very much like the ideal gas law, Eq. 32.6, especially if it is written in the form

$$pV = Nm\overline{v_x^2} \tag{34.12}$$

The left side is of the desired form, and the right side is proportional to the number of molecules N, which can be written as the number of moles n times Avogadro's number $N_0 = 6.023 \times 10^{23}$/mole, the number of molecules per mole. We still need a connection between the average $m\overline{v_x^2}$ and the temperature.

For any given molecule the speed v and the velocity components are related by

$$v^2 = v_x^2 + v_y^2 + v_z^2 \tag{34.13}$$

In our model of an ideal gas we assume that the motion of the molecules is random, so that no one direction of motion is preferred over any other. Under this assumption the average value of v_x^2 is the same as the average value of v_y^2 and of v_z^2. On taking the average of both sides of Eq. 34.13 we obtain

$$\begin{aligned}\overline{v^2} &= \overline{v_x^2} + \overline{v_y^2} + \overline{v_z^2} \\ &= 3\overline{v_x^2} = 3\overline{v_y^2} = 3\overline{v_z^2}\end{aligned}$$

or, equivalently,

$$\overline{v_x^2} = \tfrac{1}{3}\overline{v^2} \tag{34.14}$$

This enables us to rewrite the result Eq. 34.12 in the form

$$\begin{aligned}pV &= \tfrac{1}{3}nN_0m\overline{v^2} \\ &= \tfrac{2}{3}nN_0\bar{K}\end{aligned} \tag{34.15}$$

where $K = \frac{1}{2}mv^2$ is the kinetic energy of an individual molecule.

Equation 34.15 can now be compared with the ideal gas law, Eq. 32.6, which is

$$pV = nRT \tag{34.16}$$

in which n is the number of moles of gas and R is the universal gas constant. The two equations agree if

$$\tfrac{2}{3}N_0\bar{K} = RT \tag{34.17}$$

We need only correlate the temperature of the gas with the mean kinetic energy of the molecules composing the gas to have complete correspondence between the empirical gas law for low densities and the consequences of our model. The mean kinetic energy of a molecule of a gas at kelvin temperature T is then

$$\begin{aligned}\bar{K} &= \frac{3}{2}\frac{R}{N_0}T \\ &= \frac{3}{2}kT\end{aligned} \tag{34.18}$$

The new constant k is the gas constant per molecule, known as *Boltzmann's constant*. It has the value

$$
\begin{aligned}
k &= 1.38 \times 10^{-16} \text{ ergs/K} \\
&= 8.617 \times 10^{-5} \text{ eV/K} \qquad (34.19)
\end{aligned}
$$

At room temperature, 300 K, the mean kinetic energy of a molecule is thus about $\frac{1}{40}$ electron volt. The corresponding mean speed of a molecule of hydrogen gas is about 1.5×10^5 cm/sec.

The model of an ideal gas developed here has further consequences as a result of the connection between temperature and energy shown in Eq. 34.18. If a mole of gas is warmed up an amount ΔT, the volume being kept constant in the process so that no work is done against any external forces, an amount of energy ΔE given by

$$\Delta E = N_0 \, \Delta \bar{K} = \tfrac{3}{2} R \, \Delta T \qquad (34.20)$$

should be required. The quantity

$$\frac{\Delta E}{\Delta T} = C_V \qquad (34.21)$$

is called the *molar heat at constant volume* of the gas. It should have the value $C_V = 3R/2$ according to the model. Molar heats are most commonly specified in energy units called *calories,* 1 calorie being the energy required to raise the temperature of 1 g of water from 14.5 to 15.5°C. The calorie is related to the other energy units used here by

$$
\begin{aligned}
1 \text{ cal} &= 4.186 \times 10^7 \text{ ergs} \\
&= 2.613 \times 10^{19} \text{ eV} \qquad (34.22)
\end{aligned}
$$

The value of R is very nearly 2 cal/(mole K), so the value of C_V should be very nearly 3 cal/(mole K).

In Table 34.1 the value of C_V for some gases is given. It is seen

TABLE 34.1
Molar Heat of Selected Gases at Constant Volume

H_2	O_2	He	A	H_2O	
4.9	5.0	3.0	3.0	6.5	cal/(mole K)

that the prediction made by the model is good for He and A but is no good for the others. The values for H_2 and for O_2 are almost equal, but not what they should be, being too large. The value for H_2O is larger yet. The way in which the model fails suggests how to modify it. We assumed that the molecules of the gas are simple particles with no internal structure so that the only form of energy they can have is kinetic energy of linear motion, and the model works for those gases which are known from chemical evidence to have molecules consisting of one atom—monatomic gases such as helium and argon. Where the chemical evidence shows the molecule to have structure, the model fails.

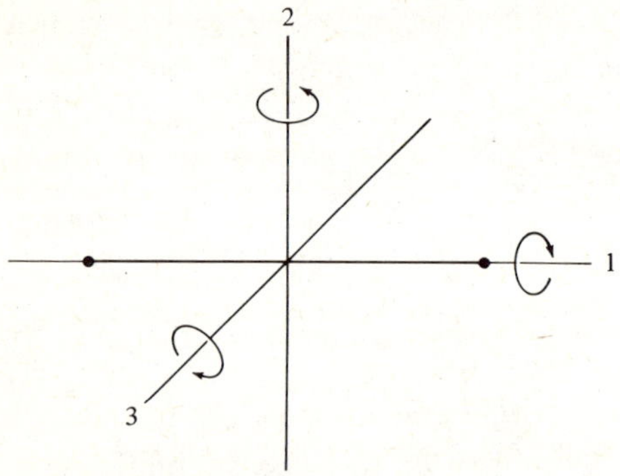

FIGURE 34.2 Rotation of the diatomic molecule shown about either of the axes 2 or 3 involves motion of the centers of mass of the two atoms and constitutes an independent kind of motion of the molecule. Rotation about axis 1 does not involve motion of the centers of the atoms and does not count as an independent motion of the molecule.

For diatomic gases like hydrogen and oxygen and polyatomic gases like water vapor the model gives too small a value for C_V, which means that in nature more energy is required to warm the gas than the model allows for. This can be accounted for by assigning new modes of motion to these more complicated molecules and by assuming that the energy of these new modes of motion also increases as the temperature is raised.

If the molecules of the gas are diatomic as in H_2 and O_2, they may be pictured as dumbbell shaped. Such a molecule can rotate about an axis through its center of mass. There are three independent axes of this kind, one passing through the two atoms and two others perpendicular to this one, as shown in Figure 34.2. Rotations about the axes labeled 2 and 3 are certainly motions of the molecule, as they involve motion of the centers of mass of the atoms constituting the molecule, though not of the center of mass of the molecule as a whole. A rotation about axis 1 does not involve motion of the centers of mass of the atoms but only rotation of the atoms about their centers of mass, and because the rotation of single atoms is seen not to contribute to C_V for monatomic gases, rotation about the molecular axis should not contribute to C_V for diatomic gases. To estimate the effect of rotation on the molar heat we must decide how much energy is to be associated with each of the two rotations that contribute to the mean energy of the molecule.

To find a reasonable value for the mean energy associated with a mode of rotation, we look at expression Eq. 34.18 for the mean kinetic energy of a monatomic molecule again and write it in the form

$$\tfrac{1}{2}m\overline{v_x^2} + \tfrac{1}{2}m\overline{v_y^2} + \tfrac{1}{2}m\overline{v_z^2} = \tfrac{3}{2}kT \tag{34.23}$$

The three quantities v_x, v_y, and v_z can vary independently of each other, but together they specify the velocity $\mathbf{v}$ of the molecule uniquely. There are, therefore, exactly three independent contributions to $\frac{3}{2}kT$ that are all equal. This suggests that an equal contribution should be made by every possible independent mode of motion, a suggestion that is actually a strict consequence of a more complete statistical theory of this model. If, then, a diatomic molecule can rotate independently about two axes, there should be a mean energy of $\frac{1}{2}kT$ associated with each of these two rotations as well as with each of the three independent linear motions. The mean energy E of a diatomic molecule that is part of a gas at temperature T then should be

$$E = \tfrac{3}{2}kT + \tfrac{2}{2}kT = \tfrac{5}{2}kT \tag{34.24}$$

and the molar heat at constant volume of this gas should be

$$C_V^{\text{diat}} = \tfrac{5}{2}R = 5 \text{ cal/(mole K)} \tag{34.25}$$

This agrees with the experimental value given in the table.

With this triumph we have not yet won everything. For a polyatomic gas whose molecules are not linear in construction, such as H_2O, there should be three rotational axes contributing to the possible modes of motion, and therefore we should have

$$C_V^{\text{poly}} = \tfrac{3}{2}R + \tfrac{3}{2}R = 3R = 6 \text{ cal/(mole K)}$$

This does not work. The value is again too low, so there must be new forms of motion. An obvious one is the vibration of atoms relative to each other within the molecule. The mean kinetic energy of a harmonic oscillator at temperature T is $\frac{1}{2}kT$, and the mean potential energy is the same, yielding a mean total energy of kT. This is too much of an increase to account for the value 6.5 cal/(mole K) for water vapor, and even worse it ruins the agreement for diatomic gases. If polyatomic molecules can vibrate, so can diatomic ones. This difficulty was recognized very shortly after the kinetic theory of gases was first proposed and stood in the way of a complete acceptance of the molecular model. This problem and closely related ones stimulated the invention of quantum mechanics, which resolves these difficulties in a simple and elegant way. We discuss this later on in this book.

The molecular model of an ideal gas developed here does much of what it was designed to do: If furnishes a model whose properties, based on particle mechanics and some statistical ideas, are the observed properties of dilute gases. The molar heat problem is the only place where difficulties arise, and for monatomic gases there is no problem even there as long as internal motions of whatever is in the atom itself are ignored. The only feature of the model needed to derive the ideal gas law is the existence of molecules. The mass of the molecules does not enter in except in the relation between the microscopic quantity $\frac{1}{2}mv^2$ and the macroscopic temperature T.

Two bodies in contact always reach the same temperature eventually, which means that if two cylinders of gas are in thermal contact the mean kinetic energies of the molecules of the two gases will be the same. This is also true if the gas in a single container consists of a mixture of molecules of different kinds. The mean energy of the molecules of any one mass m_1 will be the same as those of another mass m_2. If a container of volume V at temperature T contains n_1 moles of a gas of kind 1 and n_2 moles of a gas of kind 2, the pressure p_1 due to molecules of kind 1 is given by Eq. 34.15 to be

$$p_1V = \tfrac{2}{3}n_1N_0\bar{K}$$

while that due to molecules of kind 2 is

$$p_2V = \tfrac{2}{3}n_2N_0\bar{K}$$

Momentum is transported across a surface by the motion of molecules of either kind, and the total momentum transported is the sum of the momenta carried by the two kinds of molecules separately. Therefore the total pressure is the sum of the two partial pressures p_1 and p_2.

$$\begin{aligned} p &= p_1 + p_2 \\ &= \tfrac{2}{3}(n_1 + n_2)N_0\bar{K} \end{aligned} \tag{34.26}$$

This law of partial pressures was known very early and was one of the basic reasons for introducing the molecular model. It results from the common value of V and of $\overline{K}$ for any gas in a given container and from the fact that the molecules are assumed to have very limited interaction with each other.

In a real gas the molecules have a finite size and therefore must collide with each other from time to time. For this not to change the results of the molecular model obtained above, the fraction of the time that any given molecule spends colliding with others must be very small—otherwise we could not assign a momentum to a molecule, a momentum that remains constant while the molecule crosses the dividing surface used earlier. Also the potential energy between molecules would not be negligible, and our correlation of temperature with kinetic energy would be inaccurate for a new reason. The interactions between the molecules must be small enough so that on the average most molecules are moving like free particles. If this is so, the above derivation is good. When the interactions get stronger or the gas gets denser so that collisions are more frequent, the model gets less accurate and calculations of what does happen become very complicated. No one has yet been able to calculate, starting from the known properties of the molecules of a gas, under what pressures and at what temperature that gas will liquefy.

Problems

1. Summarize a series of experiments that lead to the numerical value of Boltzmann's constant k. Is it necessary to find the charge-to-mass ratio of the electron or of any other particle?
2. Show by an argument similar to the one leading to Eq. 34.5 that the transport of kinetic energy across the surface A of Figure 34.1 is zero when the gas is in equilibrium.
3. Find the value of $\sqrt{\overline{v^2}}$ for oxygen molecules and for electrons at room temperature.
4. A gas at temperature T fills a container of volume V and is under pressure p. Imagine that the molecules of the gas combine in pairs so that there are half as many of them but each one is twice as heavy. If the volume and temperature remain the same during this combining process, how does the final pressure differ from the initial pressure? Can you give an intuitive argument based on the molecular model to explain your answer? (How does the mean square velocity of the combined molecules compare with that of the original ones?)

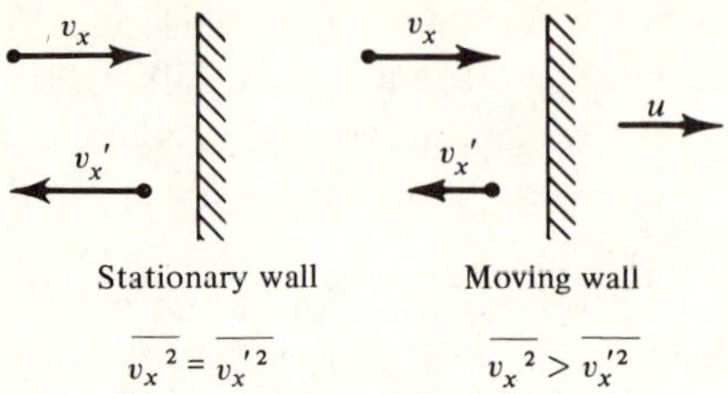

5. On the average molecules striking a wall perpendicular to the x-axis bounce off with the same distribution of x-component of velocity as they have before striking it except that the sign is reversed. Thus the average kinetic energy of the molecules is not changed. If one looks at this wall from a moving reference frame so that the wall seems to be moving away from the gas with speed u, show that on the average the kinetic energy of a molecule is less after bouncing than before. (Use the Galilean transformation Eq. 30.10, not a Lorentz transformation.)

35. THE MAXWELL-BOLTZMANN VELOCITY DISTRIBUTION

According to the molecular model of an ideal gas, a gas molecule chosen at random is equally likely to be moving in any direction. A gas is isotropic, which is a way of saying that all directions within the gas are equivalent. All speeds, however, are not equally probable. The temperature of the gas is a measure of the mean kinetic energy of the gas molecules and hence a measure of the mean square speed of the molecules. The speed distribution, therefore, must depend on the temperature of the gas.

The distribution of molecular velocities was guessed at by Maxwell in the middle of the 1800s and was derived on a firm theoretical foundation by Boltzmann later in the nineteenth century. It was not until 1947 that a good experimental verification of this distribution was made, and better ones have been made since. Maxwell and Boltzmann were correct; there had been little doubt about it in physicist's minds, but it is always worthwhile to verify such things because occasionally one is surprised by the result. The basis for Maxwell's inspired guess and for Boltzmann's derivation are mathematically too involved to describe here. All that we will do is to describe the result and to give an account of two experiments that verify it.

In deriving the relation between molecular motion and the pressure of an ideal gas, the detailed distribution of molecules with respect to some component of velocity, chosen to be v_x, was not required; its exact form did not matter. What was used was the following:

1. The total number of molecules N is the sum of the numbers of molecules in each range Δv_x of this component of velocity. In mathematical form this is stated as

$$N = \sum_{\text{all } v_x} n(v_x)\,\Delta v_x \tag{35.1}$$

2. The velocity of the gas as a whole is zero, so the mean value of the velocity is zero. For the x-component this means that

$$\bar{v}_x = \frac{1}{N}\sum_{\text{all } v_x} v_x n(v_x)\,\Delta v_x = 0 \tag{35.2}$$

3. The relation between the mean kinetic energy of a molecule and the temperature is that the mean kinetic energy associated with each independent kind of motion of a molecule in a gas at temperature T is $\frac{1}{2}kT$. Thus for motion in the x-direction,

$$\begin{aligned}\frac{1}{2}m\overline{v_x^2} &= \frac{1}{N}\sum_{\text{all } v_x}\frac{1}{2}mv_x^2 n(v_x)\,\Delta v_x \\ &= \frac{1}{2}kT\end{aligned} \tag{35.3}$$

These three conditions must be fulfilled by any distribution function, but they alone do not specify what this distribution function is.

To understand the form of the Maxwell-Boltzmann distribution one must know something about the exponential function e^x where x is the variable. The quantity e (not to be confused with the charge on an electron) is a mathematically defined number like π, and its value is

$$e = 2.71828\ldots$$

The virtue of using this number in defining the exponential function rather than a simpler number like 2 or 10 is that for small values of x one has the approximate relation

$$e^x \approx 1 + x \tag{35.4}$$

which would not hold if e were replaced by any other number. We shall have occasion to use this approximation later on. An important

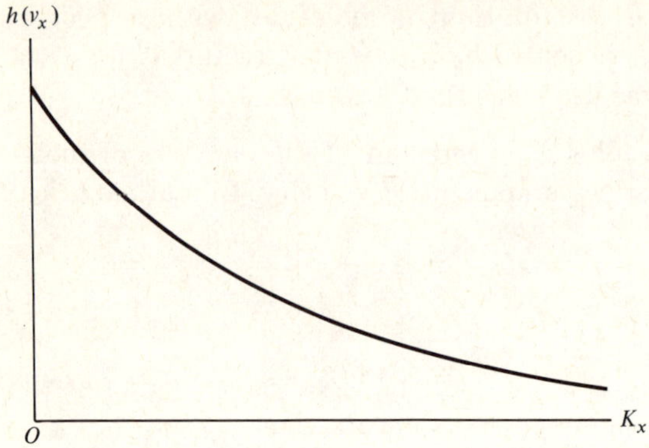

FIGURE 35.1 The distribution function for molecules of a gas at temperature T is shown as it depends on the kinetic energy associated with motion in the x-direction, rather than on the velocity component in this direction. It is an exponential function of K_x.

characteristic of the exponential function is that doubling the value of x has the same effect as squaring the function. Thus

$$e^{2x} = (e^x)^2$$

If x is positive, e^x is greater than unity and then the sequence of values $e^x, e^{2x}, e^{3x}, \ldots$ is very rapidly increasing. For example, we may choose x so that $e^x = 2$. Then we get the sequence of values

$$\begin{array}{cccccc} e^x & e^{2x} & e^{3x} & e^{4x} & \ldots & e^{10x} \\ 2 & 4 & 8 & 16 & \ldots & 1024 \end{array}$$

If x is positive, $e^{-x} = 1/e^x$ is less than unity and the sequence e^{-x}, $e^{-2x}, \ldots$ is rapidly decreasing in magnitude. With the same value of x as above we get

$$\begin{array}{cccccc} e^{-x} & e^{-2x} & e^{-3x} & e^{-4x} & \ldots & e^{-10x} \\ \frac{1}{2} & \frac{1}{4} & \frac{1}{8} & \frac{1}{16} & \ldots & \dfrac{1}{1024} \end{array}$$

If x is larger than this, the sequence decreases even faster. Figure 35.1 illustrates a decreasing exponential function.

The velocity distribution function $n(v_x)$ can now be written down.

It is

$$n(v_x) = N\sqrt{\frac{m}{2\pi kT}}\,e^{-\frac{1}{2}mv_x{}^2/2kT}$$

$$= N\sqrt{\frac{m}{2\pi kT}}\,e^{-K_x/kT} \tag{35.5}$$

if $K_x = \frac{1}{2}mv_x^2$ is the kinetic energy associated with the x-component of the velocity. The number in front is chosen so that condition 1 is fulfilled. Condition 2 is satisfied because according to Eq. 35.5, $n(v_x) = n(-v_x)$ so that both directions of motion along the x-axis are equally probable and there is no tendency for motion of the gas as a whole along the x-axis. Condition 3 is also satisfied; the proof requires calculus, and we omit it. Figures 35.1 shows $n(v_x)$ plotted against K_x; it decreases rapidly with increasing K_x, the rate of decrease depending on the value of kT. When kT is small, it does not take a very large K_x to make the exponent large in magnitude; the larger kT is, the larger K_x must be to make $n(v_x)$ small.

The dependence of $n(v_x)$ on v_x is shown in Figure 35.2 for two values of the temperature. It is a symmetrical bell-shaped curve called a "Gaussian curve." It decreases much more rapidly with increasing v_x than the curve of Figure 35.1 does with increasing K_x because v_x appears squared in the exponent.

It is also of interest to see what the distribution in speed of the molecules is, irrespective of their direction of motion. The number of mole-

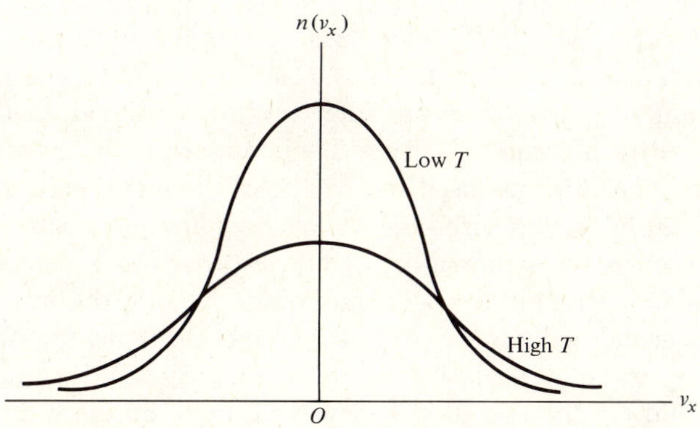

FIGURE 35.2 The distribution of molecules with respect to v_x is shown as it depends on v_x. It is a Gaussian function of v_x.

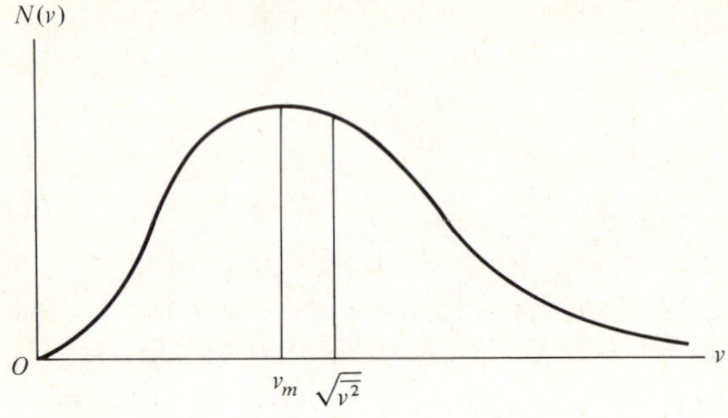

FIGURE 35.3 The distribution of molecules with respect to speed and without regard to the direction of motion. The most probable speed v_m is less than the root mean square speed $\sqrt{\overline{v^2}}$.

cules moving with speeds between v and $v + \Delta v$ in any direction is denoted by $N(v)\ \Delta v$. It can be derived from Eq. 35.5 and is given by

$$N(v) = \frac{4\pi v^2}{\sqrt{(2\pi kT/m)^3}} e^{-\frac{1}{2}mv^2/kT} \tag{35.6}$$

The graph of this distribution is shown in Figure 35.3. According to the figure, few molecules are moving very slowly and few are moving very fast. The peak of the curve gives the most probable speed v_m. Because the curve is not symmetrical the mean speed v is not the most probable one but is slightly greater. The distribution Eq. 35.6 illustrated in Figure 35.3 is called the *Maxwell-Boltzmann* speed distribution.

In an ingenious experiment I. Estermann, O. C. Simpson, and O. Stern used the known motion of particles in the earth's gravitational field to separate molecules with different velocities. Their apparatus is shown schematically in Figure 35.4. The gas used was vapor of the metal cesium or potassium. The metal was vaporized in an oven. Through a small hole in one side of the oven the vapor atoms could escape into a region where an excellent vacuum was obtained by pumping. Outside the oven was a screen with a second hole in it. Atoms leaving the hole in the oven wall followed a parabolic path like any other projectile in the earth's gravitational field. If the speed and elevation angle of the atom leaving the oven are right, the atom can go through the hole in the screen, and the distance it drops can be detected by a suitable detector. Slow atoms must follow a highly arching path, while fast atoms must follow a flatter trajectory. The distance of fall d of an atom of speed v is given by

$$d = g\left(\frac{l}{v}\right)^2 \tag{35.7}$$

At an oven temperature of 450 K and an oven screen and screen detector spacing of 100 cm, the distance of drop of an atom moving with the most probable speed is less than 0.02 cm, but this is well within the capability of the measuring apparatus to determine. By measuring the number of atoms dropping various distances d, Estermann, Simpson, and Stern obtained the velocity distribution. It should be noticed that the number of atoms with a speed between v and $v + \Delta v$ leaving the hole per unit time is proportional to $vN(v)\ \Delta v$ because in unit time fast atoms can come from deeper in the oven than can slow ones. This extra factor v also occurred in the derivation of Eq. 34.3. The results of this experiment are shown in Figure 35.5. The agreement with the Maxwell-Boltzmann formula is good except at low speeds. The discrepancy here is probably due to the effect of collisions between atoms at the hole, but quantitative corrections could not be made.

Another experiment to measure the velocity distribution of gas molecules leaving the hole in an oven was done by R. C. Miller and P. Kusch. They used potassium and thallium vapors as the gases. Instead of exploiting free fall as Estermann, Simpson, and Stern did, Miller and Kusch used a velocity selector. This consisted of a cylinder in whose

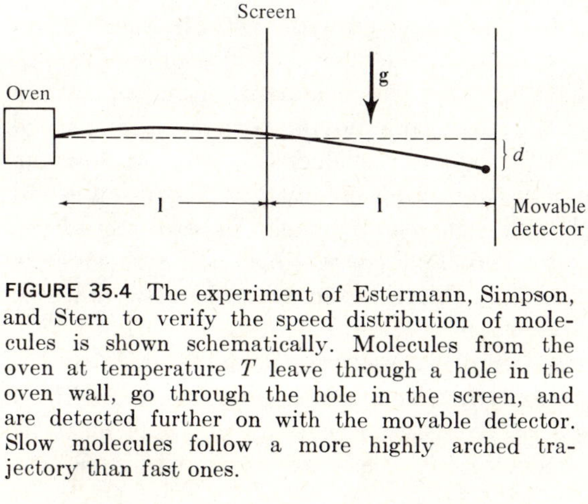

FIGURE 35.4 The experiment of Estermann, Simpson, and Stern to verify the speed distribution of molecules is shown schematically. Molecules from the oven at temperature T leave through a hole in the oven wall, go through the hole in the screen, and are detected further on with the movable detector. Slow molecules follow a more highly arched trajectory than fast ones.

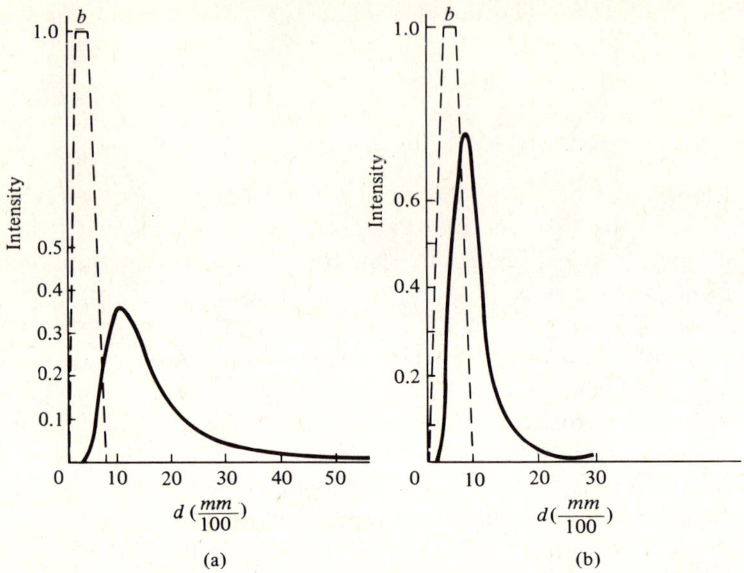

FIGURE 35.5 The number of atoms from an oven at temperature T which are detected in the Estermann-Simpson-Stern experiment illustrated in Figure 35.4 after falling a distance d is plotted as it depends on d. The curve drawn is what one would expect if the speed distribution is a Maxwell-Boltzmann one. (a) cesium atoms and (b) potassium atoms. (By permission.)

outer surface helical grooves were cut like the threads on a screw. The cylinder was rotated at an angular velocity ω about its axis. A molecule that entered one end of the slot and traveled with the right speed could go the length of the cylinder without touching the sides of the groove. Its speed would have to be that with which a nonrotating nut would travel down the rotating screw. A molecule moving with the wrong speed would strike the side of the groove and condense there. By varying the speed of rotation of the cylinder, molecules of various speeds could be selected and their number determined. A schematic drawing of the apparatus is shown in Figure 35.6, and curves illustrating the results obtained are given in Figure 35.7. Here again what is determined is $vN(v)$ rather than $N(v)$.

It seems that the statistical model of a gas, including the detailed properties of the speed distribution of molecules in a gas, is really a very accurate representation of a gas, a representation that must be taken seriously and whose difficulties such as those with the molar heat

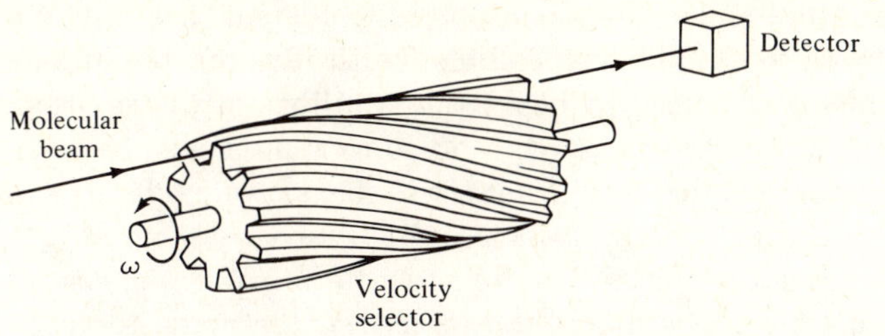

FIGURE 35.6 A schematic drawing of the Miller-Kusch experiment to verify the Maxwell-Boltzmann molecular speed distribution shows the drum with helical grooves that constitutes the velocity selector, the molecular beam, and the detector. Only molecules traveling with the right speed can go the length of the groove without being hit by a groove wall and condensing on it.

are real difficulties with mechanics as described by Newton's laws of motion.

A natural question to ask is how the molecules reach the velocity distribution corresponding to equilibrium at temperature. T. One could imagine that all the molecules in a box are started out with the same speed v and that the walls of the box are rigid so that molecules bounce off them without change of speed. Even under these conditions the molecules will soon arrive at the Maxwell-Boltzmann distribution because

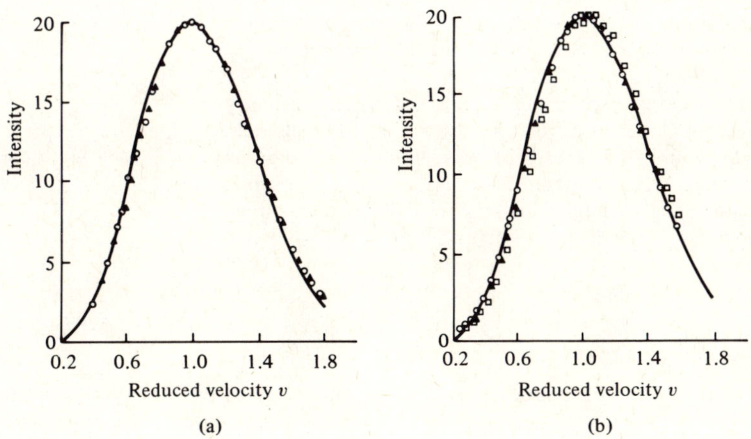

FIGURE 35.7 The speed distribution of atoms as determined by Miller and Kusch using the velocity selector as in Figure 35.6. (a) Thallium atoms and (b) potassium atoms. (By permission.)

of collisions among the molecules themselves. Collisions are random events, and after each molecule has collided with five or ten others the distribution comes very close to being the equilibrium distribution. If the gas is cool and dilute, there will be longer intervals between collisions and the approach to equilibrium will be slower. There is a time interval called the *relaxation time* required for a system to get close to equilibrium from any initial state of preparation. When measurements of distributions are to be made, they must not be made too soon after preparing the system to be studied. For hydrogen at atmospheric pressure and 0°C the relaxation time is only about 10^{-7} sec.

Problems

1. The mean kinetic energy of a molecule at temperature T is $\frac{3}{2}kT$. The root mean square speed for such a molecule is therefore obtained from $\frac{1}{2}m\overline{v^2} = \frac{3}{2}kT$ or

$$\sqrt{\overline{v^2}} = \sqrt{\frac{3kT}{m}}$$

Evaluate this speed for the cesium atoms (atomic weight 133) at an oven temperature of 450 K corresponding to an Estermann-Simpson-Stern experiment.

2. The atoms in the Miller-Kusch experiment fall in the gravitational field of the earth just as do those in the Estermann-Simpson-Stern experiment. This does not appreciably affect the Miller-Kusch experiment if the distance of fall is less than the groove depth of the velocity selector and if the beam is directly above the axis of the selector. Explain the difference between having the beam here and having it alongside this axis.
3. If the Miller-Kusch experiment were done with a velocity selector 50 cm long and rotating about a vertical axis, how much would the speed of a cesium atom moving with the typical speed given in Problem 1 change while the molecule goes through the selector? What percentage change in speed is this?

36. BROWNIAN MOTION

The molecular model of a gas requires the mean kinetic energy of linear motion of a gas molecule to be $\frac{3}{2}kT$, a value independent of the kind of molecules involved. If the gas is a mixture of various gases, these will all be at the same temperature because bodies in contact eventually reach the same temperature, and the mean kinetic energy of the different kinds of molecules will be the same. Therefore, if a very large "molecule" is introduced into a gas, a "molecule" so large that it can be seen with a microscope and yet small enough so that the speed associated with a kinetic energy $\frac{3}{2}kT$ is not completely negligible, a direct check on the model should be possible. A large "molecule" in a liquid should behave somewhat similarly.

The motion of pollen grains suspended in water was observed by the botanist Robert Brown in 1828, and even earlier observations of things that may have been essentially the same as this "Brownian motion" were made in the seventeenth century. The meaning of this motion was not understood, and it was ascribed variously to the vital force in the pollen, to gross motion of the liquid or gas, and to various other things. On the whole the motion as seen is slow; it is impossible to analyze it into the separate zigzags produced by individual molecular collisions, and therefore it is impossible to estimate the kinetic energy associated with it directly.

In 1905 Albert Einstein published a theory of this effect, attributing the motion of the grain or "molecule" to the random bombardment it undergoes by the molecules of the fluid in which it is suspended. The effect of this bombardment is equally likely to be a displacement of the grain in one direction or its opposite, so that the average displacement of a particular pollen grain as observed over many equal time intervals is zero. There are, however, fluctuations in these displacements that make the grain move first in one direction and then in another, just as an ordinary molecule follows a random path because of its collisions with other molecules. The important point of Einstein's theory is that it concentrates on what can be seen, and one thing that can be measured is the distance between the particle's starting point and its final position after the elapse of the time τ even though the twists and turns during this time cannot be followed. What Einstein calculated on the basis of molecular theory was the mean square distance the particle should travel in this time interval. To compare experiment and theory, then, the particle is observed for many time intervals, each of length τ, and the squares of the distances traveled during these intervals are averaged.

The realistic analysis of this problem is quite complicated because the effect of a collision on a particle is to change its velocity; then to find the displacement of the particle, the distance traveled between collisions must be calculated. All this can be done, but instead of doing it we consider a much simplified situation that has many of the same essential features as the real one while being easy to analyze.

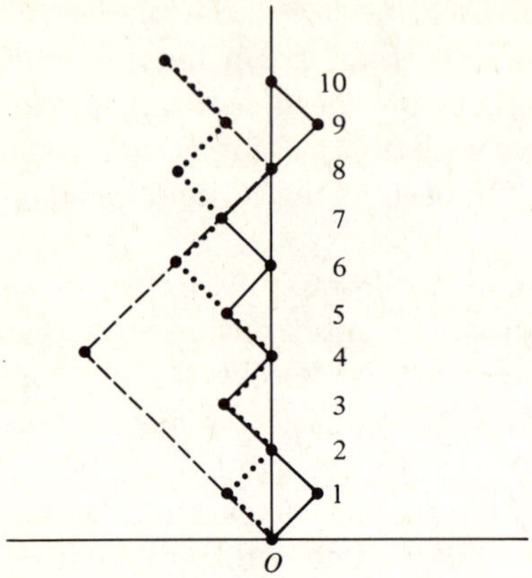

FIGURE 36.1 Three random walks of ten steps each were made by tossing a coin and moving to the right on heads or to the left on tails. The number of tails is slightly greater than the number of heads since the walks tend to end to the left, but the difference is too small to mean the coin was not balanced.

Imagine a particle restricted to move along a line, the x-axis. This particle is assumed to suffer collisions with molecules, the effect of any one collision being to move the particle a distance Δ either to the right or left, these directions having equal probabilities. If the particle starts at the origin O at time $t = 0$, what is the mean square distance of the particle from the origin after n collisions have taken place? Three possible sequences of ten displacements each are shown in Figure 36.1. The distance D_n found in any one obervation is the sum of n successive displacements.

$$D_n = \Delta_1 + \Delta_2 + \cdots + \Delta_n \qquad (36.1)$$

Because each Δ_i has a magnitude Δ and is equally likely to be positive or negative, it follows that

$$\bar{D}_n = 0 \qquad (36.2)$$

If D_n is measured many times, the average result will be zero because the particle will be found as often to the right as to the left of O, and these cancel each other. $\bar{D}_n$ is not, therefore, a particularly interesting quantity.

The value of $\overline{D_n^2}$ is much more interesting. D_n^2 is always positive and so no cancellations between positive and negative values take place in the process of averaging and a result different from zero is obtained. To evaluate $\overline{D_n^2}$, the expression for D_n must be squared. This yields

$$
\begin{aligned}
D_n^2 &= (\Delta_1 + \Delta_2 + \cdots + \Delta_n)^2 \\
&= \Delta_1^2 + \Delta_2^2 + \cdots + \Delta_n^2 \\
&+ \Delta_1(\Delta_2 + \Delta_3 + \cdots + \Delta_n) \\
&+ \Delta_2(\Delta_1 + \Delta_3 + \cdots + \Delta_n) \\
&\cdot \\
&\cdot \\
&+ \Delta_n(\Delta_1 + \Delta_2 + \cdots + \Delta_{n-1})
\end{aligned}
\tag{36.3}
$$

This is just an extension of the formula

$$
\begin{aligned}
(a + b + c)^2 &= a^2 + b^2 + c^2 \\
&+ a(b + c) \\
&+ b(a + c) \\
&+ c(a + b) \\
&= a^2 + b^2 + c^2 + 2ab + 2ac + 2bc
\end{aligned}
$$

to more than three terms. In the second line of Eq. 36.3 all the quantities Δ_i^2 are equal to Δ^2, so this line adds up to $n\Delta^2$. All the other lines average to zero.

To see why the third line of Eq. 36.3 has a zero average value over many sets of n steps each, consider how the average is found. One takes many sets of n collisions. Of these, on the average half will have Δ_1 positive and half will have Δ_1 negative. If those sets with Δ_1 positive are selected, on the average they will have $\Delta_2 + \Delta_3 + \cdots + \Delta_n$ add up to zero because each of these is equally probable to be positive or negative. The same is true for those sets with Δ_1 negative, so no matter what the value of Δ_1 the average of the sum $\Delta_2 + \cdots + \Delta_n$ in the third line is zero. The other lines average to zero for the same reason. Thus on averaging Eq. 36.3 only the first line survives because only here are there no cancellations possible and we find that

$$
\overline{D_n^2} = n\Delta^2 \tag{36.4}
$$

This can be compared with the results of observation.

If the particle is bombarded at a constant rate, then n is proportional to the time interval τ over which the particle is observed, and we write

$$D = \sqrt{\overline{D_n^2}} \propto \sqrt{\tau} \tag{36.5}$$

The "root mean square" distance is proportional to the square root of the length of the time interval. This is quite different from motion with constant velocity where the distance traveled is proportional to the time spent traveling; the difference is due to the random direction of the separate steps. The process described is called a *random walk* and is a simple example of an extremely important kind of process studied in a statistical way.

The result of Einstein's analysis of this Brownian motion is very similar to the above. It gives the value of $\overline{D^2}$ for a three-dimensional motion, and it is

$$\overline{D^2} = \frac{kT}{\pi\eta a}\tau \tag{36.6}$$

Here kT gives a measure of the violence of the individual collisions; the more violent the collision, the greater the resulting displacement. η is the viscosity of the fluid, mentioned in connection with the Millikan oil drop experiment in Section 19 and studied further in Section 37, and a is the radius of the particle, assumed to be spherical. ηa measures the resistance the fluid presents to the motion of the particle through it, and the greater this resistance, the smaller the displacement produced by a given collision.

If k is known, the entire right side of Eq. 36.6 can be evaluated experimentally and compared with $\overline{D^2}$, thereby checking the theory. On the other hand, Eq. 36.6 can be taken as correct and used to find Boltzmann's constant k. A knowledge of k is equivalent to a knowledge of Avogadro's number N_0, the macroscopically measurable gas constant R being N_0k. Thus observation of Brownian motion leads to a value of N_0. The result is consistent with values of N_0 obtained in other ways. This demonstration that the motion of a visible particle obeys the statistical laws expected from the molecular model of gases was the decisive factor in persuading many eminent physicists and chemists that atoms and molecules really exist as individual particles obeying the laws of mechanics. We are now so accustomed to this idea that it is hard to see how it could have been resisted, but the skeptics had a right to be skeptical until experimental evidence involving individual particles could be presented. The Einstein theory of Brownian motion provided one of the first pieces of this evidence.

Problems

1. Toss a coin 100 times and record each head that appears as +1 and each tail as −1. This series may be considered as 10 random walks of 10 unit steps each or as 20 random walks of 5 unit steps each. In each of these sets of walks evaluate D^2 and see how closely $\overline{D_{10}^2} = 2\overline{D_5^2}$. The result probably will be far from exact because these samples are quite small.
2. A dust speck of mass 10^{-12} g floats in the air at room temperature (300 K). What is its root-mean-square speed? Does your answer correspond to the way a dust speck is seen to move in a beam of light in a dark room?

37. VISCOSITY OF GASES

Friction in general is a very complicated phenomenon that obeys no simple laws and involves effects not at all well understood. When friction acts, the kinetic energy of a large body is partially or completely transformed into heat energy, spoiling the simple law of conservation of mechanical energy of macroscopic motion. The only case where friction is not very complicated is where the friction is that due to the low speed motion of a large object through a moderately dilute gas. The molecular model of a gas gives an account of the heat energy of the gas as mechanical energy of its molecules which is simple enough to permit an analysis of frictional effects. The discussion is more difficult than that concerning the pressure in a gas because here collisions between molecules play an essential role that was not true in the case of the pressure, and we must include the effect of these collisions.

To make the discussion as simple and specific as possible, we consider two large, parallel plates of area A at a distance h from each other. The volume between the plates is filled with gas at temperature T and of molecular density $N/V = N/Ah$. The lower plate is held fixed, and the upper plate is moved parallel to itself with velocity $\mathbf{u}$. This is illustrated in Figure 37.1. Under these circumstances it is observed that an external force parallel to $\mathbf{u}$ must be exerted on the upper plate to keep it moving. Because the plate is moving uniformly, there can be no net force on it, so the gas must be exerting a force antiparallel to $\mathbf{u}$ on the plate, a force cancelled by the external force. This is the force of *viscous friction.* Its magnitude is observed to be proportional to the area A of the plate and to the velocity $\mathbf{u}$ of the plate and inversely proportional to the separation h of the moving plate from the fixed

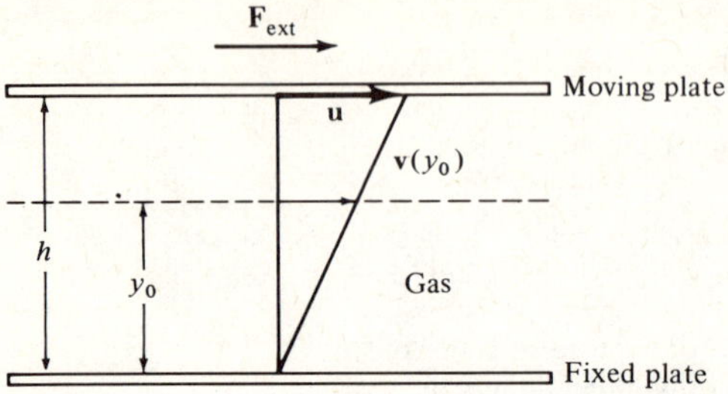

FIGURE 37.1 If the space between two parallel plates is filled with gas, a force must be continually applied to the upper plate in order to keep it moving parallel to itself. This force is due to viscosity of the gas and is quite distinct from the force necessary to prevent the plate moving away from the other plate. The latter is needed to counteract the pressure of the gas.

plate. We therefore write

$$\mathbf{F}_{vis} = -\eta A \frac{\mathbf{u}}{h} \tag{37.1}$$

The minus sign signifies that the viscous force is antiparallel to **u**. The constant η introduced here is the *coefficient of viscosity* of the gas. It is the same coefficient of viscosity that appeared in the discussion of the Millikan oil drop experiment in Section 19 and of Brownian motion in Section 36. We want to relate η to the molecular structure of the gas. As can be seen from Eq. 37.1, the units of η are g cm^{-1} sec^{-1}, these units being called "poises."

The first question to be answered is this: How can the gas exert a force on a plate in a direction parallel to the surface of the plate? If the plate were perfectly smooth so that molecules bounced off it with no change in the components of their momentum parallel to the plate, no such tangential force would appear. Collisions that do not change the tangential components of the molecular momentum do not transfer tangential momentum to the plate, and so no tangential force would be exerted on the plate by such collisions. Only the force normal to the plate, the pressure, occurs under these conditions. The existence of a viscous force shows that this is an inadequate picture of collisions between molecules and the plates, which is not surprising because it is difficult to imagine a plate made of atoms that is smooth on the scale of molecular dimensions. A much more accurate picture is that the plate is rough. This has a consequence that the direction of motion of a molecule scattered from the plate is quite independent of the direction from which the molecule struck the plate. The mean horizontal velocity of molecules that have been scattered from a horizontal plate and have not yet collided with another molecule is just the horizontal velocity of the plate. This is shown in Figure 37.2 for a plate at rest. If the molecules striking the plate also have zero mean velocity parallel to the plate, then on the average there is no tangential momentum transfer and no tangential force. On the other hand, if the incident molecules do have such a mean tangential momentum, this will be transferred to the plate and a tangential force appears. The incident molecules in Figure 37.2 do have a mean momentum to the right, so a force to the right is exerted on the plate.

When there are two plates present that are moving relative to each other as in Figure 37.1, the molecules of the gas cannot have zero mean horizontal velocity relative to them both. Molecules that have had their most recent collisions with the upper plate have mean horizontal velocity **u**. Molecules that have had their most recent collisions with the lower plate have mean horizontal velocity 0. In the interior of the gas there is a mixture of molecules that come from the upper plate with those that come from the lower plate, each kind having had many collisions with other molecules on the way. Near the bottom most molecules have hit the bottom plate more recently than they have hit the top plate, so the mean velocity is low; near the top plate most molecules have hit the top plate more recently than they have hit the bottom plate,

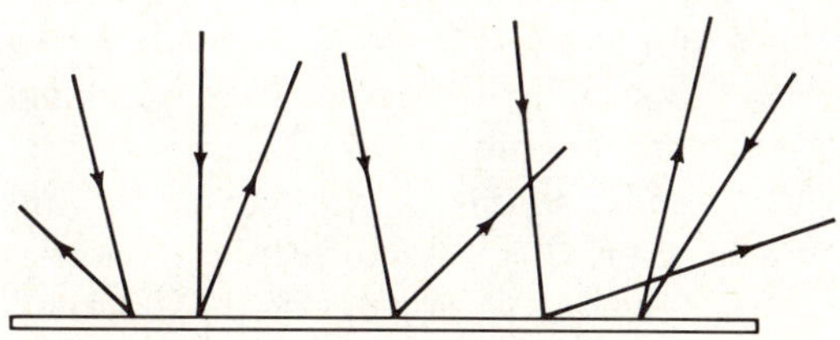

FIGURE 37.2 Gas molecules striking a rough plate are equally likely to bounce off at any angle, so the average component of molecular velocity parallel to the plate is zero if the plate is not moving. If the plate is moving parallel to itself, the mean velocity of the bouncing molecules parallel to the plate is the velocity of the plate.

so their mean velocity is near $\mathbf{u}$. Denoting by $\bar{\mathbf{v}}(y)$ the mean velocity of the molecules that had their last collision with another molecule at height y above the bottom plate, we have

$$\bar{\mathbf{v}}(y) = \frac{y}{h}\mathbf{u} \tag{37.2}$$

which has the value 0 at the lower plate where $y = 0$, and the value $\mathbf{u}$ at the upper plate where $y = h$. This mean velocity is superimposed on a Maxwell-Boltzmann velocity distribution appropriate to the temperature of the gas.

When the molecules of a gas can collide with one another, it is convenient to introduce the mean distance that a molecule goes between collisions. This is called the *mean free path* of a molecule and is denoted by λ. The probability that a particular molecule will have a collision in a given distance is proportional to the number of molecules with which it can collide, that is, to the density of molecules. The mean free path, therefore, is inversely proportional to the molecular density. It does not depend on the temperature of the gas; the effect of increasing the temperature is to speed up the molecular motion and thus to reduce the time between collisions, but this does not affect the distance between collisions. When h is much larger than λ, a molecule going between the two plates will have many collisions en route.

To find the viscous force exerted on the upper plate by the gas we calculate the transport of horizontal momentum upward through a plane parallel to the plates and a height y_0 above the lower plate. There also will be a transport of vertical momentum, but this just produces the pressure on the upper plate, which does not interest us here. Because the momentum of the gas above this plane is not changing with time, the momentum received through the plane must be passed on up, eventually reaching the upper plate and canceling the momentum provided by the external force that keeps this plate moving.

The number of molecules with speeds in the range between v and $v + \Delta v$ is $N(v)\ \Delta v$. This specifies nothing about the direction of motion or about the value of any one component of v, say v_y. If all directions of motion are equally probable, the fraction of molecules with speeds in the range between v and $v + \Delta v$ that have components v_y with values in the range between v_y and $v_y + \Delta v_y$ is

$$\frac{\Delta v_y}{2v}$$

A check on this is provided by summing over all the intervals Δv_y from the lowest possible value $-v$ to the largest possible value $+v$. The sum of the Δv_y's over this range is just $2v$, the size of the entire interval. The sum of the fractions $\Delta v_y/2v$ thus is unity, which means that all the molecules are accounted for. The number of molecules that simultaneously have speeds in the range Δv and v_y in the range Δv_y is then

$$[N(v)\ \Delta v] \times \left[\frac{\Delta v_y}{2v}\right] \tag{37.3}$$

The first factor is the total number in this speed range, and the second factor is the fraction of these in the specified component range.

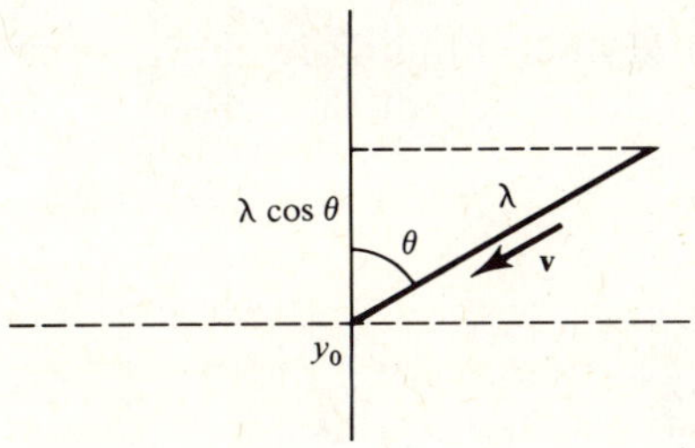

FIGURE 37.3 On the average a molecule travels a distance λ between collisions, so a molecule arriving at the plane y_0 moving in the direction shown will on the average have had its last collision at a height λ cos θ above this plane.

The number of molecules of speed v and component v_y that will cross the plane at y_0 in time Δt is the number in a layer of thickness $v_y\ \Delta t$ and area A, namely,

$$\left[N(v)\ \Delta v\,\frac{\Delta v_y}{2v}\right] \times \left[\frac{A v_y\ \Delta t}{Ah}\right] \tag{37.4}$$

On the average these will have come from a distance

$$\lambda \cos\theta = \lambda \frac{v_y}{v}$$

from the plane y_0, θ being the angle between v and the y-direction shown in Figure 37.3, and will therefore have a mean horizontal velocity given by Eq. 37.2 of

$$\frac{(y_0 - \lambda v_y/v)}{h}\mathbf{u} \tag{37.5}$$

The mean horizontal momentum transported across the plane per unit time by molecules of this kind is then

$$\left[N(v)\,\Delta v\,\frac{\Delta v_y}{2v}\frac{v_y}{h}\right] \times \left[m\,\frac{(y_0 - \lambda v_y/v)}{h}\mathbf{u}\right] \tag{37.6}$$

which is the number of molecules of this kind crossing the plane per unit time multiplied by the mean horizontal momentum of each of these molecules.

The total upward transport of horizontal momentum is found by summing first over all the intervals Δv_y, and then over all the intervals Δv. The sum over all the intervals Δv_y involves

$$\frac{m\mathbf{u}}{2vh^2} N(v)\,\Delta v \sum_{\text{all } v_y} \Delta v_y \left(y_0 v_y - \frac{\lambda}{v} v_y^2\right) \tag{37.7}$$

The sum $\Sigma v_y\,\Delta v_y$ vanishes because for each interval in which v_y is positive there is an equal interval for which it is negative and these contributions cancel. v_y^2 is positive in every interval, and its mean value we already know to be $\frac{1}{3}v^2$, so the sum $\Sigma\,\Delta v_y v_y^2$ is $2v \times \frac{1}{3}v^2$, the length of the total interval times the average value over the interval. The contribution of molecules with speeds in the range between v and $v + \Delta v$ is thus

$$\frac{m\mathbf{u}\lambda}{3h^2} vN(v)\,\Delta v$$

The sum over all the intervals Δv of $vN(v)\,\Delta v$ is just $N\bar{v}$. Finally, then, for the horizontal momentum transport per unit time across the plane y_0 we get

$$\mathbf{F}_{\text{vis}} = -\frac{m\mathbf{u}\lambda N\bar{v}}{3h^2} \tag{37.8}$$

which is to be compared with Eq. 34.1.

Comparing Eqs. 37.1 and 37.8 shows that

$$\frac{\eta A\mathbf{u}}{h} = \frac{Nm\lambda v\mathbf{u}}{3h^2}$$

or that

$$\eta = \frac{N}{Ah}\frac{m\lambda\bar{v}}{3} \tag{37.9}$$

η is proportional to the product of the molecular density, the molecular mass, the mean free path, and the mean speed. In this way the macroscopic quantity η is expressed in terms of microscopic quantities.

Two important conclusions can be drawn from Eq. 37.9. First, the coefficient of viscosity is independent of the molecular density, because while N/Ah appears, so does λ and we saw that λ is inversely proportional to the molecular density. This is a surprising result, but it is amply confirmed by observation. The second is that η should increase with increasing temperature because $\bar{v}$ increases. For a Maxwell-Boltzmann distribution $(\bar{v})^2$ is proportional to $\overline{v^2}$, which in turn is proportional to T. Hence $\eta \propto T^{\frac{1}{2}}$, again a result that is experimentally confirmed.

In this discussion the transport of momentum has been handled very carefully, but nothing has been said about energy, even though one chief characteristic of friction is that it converts mechanical energy of large objects into heat energy, or molecular kinetic energy. The external force that keeps the upper plate moving does work. This work appears as kinetic energy of the molecules, increasing the temperature of the gas. It is not hard to see how this comes about. Molecules that are about to collide with the upper plate have a mean horizontal velocity less than that of the plate, while those that leave the plate do so with mean horizontal velocity equal to that of the plate. These latter molecules are therefore moving with a slightly greater value of $\overline{v^2}$ than those that have not yet struck the plate and hence have a slightly greater mean kinetic energy. This extra energy is rapidly shared with other molecules by collisions, and the temperature rises. When $\mathbf{u}$ is small this temperature rise is very slow.

By this rather involved calculation, whose complication arises from the need to know where molecules come from as well as where they are, we have shown that one kind of friction can be described by a molecular model. Its difficulty makes it plausible that the calculation of friction involving liquids and solids starting from a molecular model is hopelessly complicated.

Problems

1. The coefficient of viscosity of air is about 1.7×10^{-4} poises at 0°C. Assuming for simplicity that a space capsule is a sphere 2 meters in diameter, what is the force of friction

on it in air when it is traveling at 20,000 mph, just a little less than escape speed, according to Eq. 19.10? (This speed is too high for the formula to be accurate. The velocity distribution of molecules after colliding with the capsule at such speeds is not given by a Maxwell-Boltzmann distribution, so no temperature can be defined in the way we have done it here. Molecules are actually ionized under these conditions, causing a radio "blackout" for a reentering space capsule.)

2. If two kinds of molecules are of the same size, so that at the same molecular density they have the same mean free path but the molecules of one are twice as massive as those of the second, which will yield a gas with the higher viscosity? How much more viscous will it be than the other at the same temperature?

38. FIRST APPEARANCE OF THE QUANTUM

In Section 34 the molecular model of an ideal gas was described. This model works beautifully to give a mechanical basis for the behavior of gases; in particular it shows why the gas law is so well obeyed, and it gives a good understanding of viscosity. The detailed predictions of the model in the form of the Maxwell-Boltzmann velocity distribution and the Brownian motion are well-confirmed by observation. The one place where it was seen not to work was in accounting for the molar heats of gases. For monatomic gases it works if the internal structure of atoms is ignored, but for diatomic gases it works only if certain types of motion are arbitrarily excluded, and for polyatomic gases it does not work at all. Newtonian mechanics together with some statistical hypothesis that are difficult to question lead to the theorem of the *equipartition of energy*. The mean value of the kinetic energy of linear motion in any direction, of the kinetic energy of rotation about any axis, and of the kinetic and also of the potential energy of any possible vibration should be $\frac{1}{2}kT$ when the motion takes place in a system at temperature T. This result is simple, and it is in accordance with observation if only certain modes of motion are admitted. It is in violent disagreement with observation if all known modes of motion are admitted, as when vibration of a diatomic molecule is considered possible. The law of equipartition of energy is too often valid for us simply to discard it, and yet it is not universally applicable. It fails when the relative motions of parts of molecules are involved but is good when the motion of the molecule as a whole is involved. This indicates that perhaps the classical mechanics of Newton does not provide a good description of motions *within* a molecule, while still being accurate for the motion of the molecule as a whole.

It was not the problem of the molar heats of gases that was first clarified but a related problem concerning the energy associated with the electromagnetic field inside a cavity in a body at temperature T. Inside a cavity there can be light waves of various wavelengths and frequencies, and the mean energy associated with each kind of wave should be equal to the mean energy of a harmonic oscillator, namely, kT. The difficulty is that there are infinitely many kinds of waves possible, there being no limit to how short the wavelength of the waves can be, so that the total mean energy of all the waves may be written, according to the ideas of Newtonian physics,

$$\bar{E} = NkT, \qquad N = \infty \tag{38.1}$$

which is infinite, but this is not directly contrary to experience because only changes in energy are directly observable. This is disagreeable enough, but things become worse if one calculates the energy needed to raise the temperature of the cavity by an amount ΔT. It is

$$\Delta\bar{E} = Nk\,\Delta T, \qquad N = \infty \tag{38.2}$$

which is also infinitely large. Our theory says that an infinite amount of energy is required to raise the temperature of a hole by one kelvin. This is an absurdity but an absurdity that is an inevitable consequence of the physics of Newton and Maxwell and some elementary statistical ideas. We shall not pursue this problem further, but instead we shall concentrate on the molar heat problem.

In the discussion given in Section 34 it appeared that the molar heat of diatomic gases is too big to be simply accounted for by the molecular model, but the actual difficulty is that it is much too small. The value $\frac{3}{2}N_0k$ for the molar heat of a monatomic gas presumes that the molecule (atom) is a structureless point that can move in three dimensions but has no internal motions with which thermal energy can be associated. We know, however, that an atom of a monatomic gas contains a nucleus and electrons, and these must be able to move relative to each other because electrons can be extracted from atoms. Why is there not an energy $\frac{1}{2}kT$ associated with each possible internal motion? Why is there no internal energy in the atoms of a diatomic molecule, and why cannot the two atoms vibrate back and forth along the line joining them, thus constituting a harmonic oscillator? These two atoms can rotate around an axis through the molecular center of mass with mean kinetic energy

$\frac{1}{2}kT$. The problem here is similar to that of the energy of light in a cavity: Why do the long wavelength waves have their share of the thermal energy as predicted by the equipartition of energy when the short wavelength ones do not?

The first step toward answering this question was taken by Max Planck in 1900. As stated before, he did this in connection with cavity radiation, but he could have done it almost as well for molecular motions. More was known about light than about molecular structure in 1900, so comparison of theory and experiment could be made much more meaningful in the case of light, and this comparison was absolutely necessary in order to convince anyone of the usefulness of the radical idea introduced by Planck. Planck's new assumption was that a physical system, in particular a harmonic oscillator, could absorb or emit energy only in discrete amounts called *quanta*. The size of the quanta that can be exchanged depend on the system, and for a harmonic oscillator of angular frequency $\omega = 2\pi\nu$ this quantum size is

$$E = \hbar\omega = h\nu \tag{38.3}$$

It is proportional to the frequency of the oscillator, the constant of proportionality being Planck's constant. The constant introduced by Planck was h; what is often called Planck's constant now is denoted by $\hbar = h/2\pi$. We will use $\hbar$ almost exclusively. Its presently accepted value is

$$\begin{aligned} \hbar &= 1.05449 \times 10^{-27} \text{ erg sec} \\ &= 6.5819 \times 10^{-16} \text{ eV sec} \end{aligned} \tag{38.4}$$

Later it emerged that $\hbar$ is also itself the quantum of angular momentum, as mentioned in Section 20.

Planck's quantum hypothesis produces the desired effect. If an oscillator is colliding with molecules whose mean kinetic energy is $\frac{3}{2}kT$ and if the atom can emit and absorb energy in quanta $\hbar\omega$ that are much smaller than kT, the fact that the energy transfer is quantized will have very little effect; the mean energy of the oscillator will be kT, as it should be on the basis of classical physics. However, if the size of these quanta $\hbar\omega$ is much greater that kT, then only very rarely will the oscillator collide with a molecule with enough energy to give it a quantum, and on the average it will have an energy much less than kT as a result.

This result can be stated quantitatively. The mean energy $\bar{E}$ of a harmonic oscillator of frequency ω is given by

$$\bar{E} = \frac{\hbar\omega}{e^{\hbar\omega/kT} - 1} \tag{38.5}$$

Here e is the number mentioned in Section 35. When $\hbar\omega$ is much greater than kT, the exponential in the denominator is large and $\bar{E}$ is small. When $\hbar\omega$ is much less than kT, the approximation

$$e^x \approx 1 + x$$

can be used with $x = \hbar\omega/(kT)$, yielding

$$\begin{aligned} \bar{E} &= \frac{\hbar\omega}{\left(1 + \dfrac{\hbar\omega}{kT}\right) - 1} \\ &= kT \end{aligned} \tag{38.6}$$

For small quantum size or for high temperature one gets the old result. Figure 38.1 shows how the mean energy approaches the classical value as the temperatures rises. The higher the oscillator frequency, the higher the temperature must get for the oscillator to approach classical behavior.

All that we need to do to apply this to the molar heat problem is

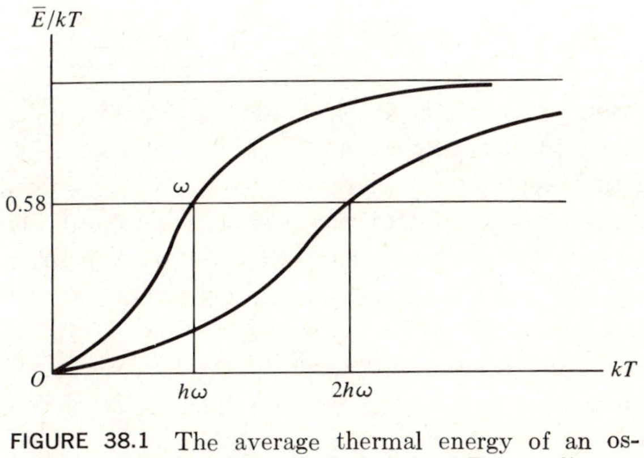

FIGURE 38.1 The average thermal energy of an oscillator at temperature T should be kT according to classical physics. The ratio of the quantum value as given by Planck's formula to the classical value is plotted for two oscillators, one of frequency ω and one of twice that frequency. The low frequency oscillator behaves almost classically at lower temperatures than the high frequency one does.

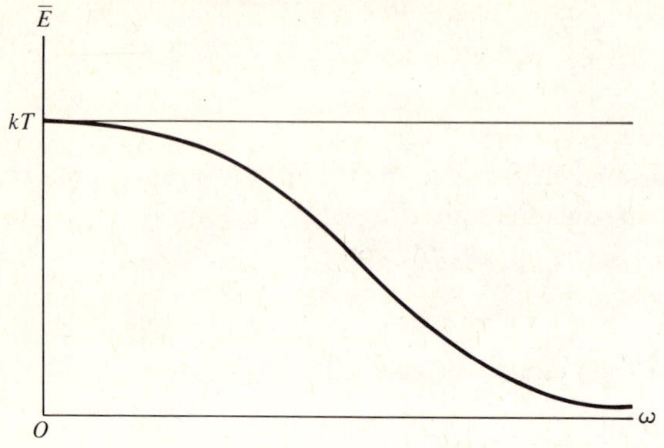

FIGURE 38.2 At a given temperature the mean energy $\bar{E}$ associated with an oscillator depends on the frequency of the oscillator. For very low frequencies it is the classical value kT, and for higher frequencies it gets less, approaching zero for very high frequencies.

to assume that the quanta associated with the internal motion of atoms and with interatomic vibration of atoms in a diatomic molecule are much larger than kT for the temperatures at which molar heats are measured. If this is so then, as can be seen from Figure 38.2, almost no energy is associated with these motions and also no contribution to the molar heat comes from them. As the temperature increases the value of $\bar{E}$ increases too until it becomes very nearly equal to kT. The molar heats of diatomic and polyatomic gases should increase with increasing temperature as the vibrational motions begin to act somewhat as they should according to Newtonian mechanics, and in fact this does occur. The quantum hypothesis makes this behavior understandable, although it does not allow one to calculate the frequencies of vibration to be expected in a molecule. Such calculations had to wait until a nearly complete theory was developed.

Another application of Planck's quantum hypothesis that was soon made was to the *photoelectric effect*. It was known that light shining on a metal surface ejects electrons from that surface. This is not hard to believe when the electromagnetic theory of light is known. The electric field in the light wave exerts a force on the electrons and can give them kinetic energy. One would expect, however, that the energy of the ejected electrons would increase as the intensity of the light increases, since an intense wave involves a stronger electric field than a dim one.

Experimentally this did not seem to be the case. The number of electrons ejected does increase with light intensity, but the energy of each electron does not. The data available in 1905 were, however, very imprecise. In that year Einstein suggested that light consists of quanta of energy $\hbar\omega$, where ω is the frequency of the light. The energy quantum emitted as light by one of Planck's oscillators travels to whatever eventually absorbs the light and is absorbed as a whole. An electron in a metal receives an amount of energy $\hbar\omega$ all at once. It requires some energy $e\varphi$ to get the electron out of the metal, and the rest of this quantum of energy appears as kinetic energy K. Conservation of energy then leads to

$$\begin{aligned} \hbar\omega &= K + e\varphi \\ K &= \hbar\omega - e\varphi \end{aligned} \tag{38.7}$$

The potential φ is called the *work function* of the metal. It varies from metal to metal and depends very strongly on the condition of the surface of the metal. The photoelectric equation, Eq. 38.7, was not accurately verified experimentally until 1916 when it was investigated by R. A. Millikan. It was not easy to make and to keep clean enough metal surfaces.

The hypothesis that light consists of quanta or photons upset the whole wave theory of light with all its detailed agreement with experiment. A theory as good as the wave theory of light is not discarded casually, so the quantum theory of light was not wholeheartedly accepted right away. It was only later when the quantum hypothesis brought order into the chaos of atomic behavior that it was recognized as inevitable. Now the quantum theory encompasses both the wave and the particle aspects of light, neither aspect giving the whole picture and neither aspect excluding the other. The particle aspect has already appeared in Section 23 and following sections, the particle being the photon γ. The application of the quantum hypothesis to atoms is described in Section 39.

Problems

1. At room temperature, 300 K or 27°C (80°F, a warm room), the value of the thermal energy kT is very nearly $\frac{1}{40}$eV. At what temperature would (a) an oxygen molecule, mass about 32 proton masses, (b) an electron have a thermal energy 1 percent of its rest energy?

2. A water molecule consists of two hydrogen atoms and one oxygen atom arranged in a boomerang-shaped structure with the oxygen in the middle. Considered separately, these three atoms can each move in three independent ways (neglecting internal motions), so there are nine independent modes of motion. Considered as a molecule, how many ways can the molecule (a) move linearly as a whole and (b) rotate as a whole and (c) how many modes are left over for vibrational forms of motion?
3. What should the molar heat of water vapor be when (a) the temperature is so low that no vibrational motion is excited and (b) the temperature is so high that the vibrational motion can be treated classically? Compare your answer with the room temperature value given in Table 34.1, and comment on where in this temperature range room temperature seems to lie.

39. ATOMIC SPECTRA

The molecular model of gases leads to difficulties in the understanding of molar heats that can be blamed on an inadequate theory of individual atoms and molecules. To resolve these difficulties we must, therefore, return to the study of individual atoms. The nuclear model of the atom was proposed by Rutherford in 1911 on the basis of experiments that were described in Section 20. These experiments were scattering experiments in which particles of known charge, mass, and energy were deflected from their original motion on going through a thin gold foil in just the way that would be expected if they were to pass at random distances from a point charge and were to be acted on by the Coulomb field produced by that point charge. The motion of the bombarding alpha particle is a large scale motion, and nothing but Newtonian mechanics and Coulomb's law is required for the analysis, so there is no escape from the conclusion that the gold atom contains a massive, concentrated charge, the nucleus. Atoms also contain electrons, since electrons can be produced as cathode rays in any gas. After Rutherford's experiments, then, the atom was known to consist of a positively charged nucleus carrying the charge $+Ze$ and Z electrons each with charge $-e$. This situation was disastrous for the classical physics of Newton and Maxwell.

The only important force acting between the nucleus and the electrons in an atom is electromagnetic. Other forces known to act are gravitation and the weak interaction, but these have undetectably small effects in comparison with the Coulomb force and some magnetic forces. The problem is to construct a model of the atom made up of the nucleus and electrons and to have it consistent with what is known about electric and magnetic interactions. The nucleus contains most of the mass of the atom, roughly 99.95 percent of it, so that it can be considered to be at rest if the atom as a whole is stationary. Then the electrons either must be stationary in equilibrium positions in the atom or must be

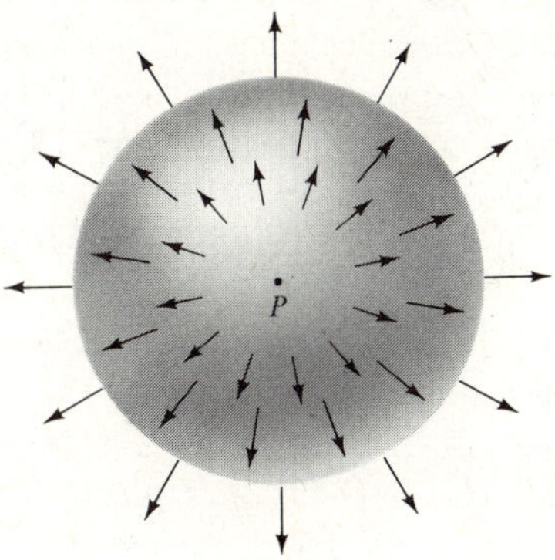

FIGURE 39.1 If point P is to be an equilibrium point for an electron, the electron must be attracted toward P no matter in what direction it is displaced from P. Therefore, there must be a net number of electric field lines leaving P, and P cannot be just a point in space but must be the site of a positive charge.

moving around in the atom, neither of which is consistent with classical physics.

It is an elementary theorem of electrostatics that a charged particle cannot be in stable equilibrium in an electrostatic field. This can be seen with the help of Gauss' theorem, as stated in Section 16. Suppose an electron is in stable equilibrium at a point P. This means that the electron is attracted to P no matter what direction it is moved away from P, so at all points of a small spherical surface drawn with P as its center there must be an electric field with a positive outward component (outward because the negative electron is attracted), as shown in Figure 39.1. There is, in other words, a number of lines that leave this sphere, and for this reason there must be a positive charge inside the sphere. Only at the nucleus, where the positive charge is concentrated, can electrons be at equilibrium, but this is impossible because the electrons' negative charges would cancel the positive nuclear charge and no alpha particle scattering would ever happen. The static model does not work.

An electron might be expected to move around the nucleus in an elliptical orbit as the planets move around the sun, the force law in each case being an inverse square law. An atom is, perhaps, a miniature solar system. This pretty model does not work because an electron moving in such an orbit would lose energy in the form of electromagnetic radiation at such a rate that the atom would collapse in less than 10^{-6} sec. The solar system also is thought to radiate energy, as gravitational energy in this case, but it does not collapse because the rate of emission of gravitational energy by the planets is too low to be detected mainly because Newton's constant of gravitation G has the small value 6.7×10^{-8} dyne cm^2/g^2. The electromagnetic radiation emitted by a planetary electron should have a fundamental frequency equal to the period of revolution of the electron around the nucleus, and multiples of this frequency should occur. The prediction of electromagnetic radiation by large moving charges was verified experimentally by H. Hertz in 1888 and is the basis for the operation of all broadcasting antennas, so the fault with the planetary model of the atom is basic.

The electrons in an atom can neither be at rest nor can they be in motion if the laws of classical physics apply to the internal motions of atoms. Atoms do exist, so there is little choice about giving up at least some part of classical physics when systems of atomic size are considered. This should not be too disturbing because classical physics was developed completely on the basis of large scale observation. In the microscopic world of the atom one might expect different principles to apply. One would like a scheme that gives an accurate description of both the large scale classical realm of physics and the atomic realm. Quantum mechanics provides such a scheme, and we say something about the quantum ideas of atomic structure here.

When atoms emit light, as in the gas discharge used by Thomson as a source of electrons, the emitted light does not consist of multiples of a few fundamental frequencies or of a continuum of frequencies as from an oven, but rather it contains series of distinct frequencies. In a spectrograph, an instrument that separates light of different frequencies, these show up as distinct bright lines called *spectral lines*. Atoms of different elements have different spectra of lines, and the presence of an element can be detected by observation of the characteristic lines of its spectrum. It is in this way that the chemical constitution of stars can be studied. Atoms left to themselves do not radiate light at all.

One of the first regularities discovered in the spectrum of lines emitted by elements was the discovery by J. J. Balmer that the visible lines in the spectrum of atomic hydrogen have frequencies given by the formula

$$\nu_n = Ry\left(\frac{1}{4} - \frac{1}{n^2}\right) \tag{39.1}$$

Here ν_n is the frequency of a line in the series and n is an integer greater than two. This series of lines is illustrated in Figure 39.2. The constant Ry (the only physical constant denoted by two letters) is a

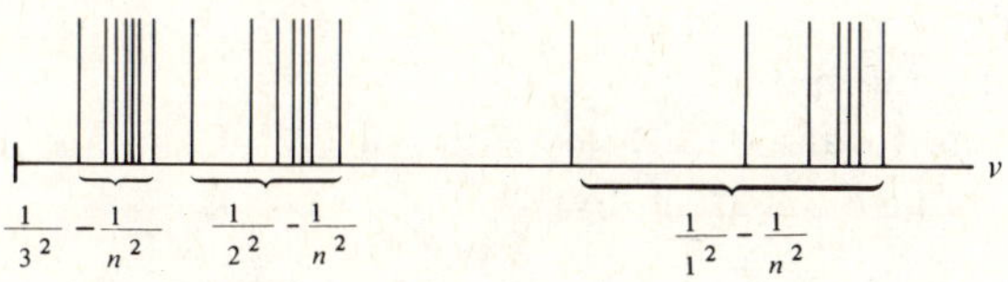

FIGURE 39.2 Three series of lines in the spectrum of hydrogen are shown. The one on the right is called the Lyman series and contains the highest frequency lines in the spectrum—in the ultraviolet and therefore invisible to the naked eye.

frequency characteristic of the hydrogen atom, called the Rydberg, and it has the value

$$1Ry = 3.28985 \times 10^9 \text{ Hz} \tag{39.2}$$

Later other series of lines with frequencies given by

$$Ry\left(\frac{1}{1^2} - \frac{1}{n^2}\right), \qquad Ry\left(\frac{1}{3^2} - \frac{1}{n^2}\right), \qquad \text{and so on}$$

also were discovered. Equation 39.1 is extraordinarily accurate, having a disagreement with observation of less than one part in 100,000. In this way the many lines in the spectrum have frequencies given by the differences between a smaller number of *terms,* as the quantities Ry/n^2 are called.

The existence of terms was found to be common to all spectra; the frequencies of the lines emitted by the atoms of any element can be expressed as the differences of terms, many lines having frequencies specified by relatively few terms. The general equation is written

$$\nu_{nm} = T_n - T_m \tag{39.3}$$

where T_n and T_m are two terms belonging to the atoms of that element. In cases other than hydrogen, no simple, accurate formula for the terms can be given. This general existence of atomic terms was noticed by Walther Ritz in 1908, and Eq. 39.3 is called the *Ritz combination principle.* It has no physical interpretation in any classical model of the atom.

A large measure of order was brought into this subject by Niels Bohr starting in 1913. Bohr combined the ideas of Planck's quantum hypothesis, Einstein's light quantum or photon, and Ritz's terms to produce the model of the atom that bears his name. This model is semiclassical, using the ideas of classical physics except at certain crucial points where quantum ideas are introduced in an *ad hoc* way. It now has a firm basis in the theory of quantum mechanics.

One of Bohr's big steps was to associate a certain energy of the atom with each of Ritz's terms, so that E_n is given by

$$E_n = -2\pi\hbar T_n \tag{39.4}$$

The negative value of E_n means that the electron does not have enough energy to escape from the atom but is in a bound state. Bohr also assumed, in contradiction to classical physics, that only those energy values corresponding to atomic terms are possible for an atom and that an atom with one of these allowed energies would not radiate light. The emission or absorption of light of frequency ν involves a *change* in the atom's energy by an amount $2\pi\hbar\nu = \hbar\omega$ according to Einstein's photon idea, and this is possible only if this energy just matches the energy difference between two allowed energies according to Ritz's combination principle. The idea of allowed energy values for an atom thus unifies several distinct notions in a way agreeing with experiment.

The hydrogen atom is the proving ground for ideas in atomic physics because it is so simple, containing only one electron. The next simplest atom is helium, containing two electrons. In the helium atom the repulsion between the electrons is half as big as the attraction of the doubly charged nucleus for either electron when the distance is the same, and this complicates the situation so much that calculations on helium are much less accurate than calculations on hydrogen. Bohr used hydrogen as the testing ground for his ideas and showed that they work well.

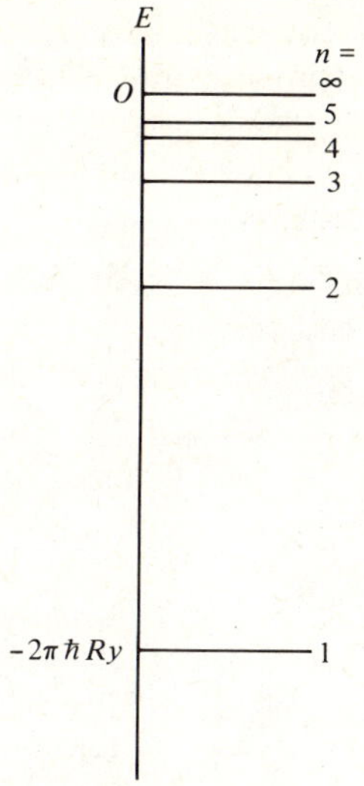

FIGURE 39.3 The allowed energy levels of an electron in a hydrogen atom according to the rules of Bohr. The lowest level has the principal quantum number $n = 1$. The energies are negative because the electron is bound in the atom, and the zero of energy is chosen to be that of an electron at rest very far from the proton.

The terms for hydrogen are given by Balmer's formula, which is rewritten using the energy as

$$E_n = \frac{-2\pi\hbar Ry}{n^2} \tag{39.5}$$

Figure 39.3 shows a diagram of the discrete allowed energy levels of hydrogen. The energy is negative because the electron is bound to the nucleus, and the zero of energy corresponds to the electron being very far from the nucleus and at rest. This negative total energy was encountered in discussing satellite motion in Section 13. As n, the so-called *principal quantum number*, gets large, the energy gets small in magnitude and the electron gets far from the nucleus. For large quantum numbers the orbits are large and, therefore, classical results should be the same as the new quantum results. Bohr used this idea, known as the *correspondence principle,* to find a connection between the Rydberg and other, previously known quantities.

When a hydrogen atom makes a transition from an energy level with principal quantum number $n + p$ to a level with principal quantum number n, the frequency of the light emitted is given by the Balmer formula Eq. 39.1 or one of its variations with $1/2^2$ replaced by the inverse square of some other integer.

$$\begin{aligned}\nu &= Ry\left(\frac{1}{n^2} - \frac{1}{(n+p)^2}\right)\\ &= Ry\left(\frac{(n+p)^2 - n^2}{n^2(n+p)^2}\right)\end{aligned}$$

Now if n is taken to be very large and p is taken to be much less than n so that n and $n + p$ are almost equal, the above expression for the frequency ν simplifies:

$$\begin{aligned}\nu &= Ry\,\frac{2np + p^2}{n^2(n^2 + p^2)}\\ &\approx Ry\,\frac{2np}{n^4} \qquad (39.6)\\ &= \frac{2Ry}{n^3}\,p\end{aligned}$$

For various values of the integer p these frequencies are all multiples of a fundamental frequency ν_1 where

$$\nu_1 = \frac{2Ry}{n^3} \qquad (39.7)$$

This is just what classical physics would predict if the fundamental frequency ν_1 is the frequency of revolution of the electron in its orbit around the nucleus. Bohr required ν_1 to be this orbital frequency and thereby found what the Rydberg is.

Classically the energy of a particle moving in a circular orbit in an inverse square force is just half the potential energy at the radius of the orbit, as was shown in Eq. 13.21. In a hydrogen atom, therefore, we may write the energy E as

$$E = -\frac{1}{2}\frac{e^2}{r}$$

$-e^2/r$ being the potential energy of the electron when it is a distance r from the nucleus of charge $+e$. If this energy is put equal to one of the allowed energy values of Bohr's model we obtain

$$-\frac{2\pi\hbar Ry}{n^2} = -\frac{1}{2}\frac{e^2}{r}$$
$$r = \frac{n^2e^2}{4\pi\hbar Ry} \tag{39.8}$$

This shows that the radii of the various possible circular orbits increase

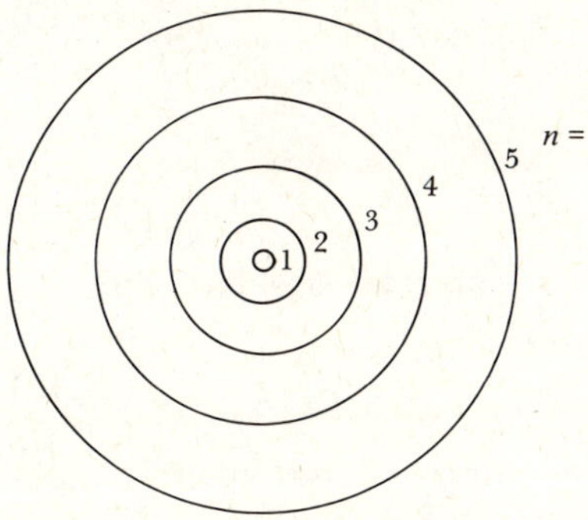

FIGURE 39.4 The circular orbits of the original Bohr model of the atom have radii proportional to the square of the principal quantum number. For high quantum numbers the orbits are large, and when large enough they should be describable by classical ideas.

with the square of the principal quantum number, as illustrated in Figure 39.4.

The centripetal force needed to make a particle move in a circular orbit is provided in the hydrogen atom by the Coulomb attraction of the nucleus for the electron, so

$$\frac{mv^2}{r} = m\omega^2 r = \frac{e^2}{r^2}$$

The frequency of revolution of the electron, $\nu = \omega/2\pi$, is therefore given by

$$\begin{aligned}\nu^2 &= \frac{e^2}{4\pi^2 mr^3}\\ &= \frac{e^2}{4\pi^2 m}\left(\frac{4\pi\hbar Ry}{n^2e^2}\right)^3\\ &= \frac{16\pi\hbar^3 Ry^3}{n^6e^4}\end{aligned} \tag{39.9}$$

This frequency is to be the same as that found from the Balmer formula for large n, Eq. 39.7. Thus, squaring Eq. 39.7 to get

$$\nu_1^2 = \frac{4Ry^2}{n^6} \tag{39.10}$$

we equate ν^2 and ν_1^2 and obtain the equation

$$\frac{16\pi\hbar^3 Ry^3}{n^6e^4} = \frac{4Ry^2}{n^6} \tag{39.11}$$

The principal quantum number cancels out, and we are left with

$$Ry = \frac{me^4}{4\pi\hbar^3} \tag{39.12}$$

This equation states that a quantity whose value is determined by measuring the frequencies of spectral lines emitted by atomic hydrogen is equal to a combination of quantities determined by three other experiments; the Thomson e/m experiment, the Millikan oil drop experiment, and the measurement of $\hbar$ by means of the radiation emitted by a hot oven or the photoelectric effect. The equation is satisfied by the experimentally determined numbers, which is a strong indication that the underlying ideas are correct even though they do not at first fit into any general theory that tells one when classical ideas are good and when they must be replaced by new, quantum ideas.

One other consequence of Bohr's model of the atom is worth deriving. It concerns the angular momentum of the electron in its motion around the nucleus. We start from the definition of this angular momentum,

$$rmv = mr^2\omega$$

For large orbits this can be evaluated with the help of Eq. 39.8 for the value of r and Eq. 39.7 for the value of $\omega = 2\pi\nu_1$

$$\begin{aligned}rmv &= mr^2\omega\\ &= m\left(\frac{e^2}{-2E}\right)^2 \times \left(\frac{4\pi Ry}{n^3}\right)\end{aligned}$$

After using the Balmer expression Eq. 39.5 for E, this reduces to simply

$$rmv = n\hbar \tag{39.13}$$

This result must be used with caution because it depends on n being large, so large that n and $n + p$ can be treated as almost equal as in the derivation of Eq. 39.7. What it does mean is that when the atom makes a transition between two energy levels, the value of the angular momentum, or more exactly the value of any component of the angular momentum, can change only by an integer multiple of $\hbar$. This result is generally valid in quantum mechanics and has been used previously in talking about the spin angular momentum of particles.

The Bohr rules make order out of the chaos of observational data on atomic spectra. There are many effects we have not gone into here that also confirm the Bohr model in many ways. This does not make a theory out of the model; it does not tell one beforehand when to give up classical ideas and use the new ones, and the two sets of ideas are mutually contradictory. This situation was cleared up by Werner Heisenberg and by Erwin Schrödinger, who independently of each other invented two distinct but equivalent forms of quantum mechanics. The theory requires considerable mathematics for its complete description because it is rather an abstract theory, as it must be because it applies to so many situations. A little bit can be said about it here, however.

Newtonian mechanics depends in its formulation on our being able to specify the position and the velocity of a particle at any instant of time. We saw in Section 9 that the essence of Newton's second law of motion is that the future motion of a particle under given external conditions is completely determined by the position of the particle $\mathbf{r}_0$ and the momentum of the particle $\mathbf{p}_0$ (or the velocity $\mathbf{v}_0$) at an initial time t_0. This implies that the position $\mathbf{r}_0$ and the momentum $\mathbf{p}_0$ can be determined simultaneously with whatever accuracy is required. Quantum mechanics differs from Newtonian mechanics in using a description of a particle that makes it impossible to specify both the position and the momentum of the particle at a single instant. This impossibility is expressed in the *Heisenberg uncertainty principle* as

$$\Delta x \, \Delta p_x \geq \frac{\hbar}{2} \tag{39.14}$$

Δx here means a range of values of x all of which are within the uncertainty in our knowledge of x, and likewise for Δp_x. The uncertainty in the value of a coordinate of a particle times the uncertainty in the value of the momentum component in the direction of the coordinate is never less than half of Planck's constant.

The Heisenberg uncertainty principle arises from the need to assign wave properties to particles, which is complementary to Einstein's assignment of particle properties to light waves when he introduced the concept of the photon. It was Louis de Broglie who saw that the Planck-Bohr relation between energy and frequency,

$$E = \hbar\omega = 2\pi\hbar\nu$$

implies a relation between momentum and wavelength. This relation is

$$p = \frac{2\pi\hbar}{\lambda} \tag{39.15}$$

The connection between these formulas comes from the theory of relativity, where the energy of a particle in one inertial coordinate system is expressed as a function of its energy and momentum in another coordinate system by a Lorentz transformation such as Eq. 22.15. We need not go into the details of this derivation. The wavelength

$$\lambda = \frac{2\pi\hbar}{p} \tag{39.16}$$

is called the *de Broglie wavelength* of the particle.

The physical existence of this wave aspect of particle behavior was confirmed experimentally by C. J. Davisson and L. H. Germer and by G. P. Thomson, the son of J. J. Thomson of the e/m experiments. The wavelength is made long by making the momentum small, which is easier to do with light particles such as electrons than with heavy particles such as atoms. Davisson and Germer observed the diffraction pattern formed when slow electrons are reflected from a single crystal of nickel. The crystal surface acts like a diffraction grating with small enough spacing to produce observable diffraction. Thomson observed the diffraction pattern formed when electrons of a given momentum go through a thin metal foil whose atoms again act like a kind of diffraction grating. Now the wave aspect of electrons is used in the electron microscope where the short wavelengths that can be obtained enable one to observe smaller structures than can be resolved with visible light.

It must be stressed that according to present ideas an electron is neither a classical particle nor a classical wave but is an entity that can display particle aspects or wave aspects depending on how it is observed. The need to assign both particle and wave aspects to objects such as electrons has the Heisenberg uncertainty principle as a consequence. In order to localize a particle accurately, the wave associated with it must have a large amplitude in only a very small region of space where the particle is known to be. In order to specify the momentum of this particle accurately, the wavelength of the wave must be well-defined, which in turn requires that many complete cycles of the wave are present. The need for many cycles of the wave is inconsistent with the need for a wave differing from zero in a small region, and this inconsistency is expressed in the uncertainty principle. Both the position of a particle and its momentum cannot be specified at a given time more accurately than allowed by the uncertainty principle. Several "wave packets" of different degrees of localization and of definition of wavelength or momentum are shown in Figure 39.5.

When large particles are involved, the momentum associated with even very slow motion is large because of the large mass, and then the restriction imposed by the uncertainty principle is smaller than experimental inaccuracy, so it is unimportant. In other words, for large particles the de Broglie wavelength is so short that many wavelengths fit into a very small region Δx, so both the coordinate x and the momentum p_x can be specified exactly enough to permit the application of Newton's laws of motion.

What replaces Newton's second law of motion when the de Broglie wavelength of the particle involved is not negligible is an equation governing the de Broglie waves associated with the particle, the Schrödinger equation. Solutions of this equation that correspond to a wave of definite frequency ν associated with an electron in an atom and describe a well-defined wave at each point of space do not exist for every frequency ν but only for a discrete set of frequencies. These solutions, then, describe the allowed energy states of the atomic system involved. An analogous situation arises when the vibrations of a string such as a guitar string are studied. With a string only certain frequencies of vibration are possible: the fundamental or lowest frequency and all its multiples or harmonics. The controlling condition here is that the two ends of the string always must be at rest. If one point of a vibrating string is at rest, so is a point one-half wavelength from it, and so if both ends of a vibrating string must be at rest, they must be an integral number of half wavelengths apart. We are led, then, to the condition that

$$L = \frac{n\lambda}{2}$$

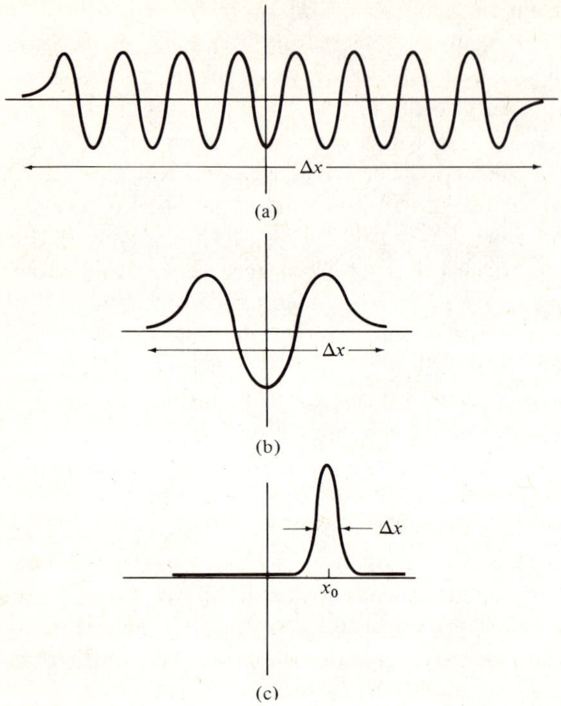

FIGURE 39.5 Three possible wave packets describing a particle. In (a) the wavelength, and therefore the momentum, is fairly well defined but the position is not. In (c) the position is fairly well defined but the wavelength is not. Situation (b) is intermediate, the wavelength being better defined than in (c) but not as well as in (a), while the position is better defined than in (a) but not as well as in (c).

or, in terms of frequency,

$$\omega = 2\pi\nu = \frac{2\pi V}{\lambda}$$

$$= \frac{\pi n V}{L}$$

Here the allowed frequencies are multiples of the fundamental $V/2L$. In a hydrogen atom this formula is replaced by the Balmer formula, Eq. 39.5, which also gives a discrete set of frequencies, but these frequencies are not quite as simply related as those for the stretched string. The energy of these states is related to the frequency of the vibration by the Bohr relation:

$$E = \hbar\omega = 2\pi\hbar\nu$$

An atom left alone will make transitions to lower energy states, emitting photons in the process, until it finally gets to its lowest allowed energy level, its ground state. In an atom containing more than one electron, one might then expect all the electrons to get down to the same lowest state, but this is not what happens. Each state of each electron is characterized by a set of quantum numbers, among which are the angular momentum and spin of the electron. There is a rule of quantum mechanics according to which no two electrons can be in states having all quantum numbers the same; at least one quantum number must be different for any two electrons. This rule is known as the *exclusion principle,* first stated by Wolfgang Pauli, and it accounts in large measure for the observed structure of atoms.

In helium with atomic number $Z = 2$ and, therefore, with two electrons in the neutral atom, both electrons can be in states very like the lowest state in hydrogen because one electron can have a spin component $+\frac{1}{2}\hbar$ in the z-direction and the other a value $-\frac{1}{2}\hbar$ for this spin component. The states of the two electrons differ by the spin quantum number and are allowed by the exclusion principle to be occupied simultaneously. Two electrons fill this electron "shell" because there is no third spin value available for a third electron. In lithium, with $Z = 3$, the third electron in the neutral atom must occupy a higher energy state, and the third electron thus starts filling a second shell. This shell holds eight electrons, and the atoms with $Z = 3, 4, \ldots, 10$ are atoms of the elements in the first row of the periodic table of the elements. The element with $Z = 10$ is neon, a noble gas like helium with $Z = 2$. All the elements with completed outer shells of electrons are noble gases; they have no energetically easy way of sharing electrons with other atoms and are, therefore, chemically inert. Elements whose atoms have one electron outside a closed shell are very active chemically, being the alkalis lithium, sodium, potassium, and so on. Elements whose atoms lack only one electron from a closed outer shell are also very active chemically, being the halogens fluorine, chlorine, and so on.

Problems

1. The electric field at a point midway between two equal electric charges is zero. Such a field is shown in Figure

16.3. Imagine a particle of charge opposite in sign to those producing the field to be located at the midpoint. For what directions of displacement away from the midpoint would the force be directed toward it, and for what directions would the force be directed away from it? What would happen if the particle were displaced in some other direction and then released from rest?

2. If an atomic system has five terms in the meaning of the Ritz combination principle, how many spectral lines might exist? (Not all pairs of terms necessarily give rise to spectral lines; there are certain "selection rules" involving the physical quantities characterizing the terms that must be satisfied if a spectral line is to appear.)
3. An atom can be excited from its lowest energy state to a higher energy state in a collision with a charged particle such as an electron. What is the threshold energy for an electron to excite a hydrogen atom at rest in its lowest energy state? (Neglect any center of mass effects; even a hydrogen atom is so much heavier than an electron that the center of mass motion is negligible.)
4. At about what temperature would the atoms of a gas made of hydrogen atoms (rather than molecules) begin to have an appreciable chance of being in the first excited state just as a result of interatomic thermal collisions? Express the temperature both in kelvins and in the energy equivalent kT measured in electron volts.
5. Is it more practical to excite hydrogen atoms by the mechanism described in Problem 3 or by that described in Problem 4 if one wants to produce some of the spectral lines of atomic hydrogen?
6. Show that according to the Bohr theory the speed of the electron moving in the smallest possible circular orbit in a hydrogen atom is given by

$$v = \frac{e^2}{\hbar} = \frac{e^2}{\hbar c} c$$

The quantity $e^2/\hbar c$ is called the "fine structure constant." It is a dimensionless number whose value is very close to $\frac{1}{137}$.

CHAPTER 8 Electrons in Solids

40. ELECTRICAL CONDUCTIVITY

With this section we start on the physics of solids. This is an important subject in physics both from the point of view of intrinsic interest and from the point of view of technical application of the results. The methods used to study solids are various, depending both on the behavior of individual particles and on the statistical behavior of large numbers of particles. We shall concentrate most of our attention on the electronic properties of solids taking for granted the fact that solids exist, most frequently in the form of crystalline structures.

One of the most noticeable characteristics of solids is their ability or inability to conduct electricity. In the discussion of electrostatics in Section 15 it was remarked that solids fall into two large classes, conductors and insulators, though the division between these classes is not sharp and depends on how much electricity is to be conducted how quickly for its definition. Here and in the following sections we want to describe experiments and theory that show how electricity is conducted in most solids and how the differences between insulators, semiconductors, and conductors can be understood on the basis of the structure of individual atoms.

Electric current is the transport of electric charge. A beam of particles of a given charge moving through vacuum constitutes a current, but only rather small currents can be carried in this way because the similarly charged particles in the beam repel one another, and a beam containing many such particles will diverge under the action of these repulsions. If charged particles can move through a region occupied by other, stationary, particles with charges of the opposite sign so that the total charge per unit volume is very small or zero, these repulsions are cancelled by the attractions between unlike charges and higher currents are possible. This situation is encountered in the condition of electricity through gases and in the conduction of electricity through metals. The latter is what concerns us now.

In our study of electrostatics, a conductor was defined as a body

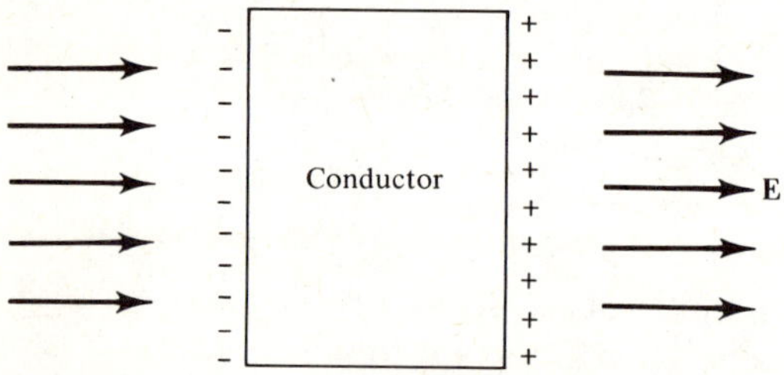

FIGURE 40.1 A conducting slab in an external electric field accumulates charge on the surface until there is no field inside. If the surface charge is continually removed, a current will flow through the conductor in the attempt to maintain this charge.

on or through which charge is free to move and inside which there can be no electrostatic field. If a conducting slab is placed in an external electric field as in Figure 40.1, its surfaces will acquire charges, and these charges produce electric fields that precisely cancel the externally applied field within the conductor, although outside the conductor the field due to these charges do not cancel the external field. When the slab is first introduced into the field there is a flow of charge in order to accumulate the needed surface charges, but this flow lasts only a very short time and then stops when the field inside has been made zero.

If somehow the surface charges are removed from the conducting slab, new charges will accumulate through the flow of charge in the slab, and if the surface charge is continually removed, there will be a continuous flow of charge through the slab. In order to remove the charge, the slab must be connected to something else by electrical conductors and must, therefore, be part of an *electric circuit*. When a circuit is present there is no contradiction in having an electric field inside a conductor because then the charges that move have someplace to go and will not necessarily accumulate on the surface until the field inside is annulled. When an electric field does exist inside a conductor, a current flows in the conductor.

Current is the flow of charge, and it is measured in charge per unit time. If the unit of charge is the statcoulomb and the unit of time is the second, the corresponding current unit often is called the statampere or esu of current. The common unit of current is the *ampere*, a flow of 1 coulomb/sec:

$$\begin{aligned} 1 \text{ amp.} &= 2.9979 \times 10^9 \text{ statamp} \\ &\approx 3 \times 10^9 \text{ statamp} \end{aligned} \tag{40.1}$$

We shall follow custom and give numerical values of current in amperes and of potential difference in volts. Of greater interest to us than the total current I flowing in a conductor is the *current density* $\mathbf{j}$, the current per unit area. $\mathbf{j}$ is a vector in the direction of current flow. If the current consists of a flow of positively charged particles, $\mathbf{j}$ is parallel to the mean velocity of these particles $\overline{\mathbf{v}}$. If the particles are negatively charged, $\mathbf{j}$ is antiparallel to $\overline{\mathbf{v}}$.

There is a very simple relation observed between the electric field $\mathbf{E}$ inside a conductor and the resulting current density $\mathbf{j}$, a relation known as *Ohm's law*. We write it in the form

$$\mathbf{j} = \sigma \mathbf{E} \tag{40.2}$$

The current density is proportional to the electric field, and the constant of proportionality σ is called the *electrical conductivity* of the material.

Ohm's law takes on a perhaps more familiar form if it is applied to a uniform conductor of cross-sectional area A and length L as in Figure 40.2. The total current I is then jA, and EL is the *potential difference* V between the ends of the conductor. Then

$$\begin{aligned} I &= A\sigma E \\ &= \frac{A\sigma V}{L} \end{aligned}$$

This is usually written as

$$\begin{aligned} V &= \frac{L}{A\sigma} I \\ &= RI \end{aligned} \tag{40.3}$$

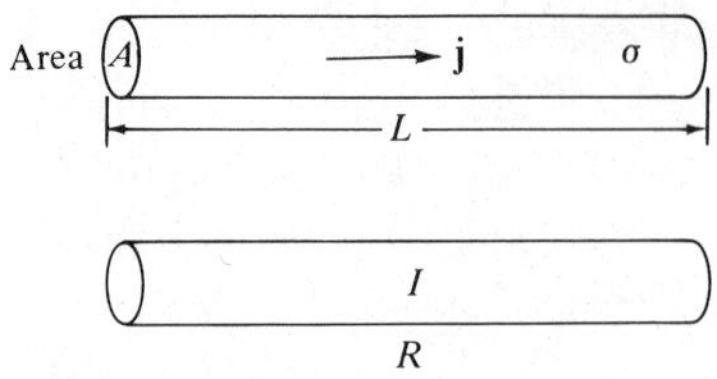

FIGURE 40.2 A piece of wire of length L and cross-sectional area A made of material of conductivity σ carries a current with density **j**. The same situation can be described by giving the resistance R of the piece of wire and the total current I carried by it. The former is more convenient when studying the material itself; the latter, when studying the use of the wire in a circuit.

The quantity $R = L/A\sigma$ is the *resistance* of the conductor. The resistance of a conductor depends on the material of which it is made through σ and on its shape through L and A. Clearly a long, thin conductor has higher resistance than a short, thick one. An advantage of σ over R is that it depends only on the material of which the conductor consists; it does not depend on the shape of the conductor, although the conductivity varies with the temperature of the material. If V is measured in volts and I in amperes, R is measured in *ohms* and σ in ohm^{-1} cm^{-1}.

If electric current consists of a flow of charged particles, it must be related to their density n (number per unit volume), their charge e, and their mean velocity $\bar{\mathbf{v}}$ by

$$\mathbf{j} = ne\bar{\mathbf{v}} \tag{40.4}$$

because this gives the amount of charge passing through unit area per unit time. Thus, using Ohm's law, we see that

$$\sigma\mathbf{E} = ne\bar{\mathbf{v}}$$

or the externally applied force on one charge carrier is given by

$$e\mathbf{E} = \frac{ne^2\bar{\mathbf{v}}}{\sigma} \tag{40.5}$$

A steady current corresponds to a steady, unaccelerated mean motion of the charge carriers, so on the average there is no force acting on these carriers even when a current is flowing. There must be a force equal in magnitude to but opposite in direction to the externally applied force $e\mathbf{E}$ acting on the carriers. Denoting this force by $\mathbf{F}_{\text{drag}}$, we see that

$$\begin{aligned} \mathbf{F}_{\text{drag}} &= -\frac{ne^2\bar{\mathbf{v}}}{\sigma} \\ &= -r\bar{\mathbf{v}} \end{aligned} \tag{40.6}$$

This force is proportional to the mean velocity $\bar{\mathbf{v}}$ and, therefore, has the character of a viscous frictional force. It should be present whenever the charge carriers move through the conductor no matter what the cause of that motion is.

Ohm's law demonstrates the existence of a drag force on the charge carriers proportional to their mean velocity but does not tell us what the carriers are. σ is directly measurable, but n, e^2, and $\bar{\mathbf{v}}$ are not, so the forces acting on the individual carriers cannot be determined from measurement of the conductivity alone. The mass and charge of these particles can be determined only by doing additional experiments. Some experiments that determine properties of the carriers are described in the next few sections. That there is much to learn is indicated by the enormous range of values the conductivity of solid materials has. Copper is a good conductor, and quartz is a good insulator. For these two materials we have

$$\begin{aligned} \sigma_{\text{copper}} &= 0.57 \times 10^6 \text{ ohm}^{-1} \text{ cm}^{-1} \\ \sigma_{\text{quartz}} &< 2 \times 10^{-19} \text{ ohm}^{-1} \text{ cm}^{-1} \end{aligned} \tag{40.7}$$

values that differ by a factor of almost 10^{25}. This must reflect some dramatic difference in the internal structure of these materials, a difference whose basis we shall see lies in the separate atoms composing the materials.

Problems

1. The fact that Ohm's law is followed by most conductors shows the existence of a viscous drag on the charge carriers. Suppose that an electric field were applied starting at time $t = 0$ in a conductor where there is *no* viscous force on the charge carriers. How would these carriers move, and what current would flow? Would the current be steady or would it depend on time?
2. Equation 40.4 is correct provided all the current is carried by the same kind of charged particle. How should it be written if there are two distinct kinds of charge carrier

present with different densities and moving with different velocities under the influence of the same electric field?

3. Consider two conductors connected "in parallel" so that the same potential difference V is applied to them and so that the total current carried by the two is the sum of the currents carried by them separately. Show from Eq. 40.3 written as $I = V/R$ that the resistance of the combination R is related to the resistances R_1 and R_2 of the separate conductors by the equation

$$\frac{1}{R} = \frac{1}{R_1} + \frac{1}{R_2}$$

4. Consider two conductors connected "in series" so that the same current flows through each one and so that the total potential difference across the two is the sum of the potential differences across them separately. Show from Eq. 40.3 that the resistance of the combination R is related to the resistances R_1 and R_2 of the separate conductors by the equation

$$R = R_1 + R_2$$

41. THE INERTIA OF CHARGE CARRIERS

Ohm's law relating the current density $\mathbf{j}$ to the electric field $\mathbf{E}$ in a conductor can be understood, as we saw in the previous section, by assuming the current to be carried by charged particles. These charged particles are accelerated by the force due to the electric field until the viscous drag proportional to their speed through the conductor just balances this force; from then on they move with constant mean velocity. The time required for this acceleration is very short if the charge carriers are light, like electrons for example. If this picture is correct it should be possible somehow to find out what the mass of these particles is. There can be little doubt that their charge is the electron charge e in magnitude. All known charged particles have a charge that is an integer multiple of e in magnitude, and the electrical neutrality of bulk metal shows that the charge of an electron in the metal is the same as the charge of one in an isolated metal atom. The charge carriers could have charge $2e$, of course, and this actually happens in superconductors, materials in which a current once started continues to flow without an applied electric field, so that the conductivity is infinite or the resistance is zero. This is a very special situation occurring only in certain metals at very low temperatures; we restrict ourselves to the case of normal conductivity.

A current flows through a conductor when the conductor as a whole is at rest and the current carriers move with mean velocity $\mathbf{v}$, or when the current carriers on the average are at rest and the conductor as a whole moves with the velocity $\mathbf{u} = -\mathbf{v}$. In each case we write the current density as

$$\mathbf{j} = ne\mathbf{v} \tag{41.1}$$

where $\mathbf{v}$ is the velocity of the current carriers relative to the conductor. The usual case is the one where the conductor is at rest and the charge carriers are caused to move by an electric field. The force on the charge carriers is proportional to their charge, the mass not entering explicitly into the relation between the drag coefficient r and the other quantities in Eq. 40.7. There is another way to produce relative motion of the charge carriers and the conductor; if the conductor as a whole is given an acceleration $\mathbf{a}_c$ by mechanical forces that do not produce an electric field inside it, the carriers will tend to be left behind. They are accelerated only by the drag force of Eq. 40.7 produced by the conductor moving past them, and only when the relative velocity of the conductor past the carriers $\mathbf{u}$ is so great that

$$r\mathbf{u} = m\mathbf{a}_c \tag{41.2}$$

will the acceleration $\mathbf{a}$ of the carriers be the same as the acceleration $\mathbf{a}_c$ of the conductor. Here m is the mass of the charge carrier. Thus there is current to be expected in a uniformly accelerated conductor if it is a part of a circuit so that the carriers can get off the ends of the accelerating conductor. Under these circumstances the direction of motion of the carriers relative to the conductor does not depend on the sign of their charge because the accelerating mechanism is not an electric field. It depends only on the direction of $\mathbf{a}_c$.

The current produced by this inertial mechanism is

$$\mathbf{j} = -ne\mathbf{u} \tag{41.3}$$

because the velocity of the conductor past the carriers is the negative of the velocity of the carriers through the conductor. Using Eqs. 41.2 and 40.7 for r we see that

$$\begin{aligned} \mathbf{j} &= -ne\frac{m\mathbf{a}_c}{r} \\ &= -\frac{m}{e}\,\mathbf{a}_c\sigma \end{aligned} \tag{41.4}$$

This contains the charge-to-mass ratio of the carriers. The equation

$$\frac{e}{m} = \frac{a_c \sigma}{j} \tag{41.5}$$

expresses this ratio in terms of observable quantities and thus provides a direct way of measuring e/m for the carriers. The direction of the current also shows directly the sign of the charge on the carriers.

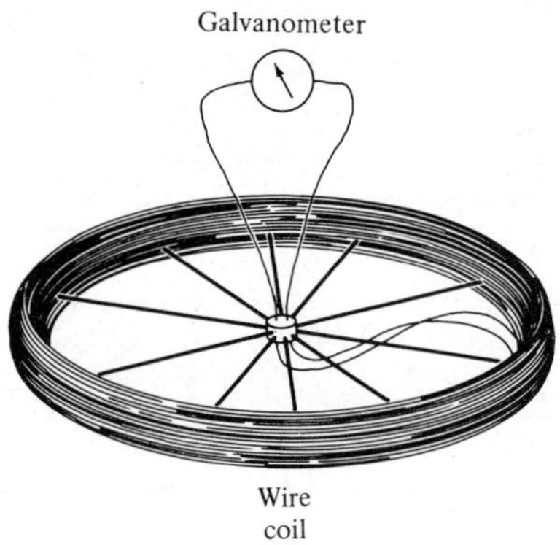

FIGURE 41.1 The inertia of the charge carriers in a wire can be measured by observing the current that flows through the galvanometer when the spinning coil is suddenly slowed down. The charge carriers tend to keep moving, thus producing the current.

An experiment along these lines was performed by T. D. Stewart and R. C. Tolman in 1916 and repeated with refinements by Tolman and co-workers later. The first experiment used a wire wrapped on the rim of a wheel and connected by long wires free to twist to a galvanometer, as shown in Figure 41.1. The wheel was spun at high speed, about 5000 revolutions per minute, and then stopped quickly. All points on the rim where the wire was experienced the same magnitude of acceleration; the bending of the wire into a circle had almost no influence on the result. The electrons inside the wire tended to keep going after the wire was stopped and a current went through the galvanometer.

The experiment was repeated with several metals and with various speeds and rates of braking. The result was that the charge-to-mass ratio was nearly that of the electron, and the direction of the inertial current showed that the charge carriers were negative. The discrepancy of e/m as measured from that of the free electron was once about 18 percent on the low side and another time about 8 percent on the high side. With the instruments available at the time this was as well as could be done. In the next section a much more precise measurement will be described, but one that does not give the sign of the charge on the carriers. The Tolman experiments show very beautifully that the charge carriers in metals are really electrons.

Recently (1967) another experiment related to the Tolman experiments was done by F. Witteborn and W. Fairbank. Here the inertial effect of the acceleration of the conductor is replaced by the acceleration of gravity. If the electrons in a conductor are free to move, why do they not fall to the bottom of the material? They do until there is a sufficient accumulation at the bottom to repel any others that approach; in other words an electric field is set up in the conductor that just cancels the gravitational field on the electrons so that they do not tend either to fall or to rise but are in equilibrium. The magnitude of this field is given by the equilibrium condition:

$$eE = mg, \qquad E = \frac{mg}{e} = 1.86 \times 10^{-15} \text{ dyne/statcoul}$$
$$= 5.58 \times 10^{-12} \text{ V/cm} \tag{41.6}$$

The magnitude of E is too small to be directly measured.

The Witteborn-Fairbank experiment used the fact that the electric field parallel to the boundary of the material must be the same inside and outside the material. If it were not, one could violate the law of energy conservation in a simple way. Suppose that there is an electric field like that shown in Figure 41.2, weak inside the conductor and strong outside. By moving an electron clockwise around the dotted path in the figure one would get more work going down the right side than one would have to do to go up the left side, and energy would be received for nothing. This does not happen, so the fields inside and outside are equal. Note that this argument works only for the field parallel to the surface of the material. The perpendicular component need not have the same value inside and outside. Witteborn and Fairbank measured the acceleration of electrons in free space inside a long tube of copper placed vertically in the earth's gravitational field, as shown in Figure 41.3. According to the arguments just given, inside the copper there is a downward electric field $E = mg/e$ that also must exist in the interior

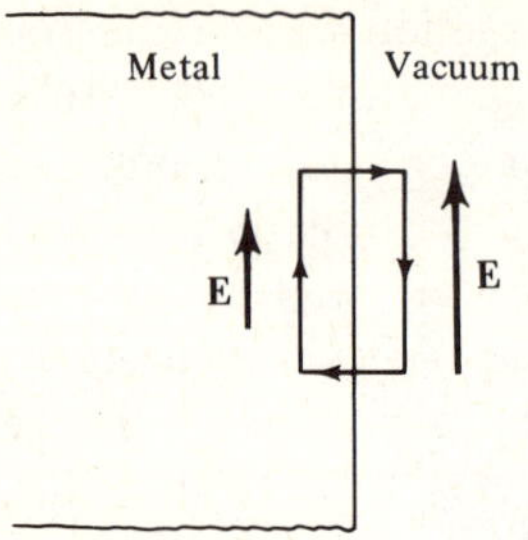

FIGURE 41.2 If the components of the electric field parallel to the surface of a conductor are not equal just inside and just outside the surface, then one can get work done for nothing by moving an electron around the path shown. (Recall that the force on an electron is antiparallel to the field.) Because no such free sources of work are known, the two fields are equal.

space surrounded by the copper. If this is so, the free electron in the tube should not fall, and its acceleration was determined to be less than $0.09g$. The technical difficulties in measuring such small accelerations of electrons are formidable.

An electric field strong enough to hold up an electron is not strong enough to affect the motion of a proton very much; it would increase its downward acceleration from

$$1 \times g \text{ to } \left(1 + \frac{m^e}{m_p}\right) \times g = (1.000545)\ \text{g}$$

The effect also exists in the Millikan oil drop experiment, but because of the enormous mass of the drop as compared with that of the electron, it is utterly undetectable. A field this strong can affect a positron, the antiparticle of the electron, and an interesting experiment using this has been proposed. It is not quite certain that antiparticles are attracted gravitationally to the earth, no direct observation of this attraction ever having been made. It is conceivable that they are repelled instead. If a positron is attracted gravitationally by the earth, it ought to fall down the Fairbank tube with an acceleration of $2g$, $1g$ being provided

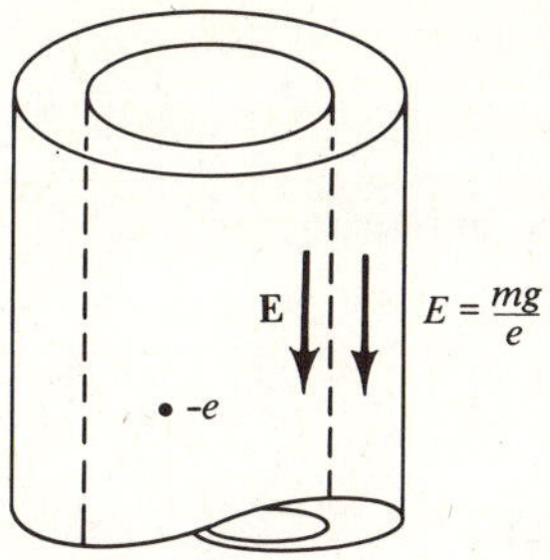

FIGURE 41.3 The electrons inside a conductor do not fall out the bottom even though they are free to move through the metal. If they are kept in the metal by an electric field that cancels the force of gravity on them, then this field also must act on an electron just outside the metal and it will not fall either. This seems to have been confirmed by Witteborn and Fairbank.

by the earth and $1g$ by the electric field holding up the electrons. If the positron is repelled by the earth, it ought to be unaccelerated in the tube. So far the experiment with positrons has not been technically feasible, so the result is not known.

It could also be that free electrons and positrons are not acted on by a gravitational field at all, even though bound electrons are. It is hard to see how to test this because to measure the force of gravity on a free electron the presence of an electric field must be avoided, and it is not obvious how to do this without using conductors as shields. If a conductor is used as a shield, one is right back to the experiment just described unless a conductor can be found in which electrons are not free to move, and no such conductor is known. If the positron experiment is done and the positron's acceleration is $2g$, the question will have been answered. If the acceleration is zero, it could be because the positron is repelled by the earth while electrons are attracted or because neither is affected by gravity. A result different from both $2g$ and zero would really provide a puzzle.

The interpretation of the experiment given has assumed that the effect of the gravitational compression of the copper crystal is negligible. The bottom part of the tube supports the upper part and is therefore more compressed than the top. There are, then, forces acting on the copper ions that hold them up. If these forces are also electrical in nature, how do the associated electric fields affect the electrons? The conclusions stated above are not beyond controversy and the meaning of the experimental result is not certain.

Problems

1. Suppose the Tolman-Stewart experiment were done with a wheel 30 cm in radius which slows down from 5000 revolutions per minute to rest in 2 sec.
   (a) What is the average acceleration of a point on the rim of the wheel?
   (b) What value of electric field is needed to produce this acceleration of a free electron?
   (c) What current density j would be caused to flow in copper by such a field?
2. In the Tolman-Stewart experiment there is an electric current flowing through the wire around the rim of the wheel. Can one apply Ohm's law to this current and find the value of electric field inside the wire? If so, what is the source of this field, and if not, why not?

42. CYCLOTRON RESONANCE IN METALS

The inertial method of determining the charge-to-mass ratio of the charge carriers in metal that was described in the last section is of limited precision. It is difficult to produce large mechanical accelerations of conductors, and various extraneous effects such as the induction of unwanted electric currents by magnetic fields are hard to avoid. Purely electromagnetic methods are therefore advantageous. The ratio e/m for free electrons is determined by deflection experiments using electric and magnetic fields. There are other methods using both electric and magnetic fields that can be applied to the charge carriers in solids.

A device like the cyclotron operates differently for particles with different values of e/m. In a cyclotron charged particles move in circular orbits in a uniform magnetic field B, as described in Section 24. The centripetal force needed to produce the circular motion is provided by the magnetic force:

$$\frac{mv^2}{r} = \frac{evB}{c} \tag{42.1}$$

or

$$\omega = \frac{v}{r} = \frac{eB}{mc} \tag{42.2}$$

This angular frequency ω is called the *cyclotron frequency* of the particle in the field B. For a given field B, it depends on the value of e/m for the particle. The frequency of revolution of the particle in revolutions per second is

$$\nu_p = \frac{\omega}{2\pi} = \frac{eB}{2\pi mc} \tag{42.3}$$

In the cyclotron operated as a particle accelerator as described in Section **24**, an oscillating electric field is applied across two accelerating gaps. This field has the frequency ν_p, so the electric field will have the same value each time the particle goes through a given gap, and if the particle starts off correctly it will be accelerated each time. As the particle gains speed it also gains kinetic energy. This energy must come from somewhere, and it comes, in fact, from the generator supplying the electric field across the gap. If the frequency of revolution of the particle is not the same as the frequency of the accelerating field, then sometimes the particle will be speeded up and sometimes it will be slowed down. On the average it will not gain energy, and the generator of the electric field need not supply this energy. In a cyclotron accelerator the amount of energy transferred to the ions being accelerated is very small compared with energy lost in other ways, so that while the energy consumption is increased when the two frequencies are the same and *resonance* exists, the increase is hardly noticeable. At very low energies the losses are less, and resonance can be detected by measuring the power delivered by the oscillator supplying the oscillating electric field.

A little (1-inch) cyclotron is often used in conjunction with a high vacuum system to determine what gases are present, information that is valuable in the search for leaks of various kinds. The gas is ionized by electron bombardment and is placed in a chamber containing an oscillating electric field and an adjustable magnetic field. The magnetic field is adjusted until cyclotron resonance is achieved, and the value of e/m for these ions is thus determined, affording an almost but not quite certain identification of the gas.

A similar scheme is applied to the study of the charge carriers in metals. If the metal in question is placed in a magnetic field B, the carriers should move in circles or helices around the field with a rotational frequency ν_p given by Eq. 42.3. If an oscillating electric field can be applied in an appropriate way these carriers will be accelerated, and the frequency of the electric field when cyclotron resonance occurs determines the e/m value for the particles. Here the only practical way to detect the resonance is by measuring the energy absorbed by the accelerated particles because it is impossible to get inside the solid to observe their behavior. The method used is quite simple. An electromagnetic wave is sent perpendicularly onto the surface of the conductor.

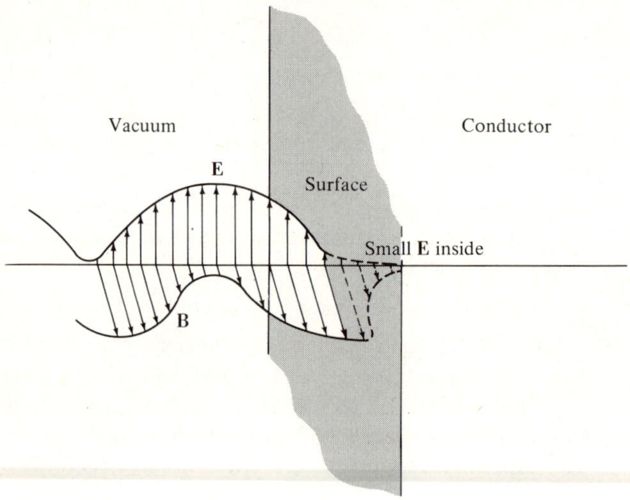

FIGURE 42.1 An electromagnetic wave incident perpendicularly on the surface of a good conductor is nearly all reflected, but the fields do penetrate a little, and some energy is absorbed by the conductor.

This wave has an electric field associated with it, the direction of this field being parallel to the surface of the conductor, as shown in Figure 42.1. The electromagnetic wave is reflected from the surface of the conductor much as a wave in a string is reflected from a fixed end. If the conductor were a perfect conductor the reflection would be complete and the electric field would be zero at the surface. If the conductivity of the material is finite, not all the energy is reflected and the electric field at the surface is not quite zero. There is thus an oscillating electric field at the surface, but the strength of this field decreases rapidly with increasing depth in the material. This is what we need to make a sort of cyclotron in the material.

A steady magnetic field **B** is applied inside the conductor, parallel to the surface, as in Figure 42.2. The charge carriers will move in circles (or helices) around the field **B**. A carrier near the surface will be accelerated a little by the electric field of the wave and will travel in a small circle, returning to the surface one cycle of its cyclotron motion later. The electric field gets small so rapidly with increasing depth that

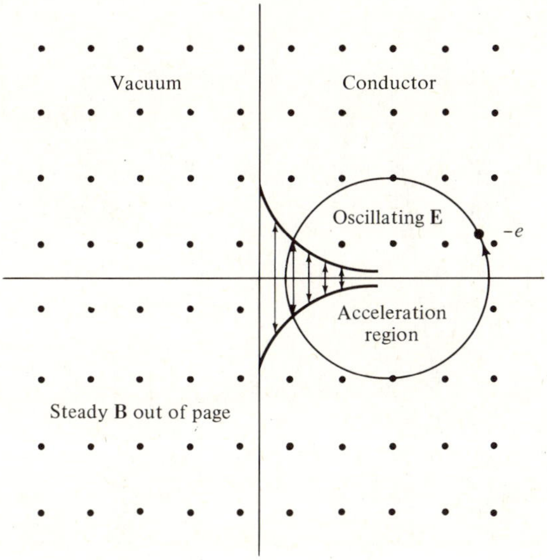

FIGURE 42.2 Electrons in the conductor moving in circular orbits as shown in the magnetic field **B** will be accelerated by the electric field near the surface but nowhere else on the orbit. The region near the surface acts like an accelerating gap in a particle accelerator, and if the frequencies match properly the electron will be speeded up on each passage, thus getting increased kinetic energy each turn.

the carrier is not affected by it except when right at the surface. If the electric field has gone through an integer number of cycles during the travel time of the carrier around the circle, the carrier will again be accelerated when it is at the surface and will move in a larger circle. It will return to the surface and be accelerated again after the same travel time, and the process will continue. This carrier absorbs energy from the electromagnetic wave when resonance occurs, diminishing the amplitude of the reflected wave, and this diminution can be detected.

The resonance condition is that during one cyclotron cycle of the carrier the electric field of the wave must have gone through one, two, or any integer number of cycles. If ν_w is the frequency of the wave, then resonance occurs when

$$\nu_w = n\nu_p \tag{42.4}$$

with integer n.

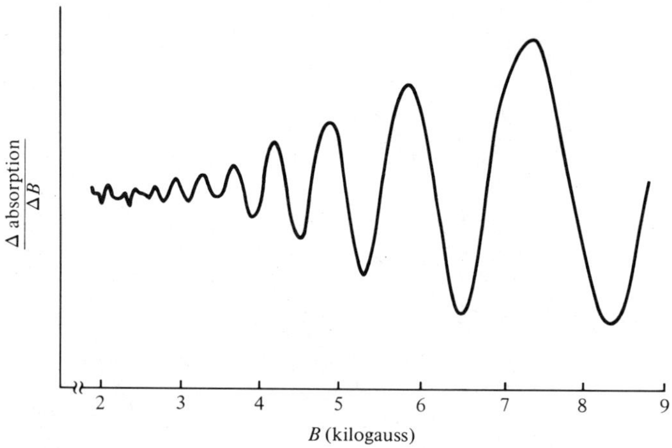

FIGURE 42.3 The cyclotron resonance curve for electrons in potassium is shown here. For technical reasons what is plotted is the rate the power absorption changes with magnetic field, not the power absorption itself, so the resonances correspond to values of B where this change is zero. The frequency of the incident electromagnetic wave was 6.62×10^{10} hertz in this experiment. (After C. C. Grimes and A. F. Kip. By permission.)

In general it is much easier to adjust the value of a static field like **B** than to change the frequency of a wave, so in this kind of experiment ν_w is held fixed and ν_p is varied by changing the magnetic field. Also it is easier technically to measure the rate at which the energy absorption changes with **B** rather than the absorption itself. Figure 42.3 shows the experimental results obtained by C. C. Grimes and A. F. Kip with potassium metal. Peaks of the absorption curve correspond to points where the curve shown crosses the axis. The different crossings are associated with different values of n. The magnetic field instead of the cyclotron frequency is plotted horizontally. The data do not extend to $n = 1$, as the magnetic field could not be made large enough to match the cyclotron frequency ν_p to the wave frequency ν_w. Larger values of n go with smaller values of ν_p and hence with smaller magnetic fields.

The results of Grimes and Kip for potassium and sodium are

$$(e/m)_{\text{potassium}} = (1.21 \pm 0.02)(e/m)_{\text{electron}}$$

$$(e/m)_{\text{sodium}} = (1.24 \pm 0.02)(e/m)_{\text{electron}} \tag{42.5}$$

The values are near that for the free electron but are not equal to it. Because no other charge carrier has a charge-to-mass ratio anywhere near this, it is clear that the carriers of charge in these two metals are electrons, although their behavior is modified by their being in the metal. This experiment does not distinguish between positive and negative carriers. These would differ only in their direction of rotation about the magnetic field, and this is not observed.

This kind of experiment not only identifies the charge carriers in the metal but also gives information about the environment of the electrons in the metal through the deviation of the value of e/m from that for free electrons. The interpretation of such data requires a detailed theory of the structure of metals and is beyond the scope of our discussion.

Problems

1. As mentioned in the beginning of this section, cyclotron resonance is used to identify residual gases in vacuum systems by ionizing some of the gas molecules and determining their cyclotron frequency, thereby determining the e/m ratio of the ion. How accurately would you have to be able to determine the cyclotron frequency to distinguish between singly charged ions of heavy hydrogen (deuterium) with molecular weight 4.028 and singly charged ions of helium with atomic weight 4.026?
2. Explain why there is no cyclotron resonance in a metal when the frequency of the electromagnetic wave is just half the cyclotron frequency of the electron while there is such a resonance when this frequency is twice the cyclotron frequency.

43. THE HALL EFFECT

It has been established experimentally that the charge carriers in some metals are electrons, for example, by inertial experiments of the Tolman type and by cyclotron resonance experiments. These experiments do not, however, establish the number of electrons that are involved in the conduction process. In copper, whose atomic number is 29, there are 29 electrons per atom while in lead there are 82, yet copper is a better conductor of electricity than lead. It is of great interest to find out how many of the electrons per atom are involved in the conduction process. The electrons that are almost free and therefore move through the metal when an electric field is applied are called *conduction electrons.*

Conductivity measurements alone cannot give the number of conduction electrons per atom, because a given electric current can be carried by many electrons moving slowly or by fewer electrons moving faster. This number can be determined if the mean velocity of the electrons carrying a known current can be measured. Because

$$\mathbf{j} = ne\bar{\mathbf{v}} \tag{43.1}$$

measurement of $\mathbf{j}$ and of $\bar{\mathbf{v}}$ determines n, the electron charge being known beforehand. The problem of measuring the speed of electrons has been met before. It faced Thomson in his measurement of e/m for free electrons, and he solved it by using a combination of electric and magnetic fields. An electric field exerts a force on a charged particle that is independent of the motion of the particle, and a magnetic field exerts a force that is proportional to the particle's speed. If, somehow, the two forces can be compared, the particle speed can be inferred. Thomson compared the forces by comparing the deflections of the electron beam produced by each field. A somewhat similar device can be used in metals, as was first shown by E. H. Hall in 1879, long before the electron was identified.

The conduction electrons in a wire are constrained to remain in the wire by strong electric fields at the surface, but these fields are restricted to a very thin layer at the surface so their only effect is to keep the electrons inside. When a current is flowing in the wire, the mean velocity of the electrons must be parallel to the wire. If some kind of sideways force is exerted on the electrons, they will move sideways only until enough have accumulated on one side of the wire to repel others so strongly as to cancel the applied sideways force. In other words, an electric field is set up across the wire, a field whose strength is automatically just that needed to cancel the applied sideways force. A measurement of this electric field then gives the strength of the applied force.

In the Hall effect experiments the sideways force is furnished by a magnetic field perpendicular to the wire and, therefore, perpendicular to the mean velocity of the electrons carrying a current in the wire.

The average force exerted on these conduction electrons is just the magnetic force whose magnitude is

$$F_B = \frac{e\bar{v}B}{c} \tag{43.2}$$

This force must be canceled, because the electrons are not accelerated out of the wire, by an electric force of equal magnitude

$$F_E = eE_H \tag{43.3}$$

where E_H, the Hall field, is the field automatically set up in the way

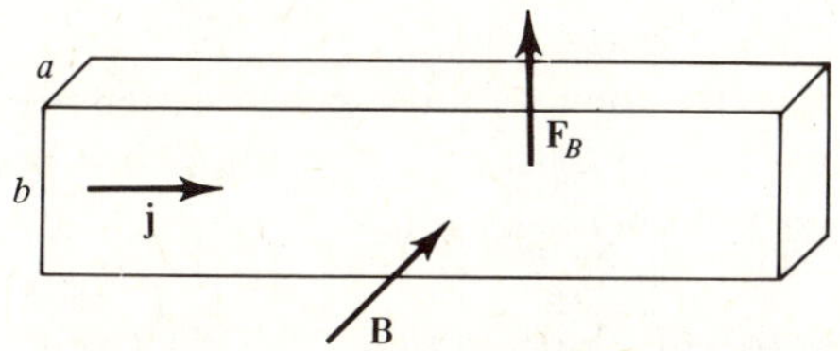

FIGURE 43.1 A current traveling to the right through a magnetic field into the page experiences an upward force $\mathbf{F}_B$, whether the current is carried by positive charges moving to the right or by negative charges moving to the left.

described above. If B is known and E_H is measured, the mean speed of the conduction electrons can be found from

$$\bar{v} = \frac{E_H}{B}c \tag{43.4}$$

Experimentally one usually uses a rectangular wire as shown in Figure 43.1. The magnetic field is applied perpendicular to two faces of the wire, and the potential difference between the other two faces is measured. This Hall potential is just the product of the Hall field E_H and the thickness of the wire in the direction of this field.

Even though the Hall field is usually detected only in experiments designed for its measurement, its existence is vital to the operation of electrical machinery of various kinds, including electric motors. These machines depend for their operation on the force exerted on a current-carrying wire in an external magnetic field. The magnetic field does not exert this force directly on the ions of the copper or whatever else the wire is made of. The magnetic force is exerted on the moving electrons. To counteract this magnetic force due to the external magnetic field and to keep the electrons in the wire, a Hall field is set up. This Hall field, acting on the moving conduction electrons, just cancels the magnetic force on them. The Hall field also exerts a force on the copper ions, the copper atoms minus the electrons they have contributed to the supply of conduction electrons. This force is just opposite to the force the Hall field exerts on the conduction electrons because the charge on the ions is equal in magnitude but opposite in sign to the charge on the conduction electrons collectively and because the force exerted by an electric field is independent of the speed of the charges acted on. The end result is, then, a force on the ions equal in magnitude and direction to the magnetic force on the electrons.

The electrons in a metal furnish a convenient handle with which to exert forces on the metal because they are constrained to remain inside the metal, just as a doorknob is a convenient handle with which to exert force on a door because it is constrained to remain on the door. If excessive forces are applied either to the doorknob or the electrons, the constraint is broken; the doorknob comes off the door, or the electrons leave the conductor. In either case considerable damage may be done.

The number of conduction electrons per unit volume n now can be expressed in terms of measurable quantities. Thus using Eq. 43.4 for $\overline{v}$ in Eq. 43.1, one obtains

$$j = ne \times \frac{cE_H}{B}$$

from which

$$n = \frac{jB}{ecE_H} \tag{43.5}$$

if Gaussian units are used. If a rectangular wire of sides a and b is used in the experiment shown in Figure 43.1, then

$$jab = I, \qquad E_H b = V_H$$

and the conduction electron density is expressed in the more directly measured quantities as

$$n = \frac{IB}{ecV_H a} \tag{43.6}$$

Values of n for various metals are given in Table 43.1. Of greater interest

TABLE 43.1
Results of Hall Effect Measurements on Simple Metals

Element	Number of Atoms per cm³	Number of Conduction Electrons per cm³	Number of Conduction Electrons per Atom
Lithium	4.8×10^{22}	3.7×10^{22}	0.8
Sodium	2.6	2.5	1
Copper	8.5	12	1.4
Silver	5.9	8	1.4
Gold	5.9	9	1.5

is the number of conduction electrons per atom. This is also tabulated and is seen to be not far from one per atom. At this point it is quite mysterious why one or two electrons out of a much larger number present in the atom are free to carry current. We return to this problem in Section 44.

It seems that there should be no question of the sign of the Hall field for a given current and magnetic field. The conduction is by electrons that have a negative charge $-e$. Surely the density of conduction electrons is positive, so how could a Hall field with the wrong sign appear? To be sure of the signs, suppose the charge carriers have a charge q that we can take to be either positive or negative. Then the various vectors involved in the Hall effect have the directions shown in Figure 43.2, (a) for $q > 0$ and (b) for $q < 0$. In both cases the magnetic force is upward, so the carriers tend to accumulate at the top of the conductor. When $q > 0$, the resulting Hall field E_H is downward in the conductor, repelling other positive carriers from the top. When $q < 0$, negative charge accumulates at the top resulting in a Hall field that is upward, repelling negative carriers from the top. Thus if the sign of the carriers' charge is reversed, the Hall field is reversed. If a Hall field in the direction corresponding to positive carriers is observed, can one conclude that the carriers are positive? This question is not as silly as it sounds because there are cases where this reverse Hall field is observed. The answer is that one has to conclude that either the charge is positive or that the density of carriers n is negative. We shall see in Section 45 that these two are in a sense equivalent to one another and that a relatively simple quantum picture makes the effect understandable.

The drift speed $\overline{v}$ of the electrons in metallic conductors is very

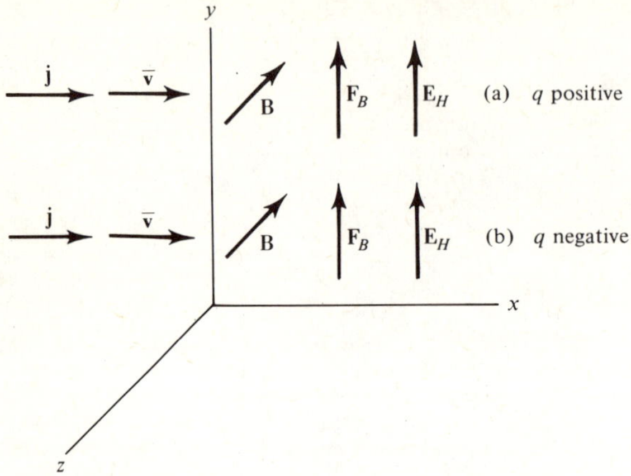

FIGURE 43.2 The directions of the vectors controlled from outside in a Hall measurement are the same for either kind of current carrier. The direction of the average velocity of the current carriers and of the Hall field $\mathbf{E}_H$ depends on the sign of the charge of the carriers.

small. In copper, for example, in a Number 16 wire carrying a current of 1 ampere, which is typical of a flexible lamp cord supplying a pair of 60-watt bulbs, the mean speed is only 4×10^{-3} cm/sec. This has nothing to do with how long it takes the light to go on after turning the switch. When electrons are pushed in one end of a wire, electrons come out the other end almost immediately. The impulse or wave along the wire caused by closing the switch travels at a speed comparable to that of light, although the electrons themselves are slow moving.

Problems

1. Equation 40.4 relates the current density in a conductor to the motion of the charge carriers if they are all alike. If two kinds of carriers are present the current density is

   $$\mathbf{j} \times n_1 e_1 \overline{\mathbf{v}_1} + n_2 e_2 \overline{\mathbf{v}_2}$$

   Show that if e_1 and e_2 are of opposite signs so that $\overline{\mathbf{v}_1}$ and $\overline{\mathbf{v}_2}$ are antiparallel, the Hall voltage may be much smaller than if only one sign of carrier is present.
2. Show that when a wire of length L is carrying a current I at right angles to a magnetic field of magnitude B, a

force given by

$$F = IBL/c$$

is exerted on it at right angles both to the wire and to **B**.

44. ELECTRON BANDS IN SOLIDS

A crystalline solid is made of atoms arranged in some regular way in space. These atoms are held together by very complicated interplays of Coulomb forces between the atomic nuclei and electrons making up the solid. We know already that Newtonian mechanics must be replaced by quantum mechanics when the behavior of electrons in a single, isolated atom is to be described, and when the atoms stick together to form a solid this remains true. The understanding of crystal structure and the mechanical properties of crystals such as strength and hardness on the basis of a knowledge of the structure of individual atoms is still beyond us; qualitatively they can be made to seem reasonable, but such properties of solids cannot yet be deduced from the properties of atoms. It is possible, however, to acquire considerable understanding of the electrical properties of solids, and to see why some are insulators, some are conductors, and some are semiconductors that combine certain properties of insulators with certain properties of conductors. In this section the band theory of solids is introduced and is used to clarify several problems that have been encountered.

To begin with, let us consider two atoms of the same element that are far enough apart so that neither is influenced by the presence of the other. Each of these atoms has a set of allowed energy levels, and enough of the lowest of these levels are occupied by electrons to make the atom neutral. Now if these two atoms are brought closer together, the electrons in each begin to feel the electric field produced by the electrons and nucleus of the other and, consequently, the energies of the levels of the two atoms together are different from the energy levels when they are apart; an atom is perturbed by the presence nearby of another atom, and it is particularly so if the two atoms are of the same kind. There is an analogy that illustrates this effect:

Imagine a pendulum suspended from an elastic support, as shown in Figure 44.1. If the pendulum is to one side, it pulls the support over to that side making the effective length of the pendulum longer than the string, as can be seen in the figure. The frequency of the pendulum

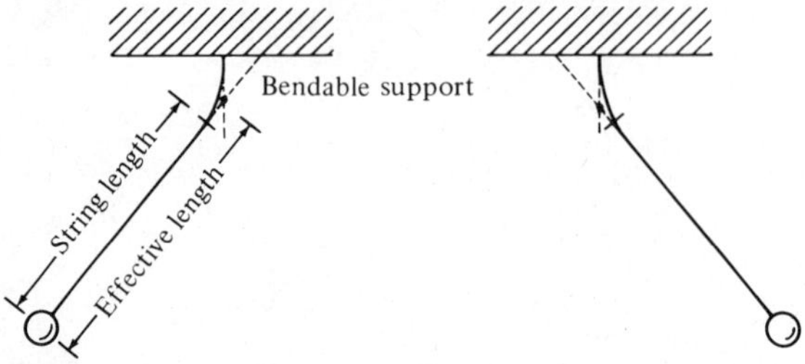

FIGURE 44.1 A pendulum hung from a bendable support swings further than one hung from a rigid support because of the yielding of the support. It has, therefore, an effective length longer than the pendulum string and oscillates more slowly than one would expect on the basis of that string length.

is inversely proportional to the square root of its length, so the elasticity of the support reduces the frequency below what it would be if the support were rigid. (The frequency ν of a pendulum depends only on its length l and the acceleration of gravity. It is given by

$$\nu = \frac{1}{2\pi}\sqrt{\frac{g}{l}}$$

A pendulum as a kind of harmonic oscillator is discussed in Section 8.)

Now imagine two similar pendulums suspended from this same support. If both pendulums are on the same side of the support, the support is pulled over twice as far as if only one pendulum were there, so the effective increase in the lengths of the two pendulums is twice as great as before and the frequency with which the two oscillate is lower than the frequency of oscillation of one pendulum by itself. This situation is illustrated in Figure 44.2.

There is no need to have the two pendulums off to the same side all the time. Their swinging could be started by pulling one to one side and the other equally far to the other side, as shown in Figure 44.3. In this case the support is pulled equally in both directions and therefore does not move in either direction but acts as though it were rigid. There is no effective lengthening of the pendulums, and they oscillate with a frequency that is higher than the frequency of either one alone. Thus two pendulums that are alike and therefore oscillate with the same frequency when alone can oscillate in two distinct ways with

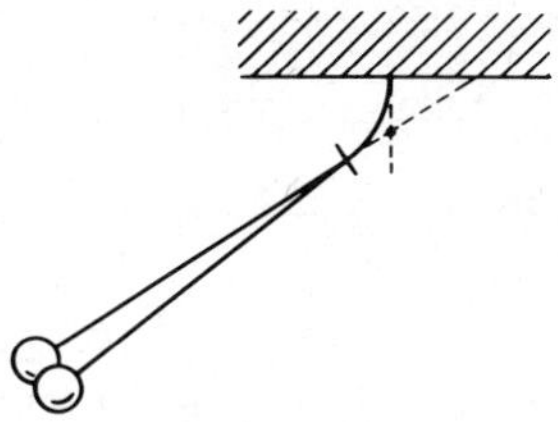

FIGURE 44.2 If a second pendulum is added to the non-rigid support of the pendulum in Figure 44.1 and if the two swing together, the support bends twice as much as before and the effective length of the pendulums is greater than when they hang alone.

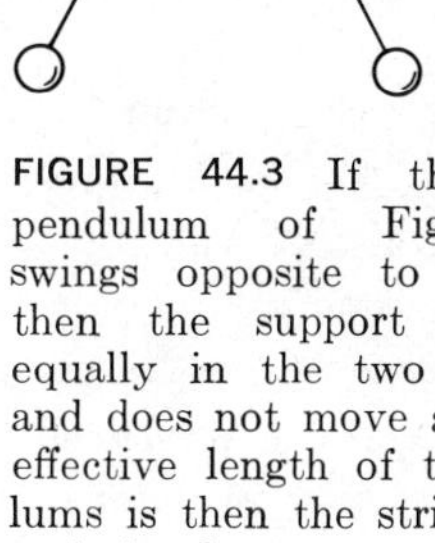

FIGURE 44.3 If the second pendulum of Figure 44.2 swings opposite to the first, then the support is pulled equally in the two directions and does not move at all. The effective length of the pendulums is then the string length, and the frequency of swinging is higher than either of the other frequencies occurring in the earlier situations.

two distinct frequencies when the motion of one can influence the motion of the other through a nonrigid support. One frequency is higher than the single pendulum frequency, and the other is lower.

The energy levels of atoms are, as we saw in Section 39, determined in much the same way as the frequencies of vibrating systems are determined, so it is not very surprising that something like the change in frequency of a pendulum in the presence of another pendulum should occur to the energy levels of an atom in the presence of another atom.

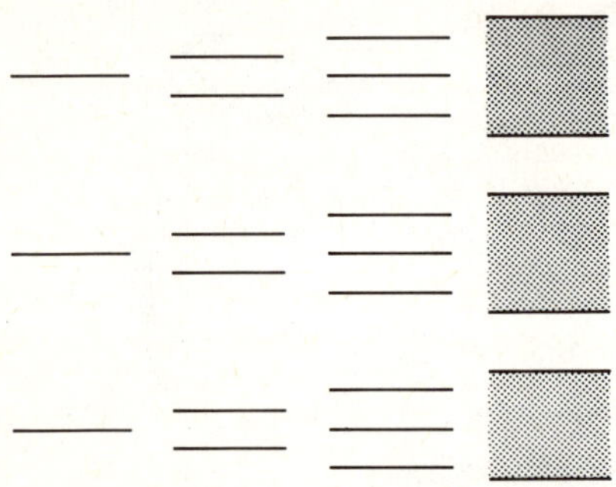

FIGURE 44.4 The splitting of three energy levels of an atom occurs when additional identical atoms are brought near. In general each level is split into as many sublevels as there are atoms involved.

In the atomic case, two identical energy levels of the separated atoms split into a pair of energy levels with slightly different energies, one higher than the energy of the level when the atoms are apart and one lower than this. The splitting gets larger the closer the atoms are to each other and may become quite large when the atoms "touch."

If three identical atoms are brought near each other, each single atom energy level is split into three levels and a fourth atom causes a four-way split. In general, if there are N atoms together, each energy level will be split into N sublevels. When N is very large, the atoms on the average cannot get very close to each other, so the splittings are small and the set of N sublevels forms an almost continuous band of energies, as shown in Figure 44.4. In general all the energy states belonging to an electron shell in an atom together form a band of allowed energies in a solid. In a solid a small cube of which may contain 10^{23} atoms, there are bands of allowed energies separated by bands of forbidden energies. In simple cases a given band comes from a single allowed electron shell of an isolated atom. In more complicated cases the bands coming from different electron shells may overlap to form wider bands. We shall not inquire into such details here; the existence of allowed energy bands for electrons is essentially all that we need.

When the electrons in a solid are at the lowest energy possible for them, the exclusion principle forbids all of them being in the lowest

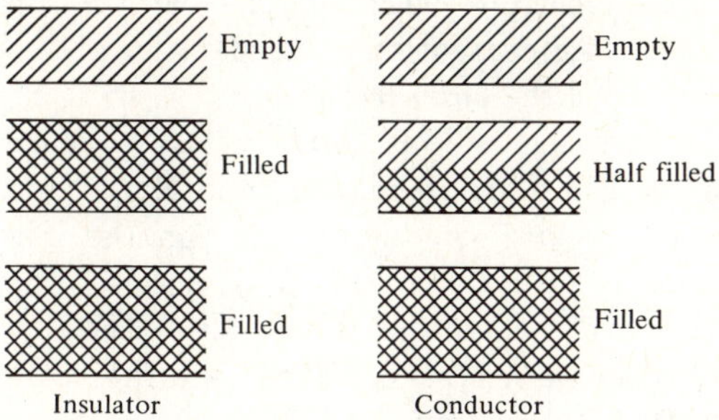

FIGURE 44.5 Whether a solid is an insulator or a conductor depends on whether the topmost energy band that contains any electrons is filled or only partly filled.

state and so the electrons fill the lowest band and then the higher bands in succession until all are placed. It is a matter of great importance whether the topmost band that contains any electrons at all is completely filled or whether it is only partly occupied. If the electron shell of an isolated atom from which this band comes is filled, then the band will be filled also, unless it overlaps some allowed energy band coming from a different atomic electron shell. If the electron shell of an isolated atom from which this band comes is not filled, then the band in the solid will not be filled either. These two situations are shown in Figure 44.5.

The electrical properties of solids are determined by the freedom of electrons in the solid to move. When the topmost occupied band of energy levels in a solid is filled, it means that an electron originally associated with one particular atom must remain associated with that atom because there is no vacant level of the same energy associated with any neighboring atom. Only if the electron can acquire enough energy to get into one of the higher, empty bands can it move from one atom to another. However, when the topmost occupied band is not completely filled, an electron associated originally with particular atom may move over to a neighboring atom because there is a vacant level there. Even though electrons in such a solid are free to move, they remain very uniformly distributed because of Coulomb forces. If many electrons tend to leave a part of a solid, they leave behind a net positive charge that then tends to attract them back to that region. Currents can flow through such solids, but very little charge accumulates anywhere under normal conditions. The partially occupied energy band of a conductor is called the *conduction band* because it is electrons with energies within this band that normally conduct the electric current that can flow in a conductor. The topmost filled band is called the *valence band*.

The result of the discussion just given is that solids whose topmost occupied electron energy bands are filled are insulators, and those with partly filled topmost occupied bands are conductors, or metals. The difference between these situations is quite clear-cut, and metals differ very much in electrical properties from insulators. There are cases that are not as clear-cut as might seem to be necessary. These are the semiconductors and are the subject of the next section.

If some of the electrons in a metal are free to move, why is it necessary to provide an electric field to keep a current flowing? Ohm's law is very accurately satisfied, so there must be some mechanism by which a current started at one time very quickly dies out unless something is done to maintain it. The damping mechanism is the scattering of electrons by irregularities in the crystal. These irregularities may be caused by impurities in the metal or by the thermal motion of the metal ions as they vibrate about their equilibrium positions with mean energy $3kT$ each. The conductivity of metals increases with decreasing temperature. Figure 44.6 shows a graph of the resistance of a sample of a metal as a function of its temperature. The decrease in resistance with decreasing temperature is rather gentle until a low temperature of a few kelvins is reached. Then the resistance of some metals drop to zero very abruptly and the metal becomes a superconductor. A current once started in a superconductor will continue to flow indefinitely without an electric field's presence. The phenomenon of superconductivity is quite complicated to explain, and we shall not attempt its explanation here.

We can now explain the failure of electrons to contribute their share to the specific heat of solids. For the electrons in the lower bands the explanation is simple: If an electron in a lower, filled band is to absorb energy, it must make a transition to a band having vacancies, and in general this requires an energy much greater than kT. Such a large

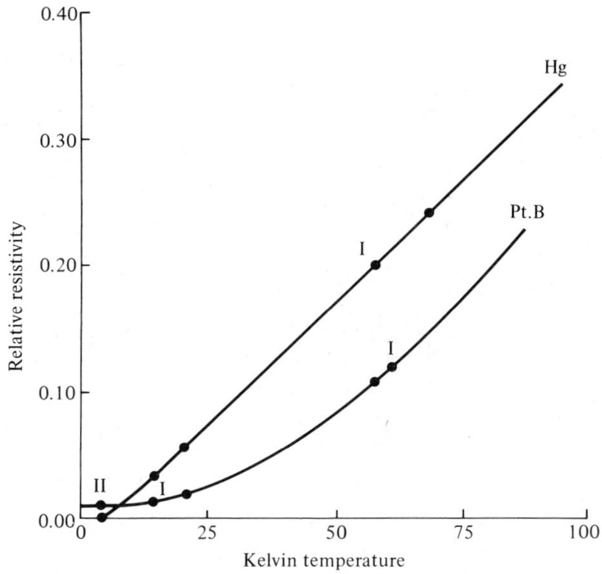

FIGURE 44.6 The resistivities of mercury (Hg) and of platinum (Pt) are shown as they vary between temperatures of almost 0 K and 100 K. At about 4 K the mercury becomes superconducting; the platinum does not. At higher temperatures the plots are both straight lines, the resistivity increasing with increasing temperature. (After K. Onnes.)

amount of energy is very seldom available all at once from the thermal motions of the system, so such transitions are extremely improbable and do not lead to much energy absorption on raising the temperature. The separation of the energy bands may be something like 1 eV, but at room temperature, say 300 K, the thermal energy kT is only $\frac{1}{40}$ eV.

For the electrons in a partially filled band the situation is different. If this band is narrow, the energy difference ΔE from top to bottom being much less than kT, then the electrons in that band are not concentrated at the bottom but are spread out randomly over all the available energy levels. An increase in temperature does not then lead to an increase of the average energy of the electrons in that band. If the partially filled band has a width ΔE much greater than kT, then the electrons will be concentrated mostly toward the bottom of the band but can make transitions up. In this case there will be a small contribution

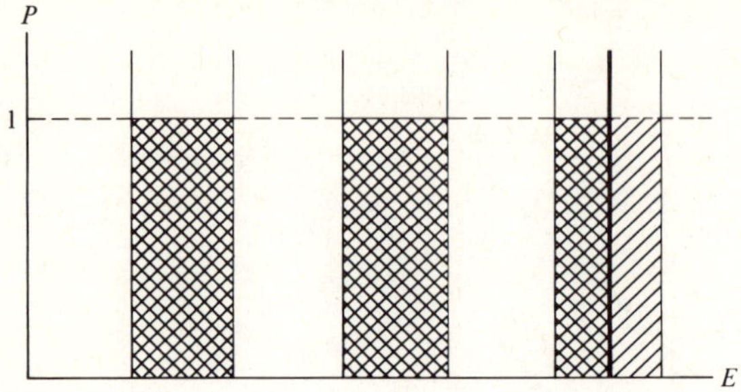

FIGURE 44.7 At absolute zero, all electrons in a metal are in the lowest possible energy states and the fraction P of levels filled is unity up to a certain energy and then drops immediately to zero. There are no thermally excited electrons at this temperature.

to the specific heat from these electrons, but even here it would not be as much as expected classically because only those electrons within an energy of about kT of the topmost filled level can make upward transitions.

The situation can be seen in the graphs of Figures 44.7, 44.8, and 44.9. Here the probability of a given energy level being occupied is plotted against the energy of the level. In Figure 44.7 the temperature is taken to be absolute zero. Two filled bands and one half-filled band are shown. At a temperature greater than zero but with kT considerably

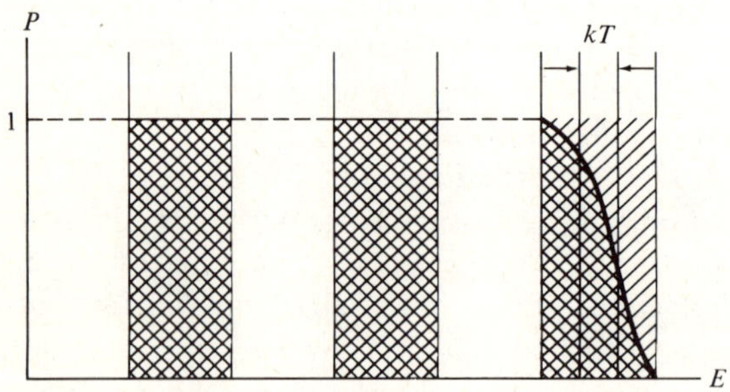

FIGURE 44.8 A little above absolute zero, when kT is still less than the width of the bands, a few electrons are thermally excited from the bottom to the top of the topmost occupied band, but not into the next band.

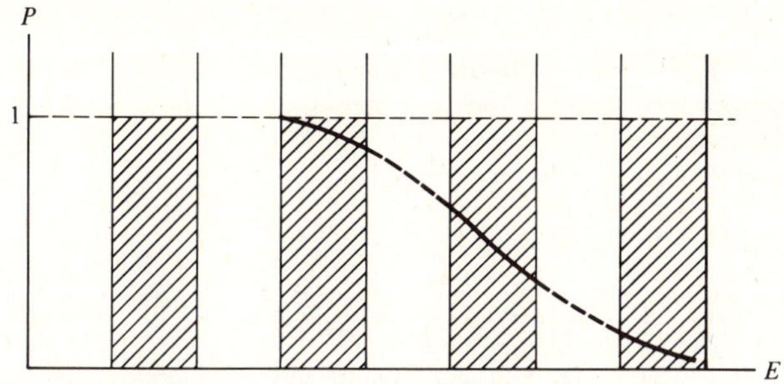

FIGURE 44.9 At a temperature much higher than that in Figure 44.8, electrons from lower bands have been thermally excited into higher bands.

less than ΔE, the distribution is that shown in Figure 44.8. A few electrons have been excited from the bottom half of the conduction band to the top half, but most of them remain near the bottom, not sharing in the thermal energy because it requires an energy $\frac{1}{2}\Delta E$ for an electron at the bottom to make any transition whatever and this energy is unlikely to be available in one lump.

At a considerably higher temperature, with kT greater than ΔE, the distribution would be that in Figure 44.9. The electrons are more uniformly distributed over the states in the conduction band, and a few have even spilled over to the next higher band, but very few.

At moderate temperatures we see that increasing the temperature slightly affects the energy of only a small fraction of the electrons in a solid, so that the average energy of the electrons changes hardly at all. For many of even the conduction electrons to be affected, kT must be comparable to the range of energies in this band that are occupied at absolute zero, so that electrons near the bottom are likely to be able to be raised to vacant levels. If very few levels at the bottom are occupied at $T = 0$, the electrons in that band will behave classically when the temperature is even slightly raised. In this way the law of Dulong and Petit, which states that the entire molar heat of solids at normal temperatures can be accounted for by the thermal vibration of the atoms considered as rigid particles, is explained. At high temperatures, when kT is about as large as or larger than the width and spacing of the allowed energy bands, the electrons can and do contribute to the molar heat, the size of their contribution depending on the detailed structure of the electronic energy bands.

Problems

1. A pendulum 1 m long is hung from a nonrigid support so that when it is pulled over 5 cm the support moves over 1 cm. What is the effective length of this pendulum? By what fraction is its period longer than it would be if the support were rigid?

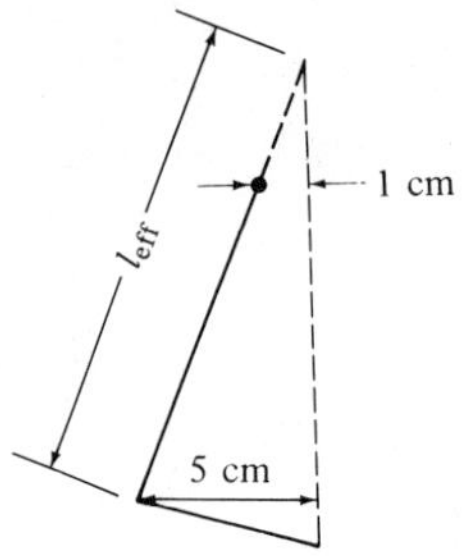

2. Suppose that electrons were bosons so that the Pauli exclusion principle did not apply to them and that all electrons in an atom could be in the lowest level. Suppose that even under these circumstances crystalline solids existed. Would the electrons contribute to the molar heats of such materials? Explain your answer with energy band diagrams analogous to those of Figures 44.7, 44.8, and 44.9.

45. SEMICONDUCTORS

In the previous section we saw how the distinction between electrical conductors and insulators arises. In conductors the uppermost electron energy band that contains any electrons is only partly filled, perhaps half the available energy levels being filled and the rest being empty. Under these circumstances an electron from a level within the band can move from atom to atom even if it has very little extra energy, and when an electric field is applied these electrons are free to move except for the drag provided by scattering from irregularities in the material. In insulators the uppermost band that contains any electrons is completely filled, so an electron that is to hop from atom to atom must get enough energy to get into the lowest empty band, and in insulators the bands are so far apart that very few electrons receive enough thermal energy to make the transition into such an energy level. Even though there are electrons in insulators, they are not free to move and, therefore, no appreciable current flows when an electric field is applied to the material.

It is possible for the topmost occupied band of electron energy levels in a solid to be filled and yet to have the next empty band of energy levels so close to the filled band that an appreciable number of electrons will receive enough thermal energy to make the transition from the filled valance band into this higher conduction band, thereby becoming conduction electrons that can carry a current if an electric field is present. Materials in which this occurs are called *intrinsic semiconductors*. These materials differ in some essential ways from ordinary metallic conductors. In general they are poor conductors when compared with metals because the number of conduction electrons is very much smaller. The number of conduction electrons increases with increasing temperature because at higher temperatures more electrons receive enough thermal energy to cross the forbidden energy band and get into the conduction band. At low temperatures the conductivity of an intrinsic semiconductor increases with increasing temperature; even though any one conduction electron is more likely to be scattered by the thermal vibration of the atoms in the warmer crystal, the larger number of electrons available more than makes up for this. In a metal the conductivity decreases with increasing temperatures. Figure 45.1 shows a plot of the conductivity of pure germanium, an intrinsic semiconductor, as a function of the temperature.

Let us consider an intrinsic semiconductor at a very low temperature so that the conduction band is empty, as shown in Figure 45.2(a), and the valence band is full. If the material is warmed slightly an electron may make the upward transition so that the situation is the one shown in (b) of the figure. Not only is there an electron free to move in the conduction band, but there is a vacant energy state in the valance band or there is a *hole* in that band. It is fairly clear that the electron in the conduction band can carry some electric current if an electric field is applied; it may not be so clear that the hole in the valance band also can carry current, but it can. An analogy may be helpful. Imagine a long traffic jam on a highway. One car pulls out of line onto the shoulder of the road; it is thereby made free to move forward, being in the "conduction lane." It leaves a vacancy in the line, which is quickly filled by the car behind. This creates a new vacancy that is itself quickly filled from behind, and the vacancy or "hole" travels down the jam in the direction opposite to that in which the cars are trying to go. Two stages in the motion of this hole are shown in Figure 45.3. This traveling of the hole causes a net transport of cars, the transport involving the small progress of many cars rather than the large progress of the single car that moved out of line initially, but the total effect can

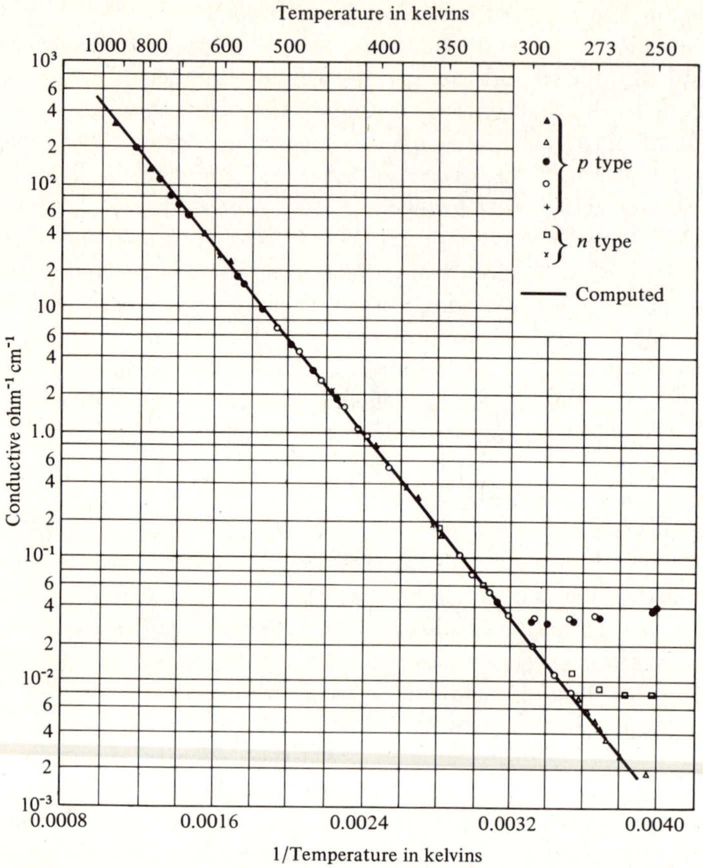

FIGURE 45.1 The conductivity of germanium is plotted against the reciprocal of the temperature, so that low temperatures are to the right and high temperatures to the left. The conductivity increases very rapidly with increasing temperature, which is the opposite of what happens with ordinary metallic conductivity. This is characteristic of intrinsic semiconductors. (After F. J. Morin and J. P. Maita. By permission.)

be about as great. It is in general easier to describe the motion of the single vacancy than it is to describe the collective motion of the cars, each of which must start and then stop in order to make the hole travel uniformly down the line.

The vacancy or hole in the valence band left by an electron that jumps to the conduction band can move through the crystal, its motion in one direction involving a complicated collective motion of the electrons

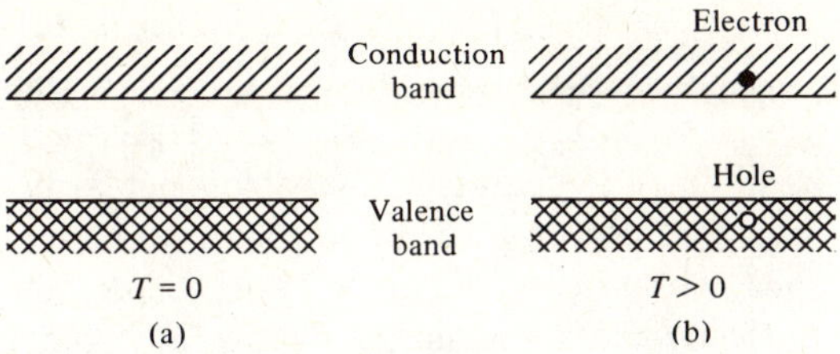

FIGURE 45.2 In (a) there is no thermal energy available, so no electrons are excited to the conduction band. In insulators the material melts before an appreciable number get there. In semiconductors the forbidden band is narrow enough to allow some excitations before the material is destroyed. In (b) an electron has been excited to the conduction band and has left a vacancy or hole in the valence band.

in the opposite direction. The hole acts like a positively charged particle, the positive charge being the charge on the ions that is not completely neutralized when the hole is nearby. The presence of this conduction by holes that are effectively positively charged shows up when a Hall effect measurement is made on an intrinsic semiconductor. According to Figure 43.2, if a current flows to the right through a magnetic field into the page, the force on the charge carriers is upward both when the carriers are positive and moving to the right and when they are negative and moving to the left. In the presence of the magnetic field, then, both the electrons and the holes carrying current in an intrinsic semiconductor will be forced in the same direction, and because they have opposite charges they will produce Hall fields of opposite signs that tend to cancel each other out. If the conduction electron and the hole are equally mobile so they move with the same average speed in

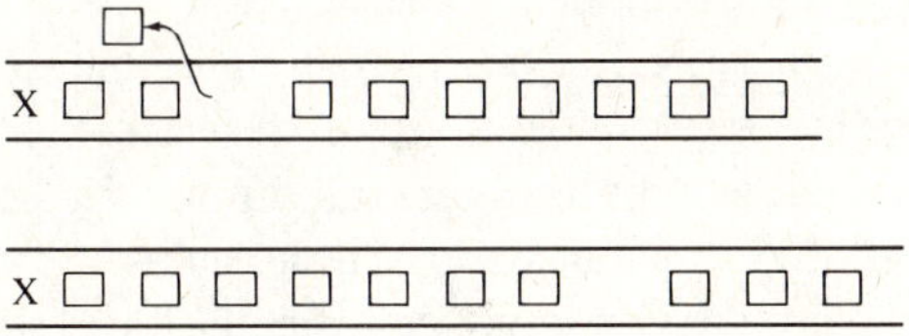

FIGURE 45.3 If a traffic lane is blocked at X, a jam will form behind the block. When a car pulls out to go round the block on the right, it leaves a vacancy which moves down the jam as one car after another goes forward one car length.

an electric field, the Hall effect will be zero. If the hole is more mobile than the electron, the Hall field will be small but in the direction corresponding to conduction by positive carriers. In most metals there is no hole conduction, and there is about one conduction electron per atom.

There is another class of semiconductors known as *impurity semiconductors*. These are of two types, either *p* type or *n* type, depending on whether the conduction of a current is primarily by positive holes (*p*) or by negative electrons (*n*). In an impurity semiconductor one of these generally dominates, unlike the intrinsic semiconductor where both are present.

An *n*-type semiconductor can be made by adding an appropriate impurity to an intrinsic semiconductor such as germanium. Germanium has atomic number 32, so every germanium atom has 32 electrons. If a small fraction of the germanium atoms in the germanium crystal are replaced by atoms of arsenic with atomic number 33, then in the vicinity of each arsenic atom there is one more electron present in the neutral crystal than there would be in pure germanium, and at very low temperature this electron is localized near the arsenic atom with its extra positive nuclear charge. This extra electron occupies a localized energy level above that of the germanium valence band because that band is already full and below the germanium conduction band because it is not free to move through the crystal but is attached to the impurity atom. This energy level is illustrated in Figure 45.4. There is one such localized level for each impurity atom present in the semiconductor provided the impurity atoms are so far apart that these levels are not split up into bands. The impurity levels lie very little below the conduction band so that at even moderate temperatures the probability is high that the electrons in these levels will be excited up into the conduction band or be "donated" to this band. Arsenic atoms are donor impurities in germanium, and the extra energy levels caused by their presence are called *donor levels*. The germanium is said to be "doped" with arsenic.

When an electron from a donor atom is excited up into the conduction band it is free to move and it will carry a current if an electric field is present. The hole it leaves behind is localized at the impurity, so there is no hole conduction caused by the motion of these holes. The conduction is thus by negatively charged carriers, and the material is an *n*-type semiconductor. At very high temperatures an *n*-type semiconductor becomes an intrinsic semiconductor because if the temperature is high enough electrons from the valence band can be excited to the conduction band, leaving mobile holes in the valence band.

A *p*-type semiconductor can be made from germanium, for example,

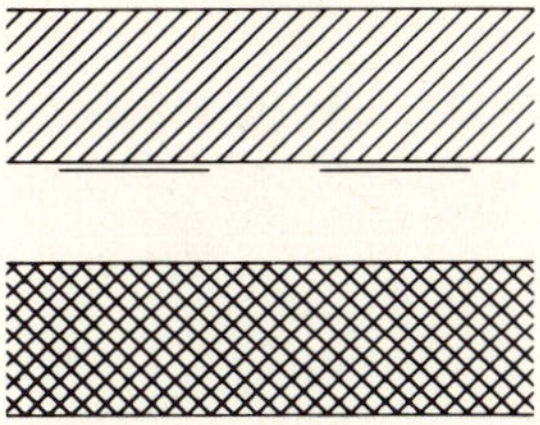

FIGURE 45.4 Energy levels of germanium doped with arsenic include localized donor levels that are occupied at zero temperature by the extra electron of the arsenic atom. Very little thermal energy is needed to excite these extra electrons into the conduction band where they can carry current.

by doping it with gallium whose atomic number 31 is one less than that of germanium. In the neutral crystal there is one less electron in the vicinity of each gallium atom than there would be in the pure germanium crystal. If an extra electron is brought near this atom, it will have an energy slightly higher than it would have in pure germanium because of the smaller nuclear charge of the gallium, so there is an energy level just above the valence band. This level is vacant at very low temperature and is called an *acceptor level,* illustrated in Figure 45.5. At moderate temperature electrons from the valence band can be excited up into these localized acceptor levels, leaving behind holes in the valence band. These holes are free to move, and we have conduction by holes or positively charged carriers only. This type of semiconductor also becomes an intrinsic semiconductor at very high temperature.

Impurity semiconductors differ from intrinsic semiconductors in three ways: (1) Conduction is either hole type or electron type but not both unless two impurities are added. (2) Even room temperature is high enough to cause almost all the donor levels to be empty and the electrons coming from them to be in the conduction band. Similarly, all the acceptor levels are filled even at room temperature, and the corresponding holes in the valence band are free to conduct a current. The number of current carriers does not increase with temperature until this gets high enough to excite an appreciable number of electrons from the valence band directly into the conduction band, making an intrinsic semiconductor. (3) The number of carriers of either kind can be controlled by the degree of doping of the basic material with impurities.

Impurity semiconductors also can be made using silicon, and various impurity elements from other columns of the periodic table can be used.

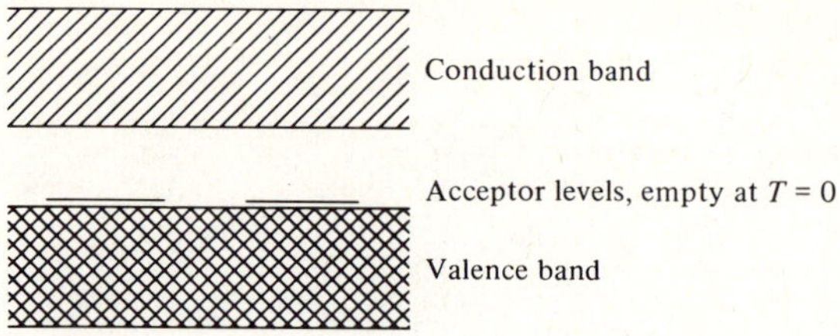

FIGURE 45.5 Energy levels of germanium doped with gallium include localized acceptor levels that are empty at zero temperature. These levels are the ones that would hold the thirty-second electron of gallium if there were one and are higher than the corresponding level for germanium because gallium has a smaller nuclear charge. Very little thermal energy is needed to excite electrons from the valence band into these acceptors, leaving behind holes that can carry current.

Figure 45.6 shows the concentration of charge carriers in several samples of n-type germanium as determined by Hall effect measurements as it depends on the temperature. At low temperatures (large values of $1/T$) the carrier concentration is very small. As the sample is warmed up, more and more electrons in the donor impurity atoms receive enough thermal energy to get into the conduction band and the concentration of carriers increases. It tends to level off at higher temperatures as all the donor atoms become ionized. At still higher temperatures, not shown, the material becomes an intrinsic semiconductor with carriers of both kinds present, and the Hall effect just about disappears. The number of carriers per cubic centimeter involved here, between 10^{13} and 10^{17}, is much smaller than that in metals. According to the Hall effect measurements given in Section 43, in simple metals this number is about 10^{22} carriers per cubic centimeter.

The great usefulness of semiconductors arises from the possibility of combining n-type and p-type semiconductors in a variety of ways to make such things as transistors and diodes. The theory of how such devices operate is based on the band picture of electrons in crystals, and we give just one simple illustration of what can be done, the rectifying p-n junction. Such a junction consists of a piece of p-type semiconductor in contact with a piece of n-type semiconductor. In Figure 45.7 the energy bands in the two materials are sketched. This figure represents the situation when no electric field is applied across the junction. Very

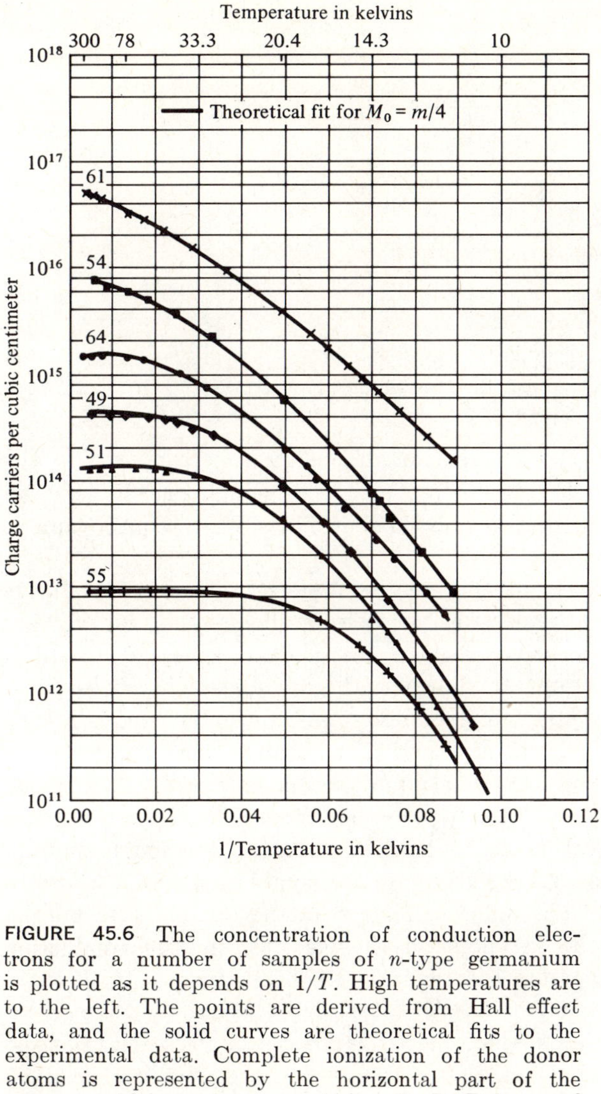

FIGURE 45.6 The concentration of conduction electrons for a number of samples of n-type germanium is plotted as it depends on $1/T$. High temperatures are to the left. The points are derived from Hall effect data, and the solid curves are theoretical fits to the experimental data. Complete ionization of the donor atoms is represented by the horizontal part of the curves at high temperature. (After P. P. Debye and E. M. Conwell. By permission.)

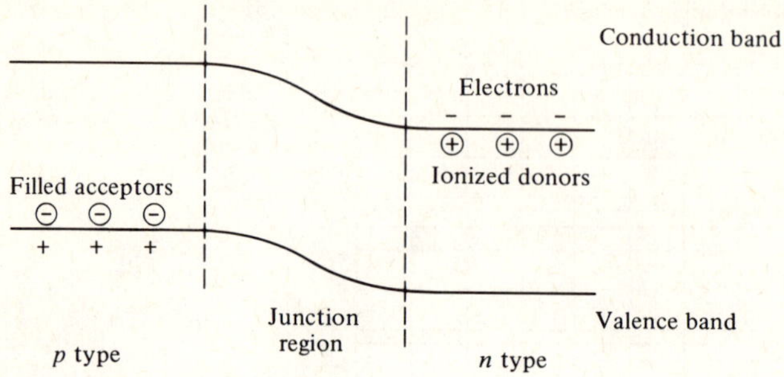

FIGURE 45.7 The energy level diagram of a *pn*-junction is shown when no electric field is applied. The levels in the two kinds of material adjust themselves relative to each other until no current flows across the junction.

few electrons in the *n*-type have enough energy to cross over to the *p*-type and very few holes in the *p*-type have enough energy to cross over to the *n*-type and, in fact, no current flows when equilibrium is reached.

If the electric potential of the *p*-type material is lowered relative to that of the *n* type, the electron energies in the *p*-type material are raised because of the negative electron charge, and an electric field is present across the junction, as shown in Figure 45.8. This field urges the electrons in the *n*-type semiconductor and the holes in the *p*-type semiconductor away from the junction, and because there is almost no supply of electrons or holes near the junction this cannot result in much steady current. What small current does flow comes from the fact that on each side of the junction a few electrons do receive enough thermal energy to get from the valence band to the conduction band, as in an intrinsic semiconductor. The holes so formed in the *n*-type material can cross the junction to the *p*-type, and the electrons in the conduction band of the *p*-type can cross the junction to the *n*-type; both of these result in a flow of current from right to left, but the size of this current depends only on the rate at which electrons are thermally excited across the forbidden energy band and not on the strength of the electric field, so this is not a normal kind of current obeying Ohm's law.

If the electric potential of the *p*-type material is raised relative to that of the *n* type, an electric field in the other direction appears across the junction and the energy levels are shifted as shown in Figure 45.9. The barrier facing the electrons on one side and the holes on the other is lower, and these particles are urged toward the barrier by the field. At the barrier the electrons fill the holes and the charge carriers are

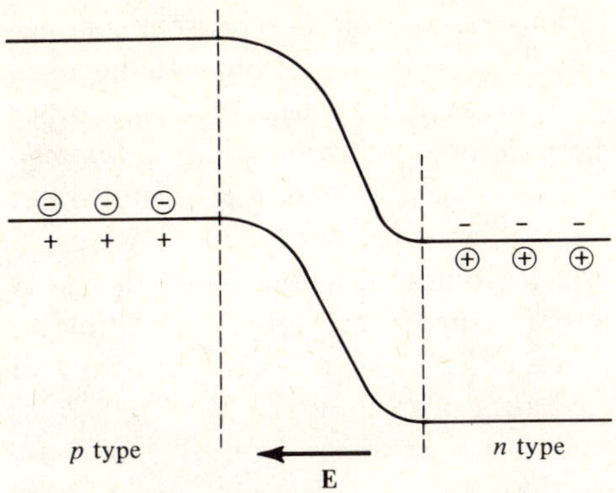

FIGURE 45.8 The situation of Figure 45.7 is changed by application of the electric field **E**. This field tends to drive the movable holes and electrons away from the junction, and since there is no source of holes or of electrons there, a current does not flow as a result of the presence of this field.

"annihilated" like particle-antiparticle pairs. As a result a relatively large current flows across the junction as long as the electric field is maintained, and this current increases with increasing electric field. The response of the junction to fields in the two possible directions is very different, and the junction acts as a rectifier allowing current to flow easily in one direction only.

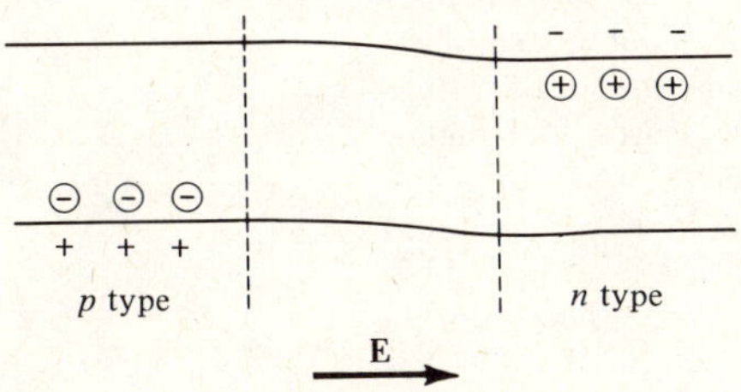

FIGURE 45.9 In this situation the applied electric field tends to drive both movable holes and electrons toward the junction, where they can disappear by the electrons filling the holes. Thus a current can continue to flow as long as there is a source of electrons and holes somewhere else in the circuit.

Other, more complicated combinations of n- and p-type semiconductors can be made. Transistors are such objects, and they can be used as amplifiers and the like in much the same way that vacuum tubes can be. The technology of semiconductor electronics has advanced rapidly and will no doubt continue to advance. The transistor and other devices of "solid state electronics" have replaced the vacuum tubes in radio and television sets and in much communications equipment. It has become possible to build entire circuits on silicon "chips" a millimeter square. Semiconductor detectors are much used in nuclear physics, and the list of applications can be extended almost ad infinitum.

We stop our discussion of semiconductors at this point. Our purpose has been to look at the principles that seem to describe the natural world and not to examine all their implications. The quantum theory of solids had to be known before solid state electronics could develop; it is almost inconceivable that any of this could have been found by experiments unguided by theory, but the theory was not developed with solid state electronics as a goal. It was pure scientific curiosity that set the stage. One cannot tell what other stages are now being set in any field of scientific investigation.

Problems

1. The creation of a particle-hole pair in an intrinsic semiconductor is analogous to the creation of a positron-electron pair in vacuum, but the energy scale is very different. What energy in the positron-electron case corresponds to the energy difference between the top of the valence band and the bottom of the conduction band in the intrinsic semiconductor? (This is the energy that must be supplied to create the electron-hole pair.) What is the temperature for which this energy is kT? Such temperatures, and higher ones, occur in the interiors of stars.
2. A bubble of a very light gas is formed in a heavy liquid and moves upward. Show by a sketch that as the bubble moves some of the liquid must move too. This motion involves kinetic energy, so the motion of the bubble has kinetic energy associated with it even if its own mass is negligible, and this kinetic energy is positive. What is the potential energy of the bubble due to gravity, and does it increase or decrease as the bubble rises?
3. If a Tolman-Stewart experiment were done with a p-type semi-conductor, would you expect the current to flow in the same or in the opposite direction as it would if an n-type were used?

CHAPTER 9 Concluding Remarks

46. ANOTHER LOOK AT THE DEFINITION OF PHYSICS

In the beginning of this book an attempt was made to define physics. Of the two definitions proposed the first was frivolous. "Physics is what physicists do" coupled with "physicists are those who do physics" leads in a circle and is uninformative. The second definition, that physics is the study of the simplest objects, is a serious one, and I want to try to justify it in this last section after the long and at times complicated-looking development that has been concluded with the discussion of semiconductors.

The central theme of this book has been the study of particles. A particle is an abstraction from experience, an object with no size or shape but with position, mass, perhaps electric charge, and some other characteristics. A particle is simple by definition; the circumstance that makes physics possible is that many objects under many conditions do behave as if they were particles subject to certain laws of motion. One of the principal arts of doing physics is to find out how a system that taken as a whole does not behave like a particle can be considered to consist of pieces each of which acts like a particle. The success of this process of analyzing systems into particles is amazing, and we have no idea of how far it can be pushed. There are indications that it will not go on forever, but the limit has not been established yet. Let us review some of the most important ideas involved in this analysis.

The motion of particles that represent large, directly observable objects is described by Newton's laws of motion. These laws have certain features that make the idea of a particle subject to them very flexible in application. Newton's third law is especially important from this point of view. The third law states that when two particles exert forces on one another, the force on one particle is the same in magnitude but opposite in direction to the force on the second. This law has the essential consequence that a pair of particles held together by any kind of binding force acts like a single particle when its motion as a whole is considered. This composite particle is not affected by the binding forces because it is influenced only by the vector sum of the forces acting on its constituent particles, and the vector sum of the binding forces is zero. The binding force on either constituent particle is not zero, and the motion of the constituent particles may be very different from the motion of the composite particle, this latter motion being a kind of average motion of the constituent particles. The structure of a composite particle need not concern us as long as we are interested in motions that do not change this structure, yet it is not inconsistent to admit the existence of such a structure when the simple particle model does not describe some observations.

For purposes of astronomy it is nearly always very accurate to assume that the sun and the planets are particles acted on by the force of gravity, even though we are certainly aware that Earth is not a point completely characterized by its mass and position but has oceans and mountains and an atmosphere and us as part of its structure. If all these features influenced the motion of Earth around the sun appreciably, the solar system would not exist as we know it and we would not have evolved to observe it. Our being here is a result of Newton's third law of motion.

Molecules can often be treated as particles. In the derivation of the ideal gas law from the molecular model it was assumed that the molecules of the gas remain unchanged even when they collide with each other or the walls of their container. The structure, and even the mass, of an individual molecule plays no part whatever in the derivation of the gas law. It is this very lack of importance of molecular characteristics that makes us accept the model, because dilute gases at ordinary temperatures do act very nearly alike. It is not always possible to treat molecules of a gas as particles. We found this out in connection with the problem of the molar heats of gases, where it turns out to be necessary to attribute some structure to at least some molecules. The molar heat of gases like hydrogen, oxygen, and nitrogen can be accounted for by making them consist of two atoms, the atoms to be treated as particles. Other kinds of molecules must be given more elaborate structures, these structures not being determined from molar heat observations alone but mainly from chemical evidence. For many purposes, then, a molecule can be treated as being made of a few particles, and this provides a simple description of the behavior of the molecule under many circumstances.

If molecules are made of atoms, then an atom must be simpler than a molecule and is therefore an object of interest. Can it always be treated as a particle? We know it cannot, because electrons can be extracted from atoms. Treating the electron as a particle, its mass and charge can be measured. Can the electron always be treated as a particle? It cannot be treated as a classical particle obeying Newton's laws of motion when it is moving in an atom, but if Newton's laws are modified to quantum mechanical laws it seems so far that the electron acts like a particle under all circumstances. It is not believed that this will always be so; there are theoretical reasons for believing that sometime the electron will be shown to have a structure of some kind, but this structure has not yet shown up in experiments. All we know about the behavior of electrons now can be accounted for by assuming that the electron is a particle with no structure, with a certain mass, with a certain electric charge, and with a spin angular momentum of $\frac{1}{2}\hbar$. Electrons are simple.

Aside from electrons we know that atoms have nuclei. For the purpose of describing atomic structure the nucleus can be regarded as a particle with a certain mass, a certain charge, a certain spin, and one or two other properties related to its electric and magnetic characteristics. Only in the most refined atomic measurements can effects be observed that require us to assign a finite size to the nucleus and thereby to ascribe some structure to it. For atomic physics, then, the atom consists of only two kinds of particles, the nucleus and electrons. Atoms are relatively simple. If the nucleus were not so small, which might be the case if the strong interaction holding its nucleons together were not nearly as strong as it is, then the atom would not be well represented by a nuclear particle and electrons; it would not be so simple, and not so much would be known about it. Under these circumstances atomic physics might well be almost impossible.

This chain of development can be followed further to the "fundamental particles" we discussed in Sections 26–28, but we do not need to follow it here. The scheme is clear: When structure appears, try to describe it in terms of a small number of new, for the moment simple, particles. The object is to find out what the properties of these various levels of "particle" are. Not all physics is of this kind.

The second part of this book was concerned not so much with the properties of individual particles as with the behavior of these particles when large numbers of them are involved. Here the particles are assumed to be known in their individual characteristics and the collective behavior is studied to see how the large scale behavior of such things as gases and solids can be accounted for. The object of such a study is partly just to understand complicated-seeming things like the electrical behavior of solids in terms of simpler things like atoms. It is also partly to learn how to control the structure of a system to obtain desired properties. Perhaps the clearest example of this control purpose is provided by the manufacture of p-type and n-type semiconductors, a process that would be very hard to discover without first knowing the band structure of crystals.

There is no sharp separation between the study of large numbers of particles and of particles individually. This we saw in connection with the molar heat of gases. In this case no classical particles could be proposed that would have the behavior needed to account for the molar heat of diatomic and polyatomic gases. The consequence was that the laws of motion of atomic particles have to be different from Newton's laws, and quantum mechanics was born. Quantum mechanics was developed fully only by the study of atoms essentially one at a time, but the need for it showed up in connection with gases.

This brief review establishes that, to a large extent at least, physics is the study of the simplest kinds of objects. In this review we have not mentioned the subject of fields, such as the electric, magnetic, and gravitational fields. Fields have been treated sketchily in this book because they are hard to describe without more mathematics than is desirable here and because there was enough to do aside from studying fields in detail. Fields are also relatively simple entities once the necessary mathematical language has been mastered, and when they are studied by the methods of quantum mechanics they have a very close connection with particles, as the existence of the photon shows. There are physicists who believe that in the end everything will be better described by fields than by particles, but this is now a matter of faith and opinion and raises a basic question that remains to be answered.

Appendixes

SUGGESTED FURTHER READINGS

References to further reading in connection with the separate chapters are given. There are a few which are cited frequently and which are worth browsing through as well as consulting for specific points of interest. Among these are:

William F. Magie, *A Source Book in Physics* (Harvard University Press, Cambridge, Mass., 1965).

Classical Scientific Papers, Introduction by Stephan Wright (American Elsevier, New York, 1965).

E. T. Bell, *Men of Mathematics* (Simon and Schuster, New York, 1952).

James R. Newman, *The World of Mathematics* (Simon and Schuster, New York, 1956).

E. T. Whittaker, *A History of the Theories of Aether and Electricity* (Harper and Row (Harper Torchbook), New York, 1960).

Chapter 1

Descartes (cartesian coordinates): Bell, Chap. 3; Newman, Part II, p. 235.

Chapter 2

Galileo: Magie, p. 1; Newman, Part V, p. 726.

Kepler: Arthur Koestler, *The Watershed* (Doubleday (Anchor Books), Garden City, N.Y.); Newman, Part II, p. 218.

Newton: Magie, p. 30; Newman, Part II, p. 254.

Young: Magie, p. 59.

Chapter 3

Coulomb: Magie, p. 408.

Oersted: Magie, p. 436.

Maxwell: Magie, p. 528. A survey of the subject of this chapter is in Whittaker, Vol. I, Chap. II.

Chapter 4

Thomson: Magie, p. 583; Wright, p. 102.

Millikan: R. A. Millikan, *Electrons, Protons, Photons, Neutrons, Mesotrons, and Cosmic Rays* (University of Chicago Press, Chicago, 1947).

Geiger and Marsden: Wright, p. 162.

Rutherford: Wright, pp. 31, 206, 218.

Chadwick: Wright, pp. 245, 246. Surveys of parts of the subject of this chapter are in David L. Anderson, *Discovery of the Electron* (Van Nostrand (Momentum Books), Princeton, N.J., 1964); Whittaker, Vol. I, Chap. XI and Vol. II, Chap. I.

Chapter 5

Einstein: Lillian R. Lieber, *The Einstein Theory of Relativity* (Holt, Rinehart and Winston, New York, 1945); Paul Schilpp, ed., *Einstein, Philosopher-Scientist* (Tudor, New York, 1951).

Accelerators and detectors: R. R. Wilson, "Particle Accelerators," *Scientific American*, March 1958; Donald A. Glaser, "The Bubble Chamber," *Scientific American*, February 1955; Gerald K. O'Neill, "The Spark Chamber," *Scientific American*, August 1962.

Elementary particles: Adolph Baker, *Modern Physics and Antiphysics* (Addison-Wesley, Reading, Mass., 1970). Kenneth W. Ford, *The World of Elementary Particles* (Blaisdell, Waltham, Mass., 1963); Luis Alvarez et al., "Search for Hidden Chambers in the Pyramids," *Science* **167,** 1970, p. 832.

Symmetry: Hermann Weyl, *Symmetry* (Princeton University Press, Princeton, N.J., 1952); Eugene P. Wigner, "Violations of Symmetry in Physics," *Scientific American*, December 1965.

Chapter 6

Young: Magie, p. 308. A survey of the topics of this chapter is in Whittaker, Vol. I, Chap. IV.

Relativity: See references for Chapter 5.

Chapter 7

Kinetic theory of gases: Magie, pp. 247, 255, 203; J. C. Maxwell, *Scientific Papers* (Dover, New York, 1952), Vol. II, p. 46.

Maxwell: Magie, p. 257.

Brown: Magie, p. 251.

Energy conservation: Magie, p. 212.

Atomic structure: A. D'Abro, *The Rise of the New Physics* (Dover, New York, 1951), Vol. II.

Chapter 8

Electron bands: J. C. Slater, "Energy Bands in Solids," *Physics Today*, April 1968, p. 61.

Hall effect: Magie, p. 541.

TRIGONOMETRIC FUNCTIONS

Degrees	Radians	Sin	Cos	Tan	Cot		
0	0.0000	0.0000	1.0000	0.0000		1.5708	90
1	0.0175	0.0175	0.9998	0.0175	57.290	1.5533	89
2	0.0349	0.0349	0.9994	0.0349	28.636	1.5359	88
3	0.0524	0.0523	0.9986	0.0524	19.081	1.5184	87
4	0.0698	0.0698	0.9976	0.0699	14.301	1.5010	86
5	0.0873	0.0872	0.9962	0.0875	11.430	1.4835	85
6	0.1047	0.1045	0.9945	0.1051	9.5144	1.4661	84
7	0.1222	0.1219	0.9925	0.1228	8.1443	1.4486	83
8	0.1396	0.1392	0.9903	0.1405	7.1154	1.4312	82
9	0.1571	0.1564	0.9877	0.1584	6.3138	1.4137	81
10	0.1745	0.1736	0.9848	0.1763	5.6713	1.3963	80
11	0.1920	0.1908	0.9816	0.1944	5.1446	1.3788	79
12	0.2094	0.2079	0.9781	0.2126	4.7046	1.3614	78
13	0.2269	0.2250	0.9744	0.2309	4.3315	1.3439	77
14	0.2443	0.2419	0.9703	0.2493	4.0108	1.3265	76
15	0.2618	0.2588	0.9659	0.2679	3.7321	1.3090	75
16	0.2793	0.2756	0.9613	0.2867	3.4874	1.2915	74
17	0.2967	0.2924	0.9563	0.3057	3.2709	1.2741	73
18	0.3142	0.3090	0.9511	0.3249	3.0777	1.2566	72
19	0.3316	0.3256	0.9455	0.3443	2.9042	1.2392	71
20	0.3491	0.3420	0.9397	0.3640	2.7475	1.2217	70
21	0.3665	0.3584	0.9336	0.3839	2.6051	1.2043	69
22	0.3840	0.3746	0.9272	0.4040	2.4751	1.1868	68
23	0.4014	0.3907	0.9205	0.4245	2.3559	1.1694	67
24	0.4189	0.4067	0.9135	0.4452	2.2460	1.1519	66
25	0.4363	0.4226	0.9063	0.4663	2.1445	1.1345	65
26	0.4538	0.4384	0.8988	0.4877	2.0503	1.1170	64
27	0.4712	0.4540	0.8910	0.5095	1.9626	1.0996	63
28	0.4887	0.4695	0.8829	0.5317	1.8807	1.0821	62
29	0.5061	0.4848	0.8746	0.5543	1.8040	1.0647	61
30	0.5236	0.5000	0.8660	0.5774	1.7321	1.0472	60
31	0.5411	0.5150	0.8572	0.6009	1.6643	1.0297	59
32	0.5585	0.5299	0.8480	0.6249	1.6003	1.0123	58
33	0.5760	0.5446	0.8387	0.6494	1.5399	0.9948	57
34	0.5934	0.5592	0.8290	0.6745	1.4826	0.9774	56
35	0.6109	0.5736	0.8192	0.7002	1.4281	0.9599	55
36	0.6283	0.5878	0.8090	0.7265	1.3764	0.9425	54
37	0.6458	0.6018	0.7986	0.7536	1.3270	0.9250	53
38	0.6632	0.6157	0.7880	0.7813	1.2799	0.9076	52
39	0.6807	0.6293	0.7771	0.8098	1.2349	0.8901	51
40	0.6981	0.6428	0.7660	0.8391	1.1918	0.8727	50
41	0.7156	0.6561	0.7547	0.8693	1.1504	0.8552	49
42	0.7330	0.6691	0.7431	0.9004	1.1106	0.8378	48
43	0.7505	0.6820	0.7314	0.9325	1.0724	0.8203	47
44	0.7679	0.6947	0.7193	0.9657	1.0355	0.8029	46
45	0.7854	0.7071	0.7071	1.0000	1.0000	0.7854	45
		Cos	Sin	Cot	Tan	Radians	Degrees

CONVERSION FACTORS

1 in.	$= 2.540$ cm	1 cm	$= 0.3437$ in.
1 ft	$= 30.48$ cm		$= 0.03281$ ft
1 mile	$= 1.609 \times 10^{5}$ cm		$= 6.214 \times 10^{-6}$ mile
1°	$= 0.01745$ rad	1 rad	$= 57.30°$
1 lb	$= 453.6$ g	1 g	$= 2.205 \times 10^{-3}$ lb
1 dyne	$= 10^{5}$ newton	1 newton	$= 10^{-5}$ dyne
1 erg	$= 10^{7}$ joule	1 joule	$= 10^{-7}$ erg
	$= 6.242 \times 10^{11}$ eV	1 eV	$= 1.602 \times 10^{-12}$ erg
	$= 2.389 \times 10^{-8}$ cal	1 cal	$= 4.186 \times 10^{7}$ erg
1 volt	$= 3.336 \times 10^{-3}$ statvolt	1 statvolt	$= 299.8$ volt
1 amp	$= 2.998 \times 10^{9}$ statamp	1 statamp	$= 3.336 \times 10^{-10}$ amp

PHYSICAL CONSTANTS

Acceleration of gravity (standard value)	g	$= 980.7$ cm/sec^2
Gravitational constant	G	$= 6.673 \times 10^{-8}$ dyne cm^2/g^2
Mass of earth	M_E	$= 5.983 \times 10^{27}$ g
Radius of earth (mean)	R_E	$= 6.367 \times 10^{8}$ cm
Radius of earth's orbit	r_E	$= 1.49 \times 10^{13}$ cm
Electron charge	e	$= 4.803 \times 10^{-10}$ statcoul
Electron mass	m_e	$= 0.911 \times 10^{-27}$ g
Electron rest energy	$m_e c^2$	$= 0.5110$ MeV
Proton mass	m_p	$= 1.673 \times 10^{-24}$ g
Proton rest energy	$m_p c^2$	$= 938.3$ MeV
Avogadro's number	N_0	$= 6.022 \times 10^{23}$
Boltzmann's constant	k	$= 1.381 \times 10^{-16}$ erg/K
		$= 6.582 \times 10^{-5}$ eV/K
Gas constant	$R = N_0 k$	$= 8.314 \times 10^{7}$ erg/K
Planck's constant	$\hbar = h/2\pi$	$= 1.055 \times 10^{-27}$ erg sec
		$= 6.582 \times 10^{-16}$ eV sec
Speed of light	c	$= 2.998 \times 10^{10}$ cm/sec

ANSWERS TO ODD-NUMBERED PROBLEMS

Section 2

1. (a) $x = 16, -4, 20$ cm
   (b) $(0, 8, 0)$ and $(0, -8, 0)$, 16 cm

Section 3

1. 5 units; $(-4, 0, 3)$
3. (a) $\left(\frac{a}{2}, \frac{b}{2}, \frac{c}{2}\right)$
   (b) $\sqrt{\left(\frac{a}{2}\right)^2 + \left(\frac{b}{2}\right)^2 + \left(\frac{c}{2}\right)^2}$
   (c) $\sqrt{\left(\frac{b}{2}\right)^2 + \left(\frac{c}{2}\right)^2}$ from x-axis
   $\sqrt{\left(\frac{c}{2}\right)^2 + \left(\frac{a}{2}\right)^2}$ from y-axis
   $\sqrt{\left(\frac{a}{2}\right)^2 + \left(\frac{b}{2}\right)^2}$ from z-axis
5. $V_x = \sqrt{3}/2\, V = 0.87$ V
   $V_y = \pm\frac{1}{2}$ V
9. $h = 5 \tan 15° = 1.34$ mi

Section 4

1. $\langle v_x \rangle = \dfrac{x(t_2) - x(t_1)}{t_2 - t_1}$ etc.
3. $v = 10^6$ cm/sec

Section 5

1. 1.609
3. (a) 3.43×10^{-4}
   (b) 3.32×10^{-4}
   (c) 1.58×10^4
   (d) 1.44×10^3

Section 6

1. 7840 cm; 0.26 sec; 6%
3. 1.21×10^3 cm/sec

Section 7

1. $a_E = 0.6$ cm/sec^2
   $a_M = 0.26$ cm/sec^2
3. 0.79×10^6 cm/sec
5. $x(t) = -r \sin \omega t$
   $y(t) = r \cos \omega t$

Section 8

1. 13.3 rad/sec
3. Stable with flatter part down, unstable with narrow part down

Section 9

1. (a) $x =$ 1, 1.1, 1.21, 1.33, 1.46, 1.61; $v_x =$ 0.1, 0.11, 0.12, 0.13, 0.15, 0.16

x	v_x
1	0.1
1.1	0.11
1.21	0.12
1.33	0.13
1.46	0.15
1.61	0.16

   (b)

x	v_x
−2	−0.2
−2.2	−0.22
−2.42	−0.24
−2.66	−0.27
−2.93	−0.29
−3.22	−0.32

3. A law giving the rate the acceleration changes; no
5. 6760 dynes; 0.00345 g

Section 10

1. 20.5 kg
3. 3.33 cm/sec^2; 10 dynes
5. 4.06×10^3 cm/sec
7. 2 mg
9. 4; $\frac{1}{4}$
11. $2k$

Section 11

1. 2.7 ft/sec; 3.3 ft/sec
3. Momentum conservation yields $(m_E + m)v = m_E v_0$ after mass m of dust has stuck.

Section 12

1. (a), (b), (c) pb into page
3. 90°W

Section 13

1. (b) maximum at ground level; minimum (zero) at y_0
3. 5×10^4 exp; 2.55×10^4 ergs kinetic; 2.45×10^4 ergs potential
5. $y = -0.735$ cm

Section 14

1. Because mass is always positive
3. 1.2×10^4 ergs; 6×10^4 ergs

Section 15

3. (a) Toward original position
   (b) Away from original position; unstable
5. It becomes charged and attracts opposite charge to the surface of the wall.
7. $\dfrac{4\,\mathrm{mg}l^2 \sin^2\theta}{qq'} = \dfrac{\cos\theta}{\sin\theta}$; yes; no

Section 16

1. 19 statcoul; 129 stat V/cm
7. 7680 V/cm; 3840 V

Section 17

1. 0.6 statcoul/cm; 1.33×10^{-9} gauss; 0.08 stat V/cm

Section 18

1. 1.25×10^8 statcoul; 1.03×10^{10} V/cm $= 0.346 \times 10^8$ stat V/cm; 0.219×10^6 V $= 0.658 \times 10^8$ stat V
3. $0.026c = 7.8 \times 10^8$ cm/sec

Section 19

1. $\dfrac{Gm_e^2}{e^2} = 0.5 \times 10^{-43}$; yes
3. 1.51×10^{-9} sec; halved
5. 1.4×10^3 cm/sec

Section 20

1. 10^{-5}
3. 4.24×10^{-11} cm
5. 3.27×10^{19} V/cm

Section 21

1. (a) 1.125 MeV
   (b) 0.844 MeV

Section 22

1. (a) $0.90c$, $0.88c$
   (b) $0.956c$, $0.368c$
   (c) $0.994c$, 0

3. $\left.\begin{aligned} K_{\text{Newton}} &= 0.125m_0c^2 \\ K_{\text{rel}} &= 0.155m_0c^2 \end{aligned}\right\}$ when $v = 0.5c$

$\left.\begin{aligned} K_{\text{Newton}} &= 0.49m_0c^2 \\ K_{\text{rel}} &= 6.07m_0c^2 \end{aligned}\right\}$ when $v = 0.99c$

5. 9.57×10^{-28} g
7. $0.976c$

Section 23

1. $E' = E\sqrt{\dfrac{1 + (v/c)}{1 - (v/c)}}$; no

5. $E'_\gamma = E_\gamma \dfrac{m_0c^2}{m_0c^2 + 2E_\gamma}$

Section 24

3. 0.956×10^{15} eV

Section 25

1. Only about $\frac{1}{50}$ of neutron energy can be transferred to a lead nucleus at rest.
3. 4.17×10^3 cm; $2.75° = 0.048$ rad

Section 26

1. yes; no
3. 0.214; smaller

Section 27

1. (a) is possible; others are impossible

Section 28

1. maximum neutrino energy, 0.8 MeV; minimum neutrino energy, 0

3. $p_\nu = p_\mu = \dfrac{m_\pi^2 - m_\pi^2}{2m_\pi} c$ (pion at rest); $p'_{\nu x} = 20p_{\nu y}$, approximately the same for $p'_{\mu x}$

Section 29

1. twenty (five in each quadrant)
3. 1.18×10^{-4} cm; larger

Section 30

1. (a) nested circles
   (b) intersecting circles

3. $\frac{4}{3}c$; no
5. $10^8\ \lambda$; 50 m for $\lambda = 5 \times 10^{-5}$ cm

Section 31

1. Reinforce between wires, cancel in regions far from wires

Section 32

1. (a) $-459°$F
   (b) $-40°$
   (c) 37°C

Section 34

3. oxygen, 0.48×10^5 cm/sec; electrons, 1.17×10^7 cm/sec
5. Reduced to half

Section 35

1. 2.90×10^4 cm/sec
3. -1.69 cm/sec; 0.0058%

Section 37

1. 2.82×10^5 dynes

Section 38

1. (a) 4×10^{12} K
   (b) 6×10^7 K
3. (a) 6 cal/K mole
   (b) 9 cal/K mole

Section 39

3. 10.4 eV
5. method of Problem 3

Section 40

1. $j = ne^2Et/m$; increase with time

Section 41

1. (a) 262 cm/sec^2
   (b) 5 stat V/cm = 1500 V/cm
   (c) 855 amp/cm^2

Section 42

1. 1 part in 2000

Section 44

1. 1.2 m; 10% about

Section 45

1. $2m_ec^2 = 1$ MeV; 1.2×10^{10} K
3. Same direction

Index